Current Topics in Microbiology and Immunology

Volume 364

Current Topics in Microbiology and Immunology

Previously published volumes

Further volumes can be found at www.springer.com

Vol. 333: **Compans, Richard W.; Orenstein, Walter A. (Eds.)**:
Vaccines for Pandemic Influenza. 2009.
ISBN 978-3-540-92164-6

Vol. 334: **McGavern, Dorian; Dustin, Micheal (Eds.)**:
Visualizing Immunity. 2009.
ISBN 978-3-540-93862-0

Vol. 335: **Levine, Beth; Yoshimori, Tamotsu; Deretic, Vojo (Eds.)**:
Autophagy in Infection and Immunity. 2009.
ISBN 978-3-642-00301-1

Vol. 336: **Kielian, Tammy (Ed.)**:
Toll-like Receptors: Roles in Infection and Neuropathology. 2009.
ISBN 978-3-642-00548-0

Vol. 337: **Sasakawa, Chihiro (Ed.)**:
Molecular Mechanisms of Bacterial Infection via the Gut. 2009.
ISBN 978-3-642-01845-9

Vol. 338: **Rothman, Alan L. (Ed.)**:
Dengue Virus. 2009.
ISBN 978-3-642-02214-2

Vol. 339: **Spearman, Paul; Freed, Eric O. (Eds.)**:
HIV Interactions with Host Cell Proteins. 2009.
ISBN 978-3-642-02174-9

Vol. 340: **Saito, Takashi; Batista, Facundo D. (Eds.)**:
Immunological Synapse. 2010.
ISBN 978-3-642-03857-0

Vol. 341: **Bruserud, Øystein (Ed.)**:
The Chemokine System in Clinical and Experimental Hematology. 2010.
ISBN 978-3-642-12638-3

Vol. 342: **Arvin, Ann M. (Ed.)**:
Varicella-zoster Virus. 2010.
ISBN 978-3-642-12727-4

Vol. 343: **Johnson, John E. (Ed.)**:
Cell Entry by Non-Enveloped Viruses. 2010.
ISBN 978-3-642-13331-2

Vol. 344: **Dranoff, Glenn (Ed.)**:
Cancer Immunology and Immunotherapy. 2011.
ISBN 978-3-642-14135-5

Vol. 345: **Simon, M. Celeste (Ed.)**:
Diverse Effects of Hypoxia on Tumor Progression. 2010.
ISBN 978-3-642-13328-2

Vol. 346: **Christian Rommel; Bart Vanhaesebroeck; Peter K. Vogt (Ed.)**:
Phosphoinositide 3-kinase in Health and Disease. 2010.
ISBN 978-3-642-13662-7

Vol. 347: **Christian Rommel; Bart Vanhaesebroeck; Peter K. Vogt (Ed.)**:
Phosphoinositide 3-kinase in Health and Disease. 2010.
ISBN 978-3-642-14815-6

Vol. 348: **Lyubomir Vassilev; David Fry (Eds.)**:
Small-Molecule Inhibitors of Protein-Protein Interactions. 2011.
ISBN 978-3-642-17082-9

Vol. 349: **Michael Kartin (Eds.)**:
NF-kB in Health and Disease. 2011.
ISBN 978-3-642-16017-2

Vol. 350: **Rafi Ahmed; Tasuku Honjo (Eds.)**:
Negative Co-receptors and Ligands. 2011.
ISBN 978-3-642-19545-7

Vol. 351: **Marcel, B. M. Teunissen (Ed.)**:
Intradermal Immunization. 2011.
ISBN 978-3-642-23689-1

Vol. 352: **Rudolf Valenta; Robert L. Coffman (Eds.)**
Vaccines against Allergies. 2011.
ISBN 978-3-642-20054-0

Vol. 353: **Charles E. Samuel (Ed.)**:
Adenosine Deaminases Acting on RNA (ADARs) and A-to-I Editing. 2011.
ISBN 978-3-642-22800-1

Vol. 354: **Pamela A. Kozlowski (Ed.)**:
Mucosal Vaccines. 2012.
ISBN 978-3-642-23692-1

Vol. 355: **Ingo K. Mellinghoff; Charles L. Sawyers (Eds.)**:
Therapeutic Kinase Inhibitors. 2012.
ISBN 978-3-642-28295-9

Vol. 356: **Cornelis Murre (Ed.)**:
Epigenetic Regulation of Lymphocyte Development. 2012.
ISBN 978-3-642-24102-4

Vol. 357: **Nicholas J. Mantis (Ed.)**:
Ricin and Shiga Toxins. 2012.
ISBN 978-3-642-27469-5

Vol. 359: **Benhur Lee, Paul Rota (Eds.)**:
Henipavirus. 2012.
ISBN 978-3-642-29818-9

Vol. 360: **Freddy Radtke (Ed.)**:
Notch Regulation of the Immune System. 2012.
ISBN 978-3-642-24293-9

Vol. 361: **Klaus Aktories; Joachim H. C. Orth; Ben Adler (Eds.)**:
Pasteurella multocida. 2012.
ISBN 978-3-642-31016-4

Vol. 362: **Marco Falasca (Ed.)**:
Phosphoinositides and Disease. 2012.
ISBN 978-94-007-5024-1

Vol. 363: **Michael G. Katze (Ed.)**:
Systems Biology. 2013.
ISBN 978-3-642-33098-8

Andreas Rummel · Thomas Binz
Editors

Botulinum Neurotoxins

Responsible series editor: Klaus Aktories

Springer

Editors
Andreas Rummel
Institut für Toxikologie
Medizinische Hochschule Hannover
Hannover
Germany

Thomas Binz
Institut für Physiologische Chemie
Medizinische Hochschule Hannover
Hannover
Germany

The cover picture illustrates the botulinum neurotoxin A shielded by its non-toxic non hemagglutinin protein (NTNHA) and was generated and kindly provided by Rongsheng Jin.

ISSN 0070-217X
ISBN 978-3-642-33569-3 ISBN 978-3-642-33570-9 (eBook)
DOI 10.1007/978-3-642-33570-9
Springer Heidelberg New York Dordrecht London

Library of Congress Control Number: 2012952294

Printed on acid-free paper

Springer is part of Springer Science+Business Media (www.springer.com)

Preface

The extremely potent substance botulinum neurotoxin (BoNT) has attracted much interest in diverse fields. Originally identified as a cause for the rare but deadly disease botulism, military, and terrorist intended to misuse this sophisticated molecule as biological weapon. This caused its classification as select agent category A by the Centers for Disease Control and Prevention and the listing in the Biological and Toxin Weapons Convention. Later, the civilian use of BoNT as long acting peripheral muscle relaxant has turned this molecule into an indispensable pharmaceutical worldwide with annual revenues >$1.5 billion. Also, basic scientists value the botulinum neurotoxin as molecular tool for dissecting mechanisms of exocytosis.

This book will cover the most recent molecular details of botulinum neurotoxin, its mechanism of action as well as its detection and application. "Genetic Diversity Within *Clostridium botulinum* Serotypes, botulinum Neurotoxin Gene Clusters and Toxin Subtypes" explains the genetics of the diverse strains of *Clostridium botulinum* and some other bacteria of the genus *Clostridium* as well as provides an overview of the diverse BoNT family currently including 35 BoNT serotypes and subtypes and their neurotoxin-associated non-toxic proteins (NAPs). "Assembly and Function of the Botulinum Neurotoxin Progenitor Complex" sheds light on the structural assembly of BoNT with its NAPs yielding stable shield complexes which allow an oral intoxication by an intact passage through the gastrointestinal tract and efficient resorption of BoNT in the gut. "Uptake of Botulinum Neurotoxin in the Intestine" describes the currently discussed three different routes for resorption of BoNT from the gut. Then the BoNT spreads via the circulation within the body. "Double Receptor Anchorage of Botulinum Neurotoxins Accounts for their Exquisite Neurospecificity" teaches that BoNT specifically bind to cholinergic motor nerve terminals via a complex dual receptor interaction which is one of the major determinant for their high potency. "The Elusive Compass of Clostridial Neurotoxins: Deciding When and Where to Go?" illustrates their subsequent endocytotic uptake which decides about the site of action of BoNT and the closely related tetanus neurotoxin. BoNT harness the synaptic vesicle recycling as a Trojan horse and "Synchronized Chaperone Function of Botulinum Neurotoxin Domains Mediates Light Chain Translocation into Neurons" exhibits the nanomodular machinery which allows the delivery of the proteolytically active domain (LC, light chain) into the neuronal cytosol. In "Clostridial Neurotoxin Light Chains: Devices for SNARE Cleavage Mediated Blockade of Neurotransmission", the extraordinary cleavage specificity of the LC for the members of the SNARE proteins is presented. "Synaptic Vesicle Proteins: Targets and Routes for Botulinum Neurotoxins" deepens the knowledge about the physiological role of the members of the SNARE proteins susceptible to BoNT hydrolysis. The unusual long persistence of LC activity within the synapses and its molecular basis are elucidated in "Persistence of Botulinum Neurotoxin Inactivation of Nerve Function". To counteract intoxications with BoNT, the major efforts of identifying LC inhibiting molecules is explicated in "Structure-Based Drug Discovery for Botulinum Neurotoxins". To allow timely and appropriate application of counter measures the most recent advanced techniques for the highly sensitive detection of BoNT are required. These are described in "Complexity of Botulinum Neurotoxins: Challenges for Detection Technology". Precise quantitation of potency of BoNT being an active pharmaceutical ingredient using modern cell-based assays is illustrated in "Progress in Cell Based Assays for Botulinum Neurotoxin Detection". Exploiting the great body of research data about BoNT to design second generation peripheral muscle relaxants as well as novel targeted secretion inhibitors is highlighted in "Transforming the Domain Structure of Botulinum Neurotoxins into Novel Therapeutics". Finally, "Botulinum Toxin: Application, Safety, and Limitations" deals with the application of BoNT as medicine and describes its extraordinary pharmacological advantages with respect to safety, treatment efficiency, and duration of action.

This book comprehensively covers all major molecular aspects of BoNT and its civilian exploitation written by experts leading in their field. We think it is appropriate for a widespread readership within the BoNT field as well as in related areas of research and medicine. We would like to thank all chapter authors for their high quality contributions and Mrs. Clauss at Springer for the smooth handling of the manuscripts.

Hannover, August 2012

Andreas Rummel
Thomas Binz

Contents

Genetic Diversity Within *Clostridium botulinum* Serotypes, botulinum Neurotoxin Gene Clusters and Toxin Subtypes 1
Karen K. Hill and Theresa J. Smith

Assembly and Function of the Botulinum Neurotoxin Progenitor Complex 21
Shenyan Gu and Rongsheng Jin

Uptake of Botulinum Neurotoxin in the Intestine 45
Yukako Fujinaga, Yo Sugawara and Takuhiro Matsumura

Double Receptor Anchorage of Botulinum Neurotoxins Accounts for their Exquisite Neurospecificity 61
Andreas Rummel

The Elusive Compass of Clostridial Neurotoxins: Deciding When and Where to Go? 91
Kinga Bercsenyi, Francesco Giribaldi and Giampietro Schiavo

Synchronized Chaperone Function of Botulinum Neurotoxin Domains Mediates Light Chain Translocation into Neurons 115
Audrey Fischer

Clostridial Neurotoxin Light Chains: Devices for SNARE Cleavage Mediated Blockade of Neurotransmission 139
Thomas Binz

Synaptic Vesicle Proteins: Targets and Routes for Botulinum Neurotoxins 159
Gudrun Ahnert-Hilger, Agnieszka Münster-Wandowski and Markus Höltje

Persistence of Botulinum Neurotoxin Inactivation of Nerve Function 179
Charles B. Shoemaker and George A. Oyler

Structure-Based Drug Discovery for Botulinum Neurotoxins 197
Subramanyam Swaminathan

Complexity of Botulinum Neurotoxins: Challenges for Detection Technology 219
Martin B. Dorner, K. Melanie Schulz, Skadi Kull and Brigitte G. Dorner

Progress in Cell Based Assays for Botulinum Neurotoxin Detection . . . 257
Sabine Pellett

Transforming the Domain Structure of Botulinum Neurotoxins into Novel Therapeutics 287
John Chaddock

Botulinum Toxin: Application, Safety, and Limitations 307
Hans Bigalke

Index 319

Genetic Diversity Within *Clostridium botulinum* Serotypes, Botulinum Neurotoxin Gene Clusters and Toxin Subtypes

Karen K. Hill and Theresa J. Smith

Abstract *Clostridium botulinum* is a species of spore-forming anaerobic bacteria defined by the expression of any one or two of seven serologically distinct botulinum neurotoxins (BoNTs) designated BoNT/A-G. This Gram-positive bacterium was first identified in 1897 and since then the paralyzing and lethal effects of its toxin have resulted in the recognition of different forms of the intoxication known as food-borne, infant, or wound botulism. Early microbiological and biochemical characterization of *C. botulinum* isolates revealed that the bacteria within the species had different characteristics and expressed different toxin types. To organize the variable bacterial traits within the species, Group I–IV designations were created. Interestingly, it was observed that isolates within different Groups could express the same toxin type and conversely a single Group could express different toxin types. This discordant phylogeny between the toxin and the host bacteria indicated that horizontal gene transfer of the toxin was responsible for the variation observed within the species. The recent availability of multiple *C. botulinum* genomic sequences has offered the ability to bioinformatically analyze the locations of the *bont* genes, the composition of their toxin gene clusters, and the genes flanking these regions to understand their variation. Comparison of the genomic sequences representing multiple serotypes indicates that the *bont* genes are not in random locations. Instead the analyses revealed specific regions where the toxin genes occur within the genomes representing serotype A, B, C, E, and F *C. botulinum* strains and *C. butyricum* type E strains. The genomic analyses have provided evidence of horizontal gene transfer, site-specific insertion, and

K. K. Hill (✉)
Los Alamos National Laboratory, Mail Stop: M 888, Los Alamos, NM 87545, USA
e-mail: khill@lanl.gov

T. J. Smith
United States Army Medical Research Institute of Infectious Diseases, 1425 Porter Street, Fort Detrick, MD, USA
e-mail: theresa.smith@amedd.army.mil

A. Rummel and T. Binz (eds.), *Botulinum Neurotoxins*, Current Topics in Microbiology and Immunology 364, DOI: 10.1007/978-3-642-33570-9_1,

recombination events. These events have contributed to the variation observed among the neurotoxins, the toxin gene clusters and the bacteria that contain them, and has supported the historical microbiological, and biochemical characterization of the Group classification within the species.

Keywords Subtype · OrfX toxin gene cluster · Ha toxin gene cluster · Horizontal gene transfer · Phylogentic tree · Clostridium botulinum genome

Abbreviations

BoNT	Botulinum neurotoxin
HA	Hemagglutinin
NTNHA	Nontoxic nonhemagglutinin
orf	Open reading frame
bp	Base pair
kb	Kilobase
Mb	Megabase
IS elements	Insertion sequence elements

Contents

1 Introduction ... 2
2 Location of the Botulinum Neurotoxin Genes Within Strains ... 6
2.1 Location of Botulinum Neurotoxin Genes Within Group I Strains ... 7
2.2 Location of Botulinum Neurotoxin Genes Within Group II and Group VI Strains ... 9
2.3 Location of Botulinum Neurotoxin Genes Within Group III Strains ... 10
2.4 Location of Botulinum Neurotoxin Genes Within Group IV and Group V Strains ... 11
3 Composition of the *HA*+ and *orfX*+ Toxin Gene Clusters ... 11
4 Botulinum Neurotoxin Diversity ... 13
4.1 Definition of a Subtype ... 13
4.2 Botulinum Neurotoxin Serotype A-G and Subtype/Variant Comparisons ... 14
4.3 Mutation and Recombination Events ... 15
5 Summary ... 16
References ... 17

1 Introduction

Modern *Clostridium botulinum* research began with the discovery of a toxin producing anaerobic spore-forming rod-shaped bacteria in 1897 by van Ermengem (1897). He named the bacteria *Bacillus botulinus*, due to frequent past cases of botulism associated with sausages, but his isolations of the organism were from

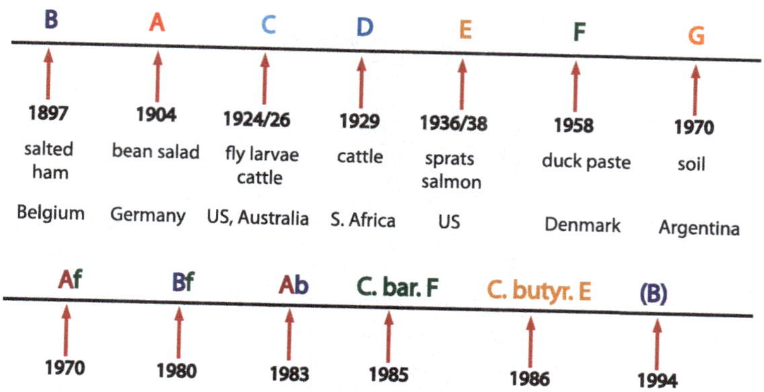

Fig. 1 Botulinum neurotoxin discovery time-line. The seven toxin serotypes were discovered over the course of 73 years, followed by the identification of bivalent strains, BoNT/F-producing *C. baratii* strains, BoNT/E-producing *C. butyricum* strains and the silent unexpressed *bont*/(B) gene. Note the time-line is not to scale

salted ham. As the time-line in Fig. 1 illustrates, six other botulinum neurotoxin (BoNT) serotypes were discovered in the next 73 years (Burke 1919; Leuchs 1910; Bengtson 1922; Seddon 1922; Meyer and Gunnison 1929; Hazen 1937; Moller and Scheibel 1960; Gimenez and Ciccarelli 1970a) followed by the discovery of bivalent (dual toxin-producing) strains (Gimenez and Ciccarelli 1970b), then BoNT/F-producing *C. baratii,* and BoNT/E-producing *C. butyricum* strains (Hall et al. 1985; Aureli et al. 1986). Lastly, the presence of silent unexpressed *bont*/(B) genes within two non-toxic *C. subterminale* strains was identified by PCR experiments in 1994 (Franciosa et al. 1994).

The identification of each new toxin type was defined by the inability of the toxin to be neutralized (or precipitated) with polyclonal antisera purified from animals immunized with culture supernatants of strains representing the other toxin types (Burke 1919; Leuchs 1910). With new toxin types being identified, efforts also focused on the biochemical and physical characteristics of the host bacteria, which were characterized using a series of microbiological and biochemical tests. A strain was then designated with the alphabetical toxin type(s) and the host bacterium was described as belonging to Group I–IV. Two additional Groups were identified—Group V which includes the BoNT/F-producing *C. baratii* strains and Group VI which represents the BoNT/E-producing *C. butyricum* strains. The attributes for each Group have been described in detail in several excellent references (Peck et al. 2011; Hatheway 1988; Smith and Sugiyama 1988; Popoff 1995). This approach provided characterization of both the bacteria and the toxin, with a systematic nomenclature to describe an isolate. For example, "Group II nonproteolytic BoNT/B-producing isolate", was terminology understood within the botulinum research community.

As molecular techniques such as DNA sequencing were developed, the *bont* gene sequences for each of the different toxin types were elucidated. Sequencing of the *16S rRNA* gene in isolates within each of the Groups was performed to

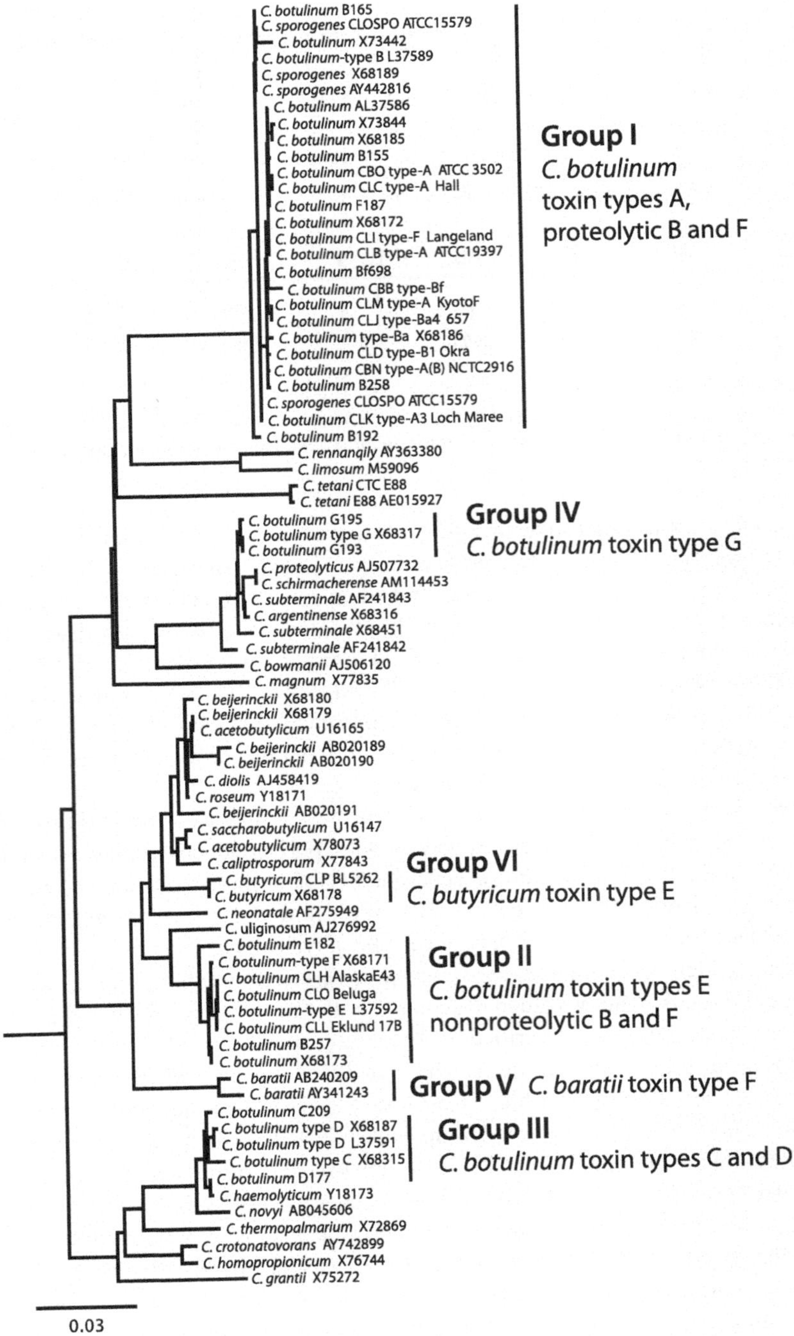

Fig. 2 A dendrogram generated from *16S rRNA* sequences representing different clostridial species illustrating the location of strains within the Group I–IV *C. botulinum*, Group V *C. baratii* type F, and Group VI *C. butyricum* type E strains. The distances between Groups illustrate the genetic diversity within the species (Hill et al. 2007)

Table 1 A list of BoNT-producing *C. botulinum* and BoNT/E-producing *C. butyricum* strain sequences and their GenBank accession numbers

Species	Group	Subtype	Strain	Sequence	Size (kb)	Accession
C. botulinum	I	A1	ATCC 3502	chr[a]	3,887	AM412317
	I			plasmid	16	AM412317
C. botulinum	I	A1	ATCC 19397	chr[a]	3,863	CP000726
C. botulinum	I	A1	Hall	chr[a]	3,761	CP000727
C. botulinum	I	A1(B)	NCTC 2916	chr[a]	4,031	ABDO02000001-49
C. botulinum	I	A2	Kyoto-F	chr[a]	4,155	CP001581
C. botulinum	I	A3	Loch Maree	chr	3,993	CP000962
	I			plasmid[a]	267	CP000963
C. botulinum	I	B5a4	strain 657	chr	3,978	CP001083
C. botulinum				plasmid[a]	270	CP001081
C. botulinum	I			plasmid	10	CP001082
C. botulinum	I	A5	H04402 065	chr[a]	3,920	FR773526
C. botulinum	I	B1	Okra	chr	3,958	CP000939
	I			plasmid[a]	149	CP000940
C. botulinum	I	B5f2	—	plasmid	4,200	ABDP01000001-70
C. botulinum	I	F1	Langeland	chr[a]	3,995	CP000728
C. botulinum	I			plasmid	18	CP000729
C. botulinum	I	F1	F230613	chr[a]	3,993	CP002011
C. botulinum	I			plasmid	18	CP002012
C. botulinum	II	B4	Eklund 17B	chr	3,800	CP001056
C. botulinum	II			plasmid[a]	48	CP001057
C. botulinum	II	E1	Beluga	chr[a]	4,000	ACSC01000001-6
C. botulinum	II	E3	E43	chr[a]	3,660	CP001078
C. botulinum	III	C	Stockholm	chr	2,660	AESA00000000
				phage[a]	55	AP008983
C. botulinum	III	C	Eklund	chr, phage[a]	2,961	ABDQ01000001-76
C. botulinum	III	CD	08-BKT015925	chr, phage[a]	2,773	CP002410
				plasmid	203	CP002411
C. botulinum	III	CD	08-V891	chr	3,139	AESC00000000
C. botulinum	III	D	1873	chr, phage[a]	2,379	ACSJ01000001-19
				plasmid	108	CP001659
				plasmid	54	CP001660
C. butyricum	VI	E4	BL5262	chr[a]	4,758	ACOM01000001-13
C. butyricum	VI	E4	BL5521	chr[a]	4,541	ABDT01000001-123

The Group designation, strain information and chromosome (chr), plasmid or phage sequence related to the GenBank accession is provided
[a] *bont* gene cluster locations

delineate the relationships among the bacteria (Hutson et al. 1993a, b; Hill et al. 2007) (Fig. 2). Gene sequences representing additional *bont*/A-G strains were released, providing information about sequence variability within and among the various bacterial Groups. As the technologies for sequencing improved, projects to sequence the entire genome of a strain were initiated.

The first completed whole genome sequence for *C. botulinum* was published in 2007 for the BoNT/A-producing strain, ATCC 3502 (Sebaihia et al. 2007). This sequence and accompanying publication described *C. botulinum* as a saprophytic organism that "relies on its toxin to rapidly kill a wide range of prey species, and to gain access to nutrient sources, it releases a large number of extracellular enzymes to soften and destroy rotting or decayed tissues". The chromosomal location of the *bont*/A was confirmed and a small 16 kb plasmid was also identified within the strain.

Additional sequencing projects for strains representing other serotypes of *C. botulinum* were initiated. The strains selected were based upon their historical recognition within the botulinum research community and included: *C. botulinum* serotype A strains Hall, ATCC 19397, NCTC 2916, Kyoto-F and Loch Maree; serotype B strains Okra, Eklund 17B and bivalent Ba strain 657; serotype C strain Eklund, serotype D strain 1873; serotype E strains Beluga and Alaska E43; *C. butyricum* type E strains BL5262 and BL5521; and serotype F strains Langeland and F230613. Most of these strains produce a single BoNT but bivalent strains, such as strain 657, produce two toxins with the capital letter designating the predominant toxin produced (Ba). Later, a genomic sequence for BoNT/A5, associated with a wound botulism case, and genomes representing serotype C and C/D strains were released (Carter et al. 2010; Skarin et al. 2011). A list of the currently available genomes is provided in Table 1.

As *C. botulinum* sequences were released, they supported the findings of the ATCC 3502 genomic analysis of an A + T-rich (~72 %) genome and in some strains the presence of toxin-encoding plasmids or phages. The genomic data supported previous findings observed using other techniques, such as pulsed-field gel electrophoresis (PFGE) (Hielm et al. 1998; Hyytia et al. 1999; Nevas et al. 2005; Nevas et al. 2006; Franciosa et al. 2009), amplified fragment length polymorphism (AFLP) analysis (Keto-Timonen et al. 2005; Hill et al. 2007), DNA melting experiments (Nakamura et al. 1977), and multilocus sequence typing (MLST) (Jacobson et al. 2008; Macdonald et al. 2011) indicating that *C. botulinum* encompassed multiple genetically diverse species. The genomic sequences provided the opportunity to understand the variation in the species and to precisely determine the location of the *bont* genes within strains, the components, and the regions flanking their toxin gene clusters.

2 Location of the Botulinum Neurotoxin Genes Within Strains

Although much of the research to date within *C. botulinum* has focused on the toxin, its structure, its mode of action, diagnostics, and therapeutics, the new sequencing platforms provide an opportunity to understand the phylogeny within the species. Determining whether the toxin is present in a strain in similar places or at different locations can provide clues as to whether the strains expressing the same toxin were different events of horizontal gene transfer and/or represent different lineages within the species. Therefore the location of the toxin gene was examined in some of the available sequences and the results are presented below.

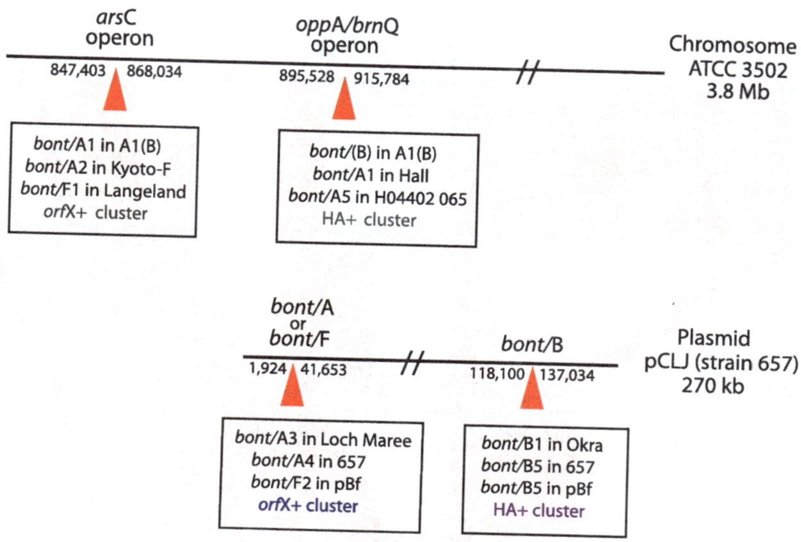

Fig. 3 Relative locations of *bont*/A, *bont*/B and *bont*/F genes in Group I strains. The chromosome of ATCC 3502 is used to illustrate the two sites (the *ars*C operon or *opp*A/*brn*Q operon) where the *bont* genes are located within other strains. The *arrows* indicate regions where toxin genes are located; associated toxin genes, strains, and cluster types are listed below each *arrow*. The plasmid pCLJ within bivalent strain 657 illustrates the two relative locations of the toxin genes and clusters in plasmids within other strains. These four regions illustrate that *bont*/A, *bont*/B, or *bont*/F are in specific locations in Group I strains (Hill et al. 2009)

2.1 Location of Botulinum Neurotoxin Genes Within Group I Strains

When several sequences of serotype A, B, and F Group I strains were available for analysis, the comparative genomics showed that the locations for the *bont*/A, *bont*/B, or *bont*/F genes within the Group I strains were not in random locations (Hill et al. 2009). The analysis showed specific locations for the *bont*/A, *bont*/B, and *bont*/F genes as shown in Fig. 3. Using the chromosome of ATCC 3502 as an example representing a Group I strain, Fig. 3 illustrates that there are two chromosomal locations for the *bont* genes in different strains.

The first location within the chromosome is referred to as the "*ars*C operon" due to its presence adjacent to the arsenate reductase gene cluster. This is the location for the *bont*/A1 gene in the A1(B) strain, NCTC 2916 (based upon the alignment of the multiple contigs representing this strain). It is also the location of the *bont*/A2 within the Kyoto-F strain and the location of the *bont*/F within the Langeland strain. This single location is the site of three different *bont* genes in the Group I strains (Hill et al. 2009).

Downstream from the *ars*C operon is the *opp*A/*brn*Q operon, which is the chromosomal location for the *bont*/A1 gene in ATCC 3502 and in two other fully

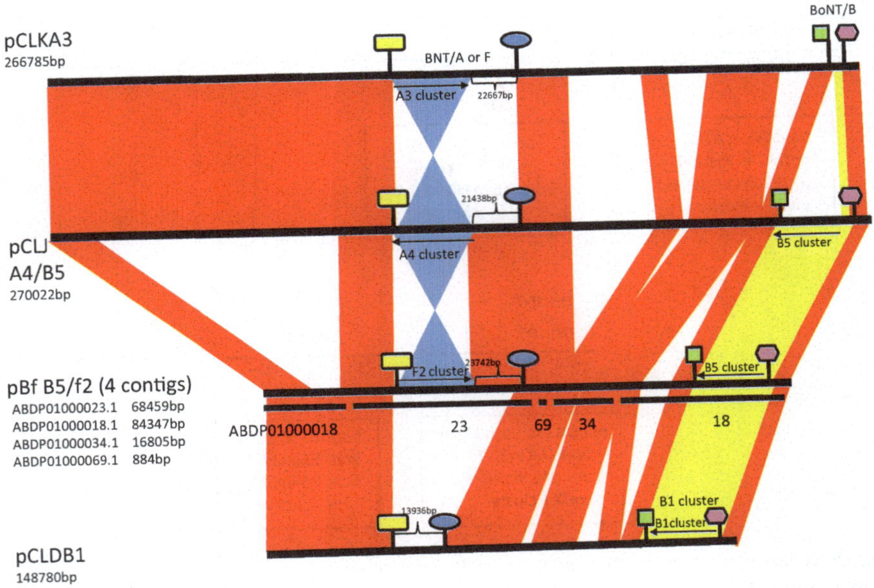

Fig. 4 Plasmid synteny and *bont* gene locations in four Group I strains. Three fully sequenced plasmids (pCLK in Loch Maree strain, pCLJ in bivalent strain 657, and pCLD in Okra) are compared to four contigs of the Bf strain. Regions of homology are shown in red, toxin gene regions containing *bont*/A or *bont*/F are shown in *blue* and *bont*/B gene is shown in *yellow*. The symbols are regions that bracket the toxin gene clusters. The comparison using sequence data show the synteny among the plasmids and the conserved location of the *bont*/A, *bont*/B and *bont*/F genes within the plasmids (Hill et al. 2009)

sequenced *bont*/A1 strains, ATCC 19397 and Hall. It is also the location of the silent *bont*/(B) within the A1(B) strain, NCTC 2916 (based upon the alignment of the multiple contigs representing this strain) and the location of *bont*/A5 in strain in H04402 065 (Carter et al. 2011). The *bont/*(B) gene does not produce toxin due to a mutation that introduces a premature stop codon at amino acid 128 (Hutson et al. 1996).

In addition to the chromosomal location of the *bont* genes, plasmids that contained *bont* genes were discovered in multiple strains. The existence of *bont* genes within plasmids had also been identified by PFGE studies and curing experiments (Eklund et al. 1988; Zhou et al. 1995; Marshall et al. 2007). The sequence data supported these findings and provided the exact size of the plasmids ranging from 148 to 270 kb. In Fig. 3 pCLJ and pBf within bivalent strains are used as examples to show the conserved locations of the *bont*/B and *bont*/A or *bont*/F, respectively, within the Group I toxin-encoding plasmids (Hill et al. 2009; Smith et al. 2007).

Comparisons of the four plasmid sequences that contain *bont* genes in the sequenced Group I strains show that the plasmids themselves share synteny as shown in red in Fig. 4. The location of the *bont* genes within the plasmid is indicated in

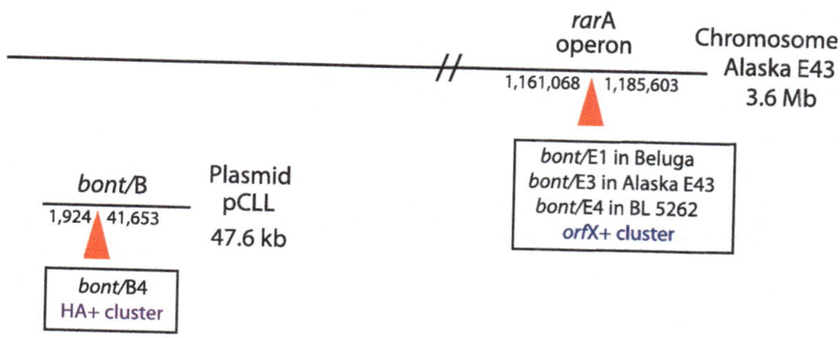

Fig. 5 Relative locations of the *bont* genes in the chromosome or plasmid in Group II and Group VI (BoNT/E-producing *C. butyricum*) strains. The chromosome of *C. botulinum* serotype E Alaska E43 strain is used to show the relative locations of the *bont*/E in Alaska E43 and its placement in the *rar*A operon in Group VI, *C. butyricum* BL 5262 strain. The 47.6 kb plasmid present in the Eklund 17B strain indicates the location of the *bont*/B4 gene (Hill et al. 2009)

either blue (*bont*/A3, *bont*/A4, or *bont*/F2) or yellow (*bont*/B1 or *bont*/B5) and regions of insertions or deletions are indicated in white. The figure shows that the plasmids appear to be dynamic with regions of insertions and deletions but do share synteny and have specific locations for either the *bont*/A, *bont*/B, or *bont*/F.

The sequence comparisons among the Group I strains identified both chromosomal and plasmid locations for the *bont* genes (A, B or F). Using PFGE, Franciosa et al. examined the location of the *bont*/B in 60 serotype B-producing Group I and Group II strains and found about half of the strains contained *bont*/B genes within a plasmid (Franciosa et al. 2009).

2.2 *Location of Botulinum Neurotoxin Genes Within Group II and Group VI Strains*

The Group II non-proteolytic strains include the *C. botulinum* strains that produce BoNT/B4 and strains that produce BoNT/E. No genomic sequences are yet available representing a type F-producing strain in the Group II *C. botulinum*. Analysis of the genomic sequences in the Group II strains shows that like the Group I strains there are plasmid and chromosomal locations for the *bont*/B and *bont*/E genes. As shown in Fig. 5, a small 47.6 kb plasmid contains the *bont*/B4 gene within the Eklund 17B strain. This plasmid does not share homology with the 16 kb or larger plasmids within the Group I strains. The location of the *bont*/E within the chromosomes of Group II strains Beluga and Alaska E43 appears to be conserved as shown in Fig. 5.

The genomic sequences in the serotype E strains revealed the association of the *bont*/E with an intact transposon-associated resolvase, called *rar*A. In these strains, the *rar*A gene appears to find a homolog (gene similar in sequence), split the

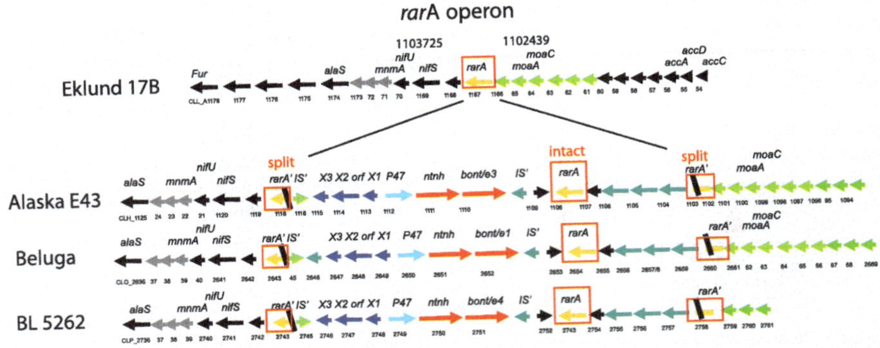

Fig. 6 Transposon-associated insertion of the *bont*/E gene between a split *rar*A gene. A portion of the chromosome of the Group II Eklund 17B strain shows the presence of an intact *rar*A operon. In serotype E-producing strains Alaska E43 and Beluga a homolog of the *rar*A is split and the toxin gene cluster containing *bont*/E and an intact *rar*A gene is inserted. BoNT/E-producing *C. butyricum* strain BL5262 at the base of the figure illustrates the same mechanism of insertion within a Group VI strain (Hill et al. 2009)

homolog, and insert the intact *rar*A gene with associated DNA (that contains the *bont*/E gene) within the split homolog (Hill et al. 2009). This results in an insertion event that contains the following: a 5' partial *rar*A, inserted DNA that includes the *bont*/E gene, an intact *rar*A gene, other associated DNA, and a 3' partial *rar*A gene. This insertion event is diagramed in Fig. 6. Additional serotype E strains were analyzed by PCR amplification and fragment sequencing to confirm the presence of both a split and an intact *rar*A gene in 41 other strains (Macdonald et al. 2011).

When the Group VI BoNT/E-producing *C. butyricum* strain BL5262 was sequenced, the same type of insertion event was observed. However the sequence of the homolog that was split differed in the two species. That is, the *rar*A homologs in the two *C. botulinum* strains were very similar in sequence, but differed from the homolog in *C. butyricum* BL5262. Because the *rar*A homologs in the two species had differing sequences, the *bont*/E insertion event in the *C. butyricum* Group VI strain would indicate that it was a separate insertion event from the Beluga and Alaska E43 *C. botulinum* strains.

2.3 Location of Botulinum Neurotoxin Genes Within Group III Strains

C. botulinum Group III strains have historically been associated with the presence of the *bont*/C or *bont*/D or chimeric *bont*/CD or *bont*/DC genes within bacteriophages. The BoNT-encoding prophage propagates within the bacterium as a large extra-chromosomal plasmid. Recently, several Group III genomes and their associated plasmids were sequenced (Skarin et al. 2011). The BoNT/C-producing Stockholm strain was sequenced and the location of the *bont*/C gene was identified within a 186 kb phage. In addition, two BoNT/CD isolates from poultry farms were sequenced. The toxin gene in one isolate, 08-BKT015925 containing five plasmids, was located on the 203 kb prophage/plasmid. The other poultry farm isolate was sequenced and had lost the BoNT prophage plasmid during cultivation.

Within the BoNT/D-producing strain 1873, genomic contigs ACSJ01000012-19 encode for phage D-1873 and the genes coding for BoNT/D are located within the phage sequences. The strain 1873 contains two additional plasmids with the larger plasmid encoding for toxin C2, which is not a neurotoxin (Sakaguchi et al. 2009).

Comparisons among the Group III strains indicate that the chromosomes of these strains are conserved, and share synteny with the non-toxic *C. novyi.* However, the phages and plasmids show a great deal of diversity (Skarin et al. 2011; Sakaguchi et al. 2005), with evidence of significant horizontal gene transfer (Skarin and Segerman 2011).

2.4 Location of Botulinum Neurotoxin Genes Within Group IV and Group V Strains

Currently, no whole genomic sequences of Group V *C. baratii* BoNT/F-producing strains or Group IV BoNT/G-producing strains have been sequenced. Group V strains are associated with infant and adult toxicoinfections (Gupta et al. 2005). Group IV strains have been isolated on rare occasions from environmental samples in Argentina (Gimenez and Ciccarelli 1970a) and Switzerland (Sonnabend et al. 1987), and from clinical (autopsy) specimens in Switzerland (Sonnabend et al. 1981). The presence of *bont*/G within large plasmids has been documented (Eklund et al. 1988; Zhou et al. 1995), but the plasmid sequences and the specific location of the *bont*/G genes within these plasmids are unknown.

3 Composition of the *HA*+ and *orfX*+ Toxin Gene Clusters

The *bont* gene is preceded by a series of other genes or predicted open reading frames (orfs) that can total about 12–15 kb in length. The two types of toxin gene clusters are the *HA*+ cluster or the *orfX*+ cluster. Other terminology such as "*HA*+/*orfX*− "or the "*HA*−/*orfX*+" has also been used to describe the clusters.

Figure 7 shows the *orfX*+ cluster and/or the *HA*+ cluster in strains representing each of the seven serotypes. In general the arrangements of the toxin clusters associated with the *bont* gene appear to have a pattern: The *bont* genes within serotype B, C, D, and G—producing strains are associated with the *HA*+ cluster, while the *bont* genes within serotype E and F—producing strains are within the

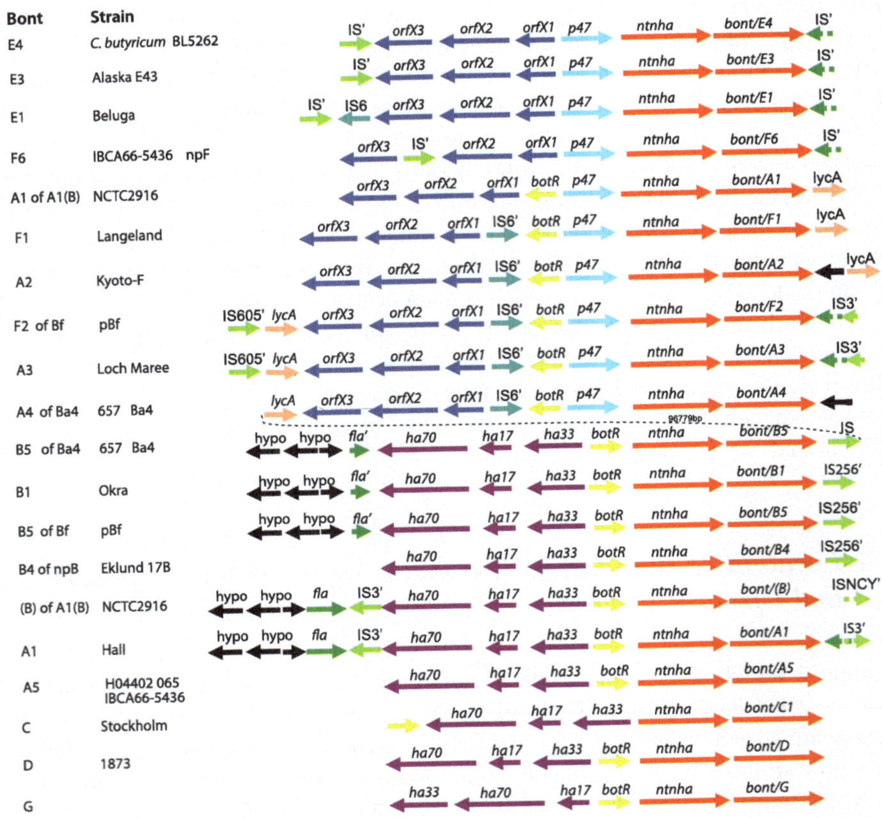

Fig. 7 The variation in the composition of the two types of toxin gene clusters in strains representing the seven serotypes. The *upper* portion of the figure illustrates the arrangements within the *orfX*+ toxin cluster in 10 strains and the lower portion shows the arrangements within the *HA*+ toxin cluster in 10 strains. The *bont* subtype or variant and strain are noted and the flanking regions containing partial IS elements, flagellin genes, or hypothetical (hypo) proteins are indicated. The *orfX*+ toxin cluster within serotype A, E and F-producing strains consists generally of the three open reading frames (*orfX1*, *orfX2* and *orfX3*), the *botR*, *p47*, *ntnha* and *bont* genes. Serotype A, B, C, D and G—producing strains generally consist of several hemagglutinin genes (*ha70*, *ha17*, *ha33*), *botR*, *ntnha* and *bont* genes [adapted from (Hill et al. 2009)]

orfX+ cluster. The bivalent strains B5a4 657 and B5f2 contain both clusters within the bacteria (the *bont*/B5 within the *HA*+ cluster and either the *bont*/A4 or *bont*/F2 within the *orfX*+ cluster).

Figure 7 shows 10 examples of the *orfX*+ toxin cluster in different strains and indicates there is some variation but *orfX*+ generally contains the three predicted open reading frames (*orfX1*, *orfX2* and *orfX3*), the *p47*, and the non-toxic non-hemagglutinin (*ntnha*) preceding the *bont* gene. The *orfX*+ toxin cluster may or may not contain the *botR* gene. The 10 examples of the *HA*+ cluster can also vary but generally consists of three hemagglutinin genes differing in sizes (*ha70, ha17, ha33*), *botR* gene and the *ntnha* gene that precedes the *bont* gene.

Variations of the composition of the toxin cluster have been identified in some strains. For example, within the toxin cluster that contains the *bont*/F6 of non-proteolytic *C. botulinum*, the cluster lacks the *botR* gene (as do the serotype E clusters) and contains an intergenic partial IS element between *orfX2* and *orfX3* (Dover et al. 2011). The toxin cluster containing *bont*/A5 has a short nucleotide deletion between the *ha33* and *botR* genes, and downstream of the *bont*/A5 in two different strains (H04402 065 and IBCA94-0214) a partial *bont*/B gene is present (Carter et al. 2011; Dover et al. 2009).

In contrast to the six other serotypes, only serotype A-producing strains have their *bont* gene within either of the two toxin clusters, the *HA*+ cluster or the *orfX*+. This is also illustrated in Fig. 3 where the location and type of toxin gene cluster is provided. The genes from the majority of *bont*/A1 strains are located within the *HA*+ toxin cluster, yet the *bont*/A1 in A1(B) strains as well as the *bont*/A2, *bont*/A3, and *bont*/A4 are within the *orfX*+ cluster. The *bont*/A5 is within the HA+ cluster. When the sequence of the *ntnha* gene was examined in the HA+ *bont*/A1 strains, a recombination event was discovered that placed the C-terminal half of the *ntnha-A* gene and the *bont*/A1 gene within the toxin gene cluster usually associated with *bont*/B strains (Hill et al. 2009; East et al. 1996).

4 Botulinum Neurotoxin Diversity

The time-line presented in Fig. 1 illustrates the discovery of the different serotypes A-G followed by the bivalent strains that express two toxins, then the presence of the toxin in different species, BoNT/F-producing *C. baratii* strains and BoNT/E-producing *C. butyricum* strains. When the seven neurotoxins, A-G, are analyzed in pairwise comparisons, they can range in difference from ~25 to 45 % (nucleotide) or ~37 to 70 % (amino acid) from each other. With more sequences being generated by PCR amplification and neurotoxin sequencing, no new serotypes have been identified but variation within the toxin of the existing serotypes has been described. Below are examples of the variation between serotypes and the subtypes or variants within a serotype.

4.1 Definition of a Subtype

Within a serotype, the differences in the amino acid sequences of the BoNTs can be minor (0.9 %) (Macdonald et al. 2011) or can be quite large (Raphael et al. 2010) (36 %). Some researchers have adopted a definition for subtype as being a difference of greater than 2.5 % in the amino acid sequence compared with other strains within that serotype (Carter et al. 2009; Smith et al. 2005; Arndt et al.

Table 2 Nucleotide and amino acid identity comparisons of the *bont* genes or proteins representing the seven serotypes

	Percent nucleotide identity						
	A1	B1	C1	D	E3	F1	G
A1	-----	60.2%	55.3%	57.8%	60.7%	62.1%	59.6%
B1	37.5%	-----	56.6%	58.0%	59.6%	59.5%	71.8%
C1	30.4%	30.8%	-----	68.3%	56.7%	56.8%	56.3%
D	31.2%	32.6%	51.4%	-----	57.7%	57.8%	57.8%
E3	37.7%	35.9%	30.9%	31.6%	-----	75.5%	59.0%
F1	38.7%	37.4%	30.9%	32.7%	62.8%	-----	59.0%
G	38.0%	57.1%	32.5%	34.0%	36.4%	36.8%	-----
	Percent amino acid identity						

Nucleotide identities are in *blue*; amino acid identities in *red*. Maximum and minimum identities are highlighted in *gray*. The GenBank accession numbers for the A-G strains used in the comparisons are provided in Table 3

2006), while others have based subtype definitions on genetic clade differences, where strains that cluster to a unique clade are considered to be separate subtypes (Hill et al. 2007; Raphael et al. 2010; Chen et al. 2007). Many of these newer subtypes have been delineated solely on the basis of genetic differences, so that the term "subtype/genetic variant" may be more appropriate than the use of "subtype", which has historically been used to designate immunological or functional differences among these strains.

4.2 Botulinum Neurotoxin Serotype A-G and Subtype/Variant Comparisons

There are seven serologically distinct botulinum neurotoxins, designated A-G. When the toxin types are compared, their nucleotide identities range from 55.3 % identity (*bont*/A1 compared to *bont*/C1) to 75.5 % identity (*bont*/E3 compared to *bont*/F1), as shown in Table 2. These nucleotide identities produce amino acid identities of 30.4–62.8 %, respectively.

BoNT variation within serotypes has been recognized, but the level of diversity became more apparent with the advent of PCR and DNA sequencing of *bont* genes, and later, whole genomes. Currently, 31 *bont* gene subtypes/variants have been identified and published within BoNT/A-F. Subtypes/variants are typically designated using an alpha-numeric code, such as A1, A2, A3, A4, and A5 within serotype A. Table 3 lists the representative strains for the different subtypes/variants and indicates the percent nucleotide identities in pairwise comparisons.

Table 3 Nucleotide and amino acid percent identities between subtypes/variants within serotypes

	Strain	Percent nucleotide identity								Accession #
		A1	A2	A3	A4	A5				
A1	Hall	-----	94.6%	92.0%	94.3%	98.6%				AF488749
A2	Kyoto-F	89.9%	-----	96.4%	93.6%	94.9%				CP001581
A3	Loch Maree	84.6%	93.0%	-----	91.6%	92.3%				DQ185900
A4	Ba 657	89.3%	88.3%	84.4%	-----	93.7%				CP001082
A5	H04402 065	97.1%	90.3%	85.0%	87.4%	-----				EU679004
		B1	B2	B3	B4	B5	B6	B7		
B1	okra	-----	97.5%	97.9%	96.2%	98.0%	97.9%	97.5%		CP000940
B2	213B	95.6%	-----	99.1%	96.2%	97.3%	99.1%	97.7%		EF028395
B3	CDC 795	96.0%	98.4%	-----	96.4%	97.7%	98.9%	97.8%		EF028400
B4	Eklund 17B [1]	93.2%	93.9%	93.7%	-----	96.0%	96.0%	96.3%		EF051570
B5	Ba 657 [2]	96.1%	95.0%	95.4%	92.7%	-----	97.3%	97.1%		EF033130
B6	Osaka05	96.1%	98.4%	98.1%	93.1%	95.4%	-----	97.3%		AB302852
B7	Bac-04-07755	94.7%	95.6%	95.7%	93.6%	94.1%	95.1%	-----		JQ354985
		C1	C/D	D	D/C					
C1	Stockholm	-----	85.3%	68.3%	76.2%					D90210
C/D	003-9	76.0%	-----	80.3%	68.8%					AB200360
D	1873	51.2%	68.7%	-----	85.9%					AB012112
D/C	5995	64.7%	51.9%	76.5%	-----					EF378947
		E1	E2	E3	E4	E5	E6	E7	E8	
E1	Beluga [1]	-----	99.4%	99.3%	98.5%	98.3%	98.5%	99.1%	97.9%	X62089
E2	CDC 5247 [1]	99.0%	-----	98.7%	98.3%	97.9%	98.2%	98.5%	98.5%	EF028404
E3	Alaska E43 [1]	98.2%	97.4%	-----	97.9%	97.6%	98.0%	98.8%	97.6%	EF028403
E4	BL5262 [1,3]	97.3%	97.0%	95.6%	-----	97.2%	98.3%	98.1%	98.0%	ACOM01000005
E5	LCL155 [1,3]	96.8%	96.3%	95.1%	95.0%	-----	97.2%	97.3%	96.9%	AY327861
E6	K35 [1]	96.9%	96.7%	95.8%	96.8%	94.7%	-----	98.2%	98.3%	AM695752
E7	IBCA97E-0192 [1]	97.8%	97.0%	97.3%	96.2%	94.7%	96.2%	-----	98.8%	JN695729
E8	Bac-02-06430 [1]	96.2%	97.0%	95.6%	96.1%	94.0%	96.6%	98.2%	-----	JN695730
		F1	F2	F3	F4	F5	F6	F7		
F1	Langeland	-----	91.2%	91.5%	96.0%	81.5%	93.1%	82.4%		X70821
F2	Bf An436 [2]	83.4%	-----	98.9%	91.9%	84.6%	95.2%	80.2%		EF028397
F3	CDC 54086	83.9%	97.0%	-----	92.1%	84.5%	95.1%	79.6%		GU213218
F4	CDC 54076 [2]	92.2%	83.5%	83.8%	-----	81.9%	93.4%	81.5%		GU213213
F5	CDC 54096 [2]	69.8%	74.0%	74.0%	69.4%	-----	83.8%	75.9%		GU213225
F6	Eklund 202F [1]	87.4%	89.8%	89.8%	86.9%	73.6%	-----	80.2%		M92906
F7	Sullivan [4]	73.7%	68.6%	68.9%	71.9%	63.8%	69.8%	-----		HM746656
		Percent amino acid identity								

Nucleotide identities are in *blue*; amino acid identities in *red*. Strains, subtype/variant designation, and their GenBank accession numbers are listed within serotypes A-F. Variants within a serotype are shaded to highlight the pairwise comparisons that show the least or the greatest similarity

[1] nonproteolytic strains

[2] bivalent toxin producing strains

[3] *C. butyricum* strains

[4] *C. baratii* strain

4.3 Mutation and Recombination Events

Sequence differences can be compared at the nucleotide or at the amino acid level. Some nucleotide changes do not affect the encoded amino acid and therefore can be silent, while others may result in dramatic differences. The "silent" *bont/*(B) gene in strains is the result of a significant mutation in the *bont/*B5 gene. In these strains a single nucleotide mutation encodes for a premature stop codon, so that the

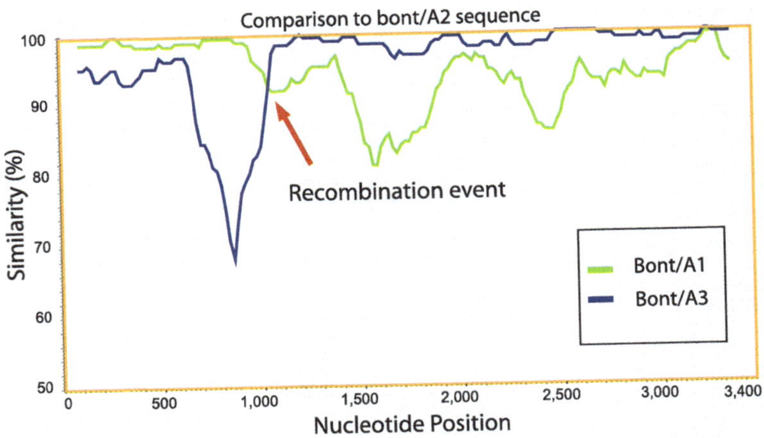

Fig. 8 Similarity plot comparing sequences of *bont*/A1 and *bont*/A3 to *bont*/A2. This plot illustrates that the *bont*/A2 is approximately 99 % identical to *bont*/A1 (*green line*) from nucleotides 1–1,146 then approximately 99 % identical to *bont*/A3 (*blue line*) through nucleotides 1,147–3,450

full-length toxin gene is present but the toxin protein is not expressed; the presence of the silent gene is noted as *bont*/(B) (Hutson et al. 1996).

Another source of genetic variation can be the result of recombination events within the *bont* gene itself or the *ntnha* gene that precedes the *bont*. Two examples of recombination within the *bont* gene are observed in the *bont*/A2 gene and the *bont*/E6 gene. These variants share regions of toxin genes that are found in other strains. For example, as the similarity plot within Fig. 8 shows, the *bont*/A2 gene has ~99 % identity with the *bont*/A1 (green line) in the first part of the gene and averages ~99 % identity with the *bont*/A3 gene (blue line) in the latter portion of the gene. In another example, regions within the *bont*/E6 in *C. botulinum* are similar to the *bont*/E4 within BoNT/E-producing *C. butyricum* strains (Chen et al. 2007).

Another example of a significant recombination event was previously described in Sect. 3, where recombination within the *ntnha* gene resulted in a mosaic *ntnha-BA* gene followed by a *bont*/A1 gene. This event placed the *bont*/A1 within the HA+ cluster. These examples illustrate how the mutations and recombination events affect toxin expression, the sequence variation within the toxin, and its presence within a toxin gene cluster.

5 Summary

The initial isolation and study of the anaerobic spore-forming toxin-producing bacteria now known as *C. botulinum* that confirmed the cause for botulism was a bacterial toxin. The characterization of the species continued with additional toxin types being discovered and the use of biochemical and microbiological methods to characterize the varied host bacteria. Six Group designations were defined for the bacteria that contained any one or two of seven serologically different toxin types. The discordant phylogeny between the toxin types and the host bacteria indicates that horizontal gene transfer of the toxin is responsible for the toxin's presence within the different bacteria.

The characterization of this species and its toxins has encompassed knowledge and skills in many disciplines. The names of many excellent historic researchers are associated with their strains and provide a continuity and reminder of the insight, dedication, and microbiological and biochemical expertise that they possessed to understand this organism. The modern molecular tools support their original observations and although the data add information they also add a deep respect for the discoveries made prior to the advent of molecular methods. The methods of microbiology and molecular and structural biology now combine to further our knowledge of this unique species.

The genomic sequences revealed what the previous researchers had already discovered—the toxin genes are within the chromosome, plasmid, or phage. The molecular analysis has added more information about the locations of the toxin genes, the components of the toxin gene cluster, evidence of insertion and recombination events that have contributed to the great genetic variation observed within the toxin and its host bacteria, and provide a challenge for diagnostics and treatment of botulism.

Acknowledgments Funding for this research was provided by the Department of Homeland Security Science and Technology Directorate contract HSHQDC-10-C-00139 and NIAID IAA 120.B18. Los Alamos National Laboratory strongly supports academic freedom and a researcher's right to publish; however the Laboratory as an institution does not necessarily endorse the viewpoint of a publication or guarantee its technical correctness. Opinions, interpretations, conclusions and recommendations are those of the authors and not necessarily endorsed by the U.S. Army, the National Institute of Allergy and Infectious Diseases, or the National Institutes of Health.

References

Arndt JW, Jacobson MJ, Abola EE, Forsyth CM, Tepp WH, Marks JD, Johnson EA, Stevens RC (2006) A structural perspective of the sequence variability within botulinum neurotoxin subtypes A1–A4. J Mol Biol 362(4):733–742

Aureli P, Fenicia L, Pasolini B, Gianfranceschi M, McCroskey LM, Hatheway CL (1986) Two cases of type E infant botulism caused by neurotoxigenic *Clostridium butyricum* in Italy. J Infect Dis 154(2):207–211

Bengtson IA (1922) Preliminary note on a toxin-producing anaerobe isolated from larvae of the green fly. Pub Health Rpts (US) 37:2252–2256

Burke GS (1919) Notes on *Bacillus botulinus*. J Bacteriol 4:555–565

Carter AT, Paul CJ, Mason DR, Twine SM, Alston MJ, Logan SM, Austin JW, Peck MW (2009) Independent evolution of neurotoxin and flagellar genetic loci in proteolytic *Clostridium botulinum*. BMC Genomics 10:115

Carter AT, Mason DR, Grant KA, Franciosa G, Aureli P, Peck MW (2010) Further characterization of proteolytic *Clostridium botulinum* type A5 reveals that neurotoxin formation is unaffected by loss of the cntR (botR) promoter sigma factor binding site. J Clin Microbiol 48 (3):1012–1013. doi:JCM.01774-09 [pii]

Carter AT, Pearson BM, Crossman LC, Drou N, Heavens D, Baker D, Febrer M, Caccamo M, Grant KA, Peck MW (2011) Complete genome sequence of proteolytic *Clostridium botulinum* type A5 (B3') strain H04402 065. J Bacteriol 139(9):2351–2352

Chen Y, Korkeala H, Aarnikunnas J, Lindstrom M (2007) Sequencing the botulinum neurotoxin gene and related genes in *Clostridium botulinum* type E strains reveals orfx3 and a novel type E neurotoxin subtype. J Bacteriol 189(23):8643–8650

Dover N, Barash JR, Arnon SS (2009) Novel *Clostridium botulinum* toxin gene arrangement with subtype A5 and partial subtype B3 botulinum neurotoxin genes. J Clin Microbiol 47(7): 2349–2350

Dover N, Barash JR, Hill KK, Detter JC, Arnon SS (2011) Novel structural elements within the non-proteolytic *Clostridium botulinum* type F toxin gene cluster. Appl Environ Microbiol 77(5):1904–1906

East AK, Bhandari M, Stacey JM, Campbell KD, Collins MD (1996) Organization and phylogenetic interrelationships of genes encoding components of the botulinum toxin complex in proteolytic *Clostridium botulinum* types A, B, and F: evidence of chimeric sequences in the gene encoding the nontoxic nonhemagglutinin component. Int J Syst Bacteriol 46(4):1105–1112

Eklund MW, Poysky FT, Mseitif LM, Strom MS (1988) Evidence for plasmid-mediated toxin and bacteriocin production in *Clostridium botulinum* type G. Appl Environ Microbiol 54(6):1405–1408

Franciosa G, Ferreira JL, Hatheway CL (1994) Detection of type A, B, and E botulism neurotoxin genes in *Clostridium botulinum* and other Clostridium species by PCR: evidence of unexpressed type B toxin genes in type A toxigenic organisms. J Clin Microbiol 32(8): 1911–1917

Franciosa G, Maugliani A, Scalfaro C, Aureli P (2009) Evidence that plasmid-borne botulinum neurotoxin type B genes are widespread among *Clostridium botulinum* serotype B strains. PLoS ONE 4(3):e4829

Gimenez DF, Ciccarelli AS (1970a) Another type of *Clostridium botulinum*. Zentralbl Bakteriol Orig 215(2):221–224

Gimenez DF, Ciccarelli AS (1970b) Studies on strain 84 of *Clostridium botulinum*. Zentralbl Bakteriol Orig 215(2):212–220

Gupta A, Sumner CJ, Castor M, Maslanka S, Sobel J (2005) Adult botulism type F in the United States, 1981–2002. Neurology 65(11):1694–1700

Hall JD, McCroskey LM, Pincomb BJ, Hatheway CL (1985) Isolation of an organism resembling *Clostridium baratii* which produces type F botulinal toxin from an infant with botulism. J Clin Microbiol 21(4):654–655

Hatheway CL (1988) Botulism. In: Balows A, Hausler WHJ, Ohashi M, Turano A (eds) Laboratory diagnosis of infectious diseases: principles and practice, vol 1. Springer, New York

Hazen EL (1937) A strain of *B. botulinus* not classified as type A, B, or C. J Infect Dis 60(3): 260–264

Hielm S, Bjorkroth J, Hyytia E, Korkeala H (1998) Genomic analysis of *Clostridium botulinum* group II by pulsed-field gel electrophoresis. Appl Environ Microbiol 64(2):703–708

Hill KK, Smith TJ, Helma CH, Ticknor LO, Foley BT, Svensson RT, Brown JL, Johnson EA, Smith LA, Okinaka RT, Jackson PJ, Marks JD (2007) Genetic diversity among botulinum neurotoxin-producing clostridial strains. J Bacteriol 189(3):818–832

Hill KK, Xie G, Foley BT, Smith TJ, Munk AC, Bruce D, Smith LA, Brettin TS, Detter JC (2009) Recombination and insertion events involving the botulinum neurotoxin complex genes in *Clostridium botulinum* types A, B, E and F and *Clostridium butyricum* type E strains. BMC Biol 7:66

Hutson RA, Thompson DE, Collins MD (1993a) Genetic interrelationships of saccharolytic *Clostridium botulinum* types B, E and F and related clostridia as revealed by small-subunit rRNA gene sequences. FEMS Microbiol Lett 108(1):103–110

Hutson RA, Thompson DE, Lawson PA, Schocken-Itturino RP, Bottger EC, Collins MD (1993b) Genetic interrelationships of proteolytic *Clostridium botulinum* types A, B, and F and other members of the *Clostridium botulinum* complex as revealed by small-subunit rRNA gene sequences. Antonie Van Leeuwenhoek 64(3–4):273–283

Hutson RA, Zhou Y, Collins MD, Johnson EA, Hatheway CL, Sugiyama H (1996) Genetic characterization of *Clostridium botulinum* type A containing silent type B neurotoxin gene sequences. J Biol Chem 271(18):10786–10792

Hyytia E, Hielm S, Bjorkroth J, Korkeala H (1999) Biodiversity of *Clostridium botulinum* type E strains isolated from fish and fishery products. Appl Environ Microbiol 65(5):2057–2064

Jacobson MJ, Lin G, Whittam TS, Johnson EA (2008) Phylogenetic analysis of *Clostridium botulinum* type A by multi-locus sequence typing. Microbiology 154(Pt 8):2408–2415

Keto-Timonen R, Nevas M, Korkeala H (2005) Efficient DNA fingerprinting of *Clostridium botulinum* types A, B, E, and F by amplified fragment length polymorphism analysis. Appl Environ Microbiol 71(3):1148–1154

Leuchs J (1910) Beitraege zur kenntnis des toxins und antitoxins des *Bacillus botulinus*. Z Hyg Infectionskr 76:55–84

Macdonald TE, Helma CH, Shou Y, Valdez YE, Ticknor LO, Foley BT, Davis SW, Hannett GE, Kelly-Cirino CD, Barash JR, Arnon SS, Lindstrom M, Korkeala H, Smith LA, Smith TJ, Hill KK (2011) Analysis of *Clostridium botulinum* serotype E strains by using multilocus sequence typing, amplified fragment length polymorphism, variable-number tandem-repeat analysis, and botulinum neurotoxin gene sequencing. Appl Environ Microbiol 77(24): 8625–8634

Marshall KM, Bradshaw M, Pellett S, Johnson EA (2007) Plasmid encoded neurotoxin genes in *Clostridium botulinum* serotype A subtypes. Biochem Biophys Res Commun 361(1):49–54

Meyer KF, Gunnison JB (1929) South African cultures of *Cl. botulinum* and *parabotulinum*. XXXVII (with a description of *Cl. botulinum* type D, N. Sp.) J Inf Dis 45:106–118

Moller V, Scheibel I (1960) Preliminary report on the isolation of an apparently new type of *Cl. botulinum*. Acta Pathol Microbiol Scand 48:80

Nakamura S, Okado I, Nakashio S, Nishida S (1977) Clostridium sporogenes isolates and their relationship to *C. botulinum* based on deoxyribonucleic acid reassociation. J Gen Microbiol 100(2):395–401

Nevas M, Lindstrom M, Hielm S, Bjorkroth KJ, Peck MW, Korkeala H (2005) Diversity of proteolytic *Clostridium botulinum* strains, determined by a pulsed-field gel electrophoresis approach. Appl Environ Microbiol 71(3):1311–1317

Nevas M, Lindstrom M, Horman A, Keto-Timonen R, Korkeala H (2006) Contamination routes of *Clostridium botulinum* in the honey production environment. Environ Microbiol 8(6): 1085–1094

Peck MW, Stringer SC, Carter AT (2011) *Clostridium botulinum* in the post-genomic era. Food Microbiol 28(2):183–191

Popoff M-R (1995) Ecology of neurotoxigenic strains of clostridia. In: Montecucco C (ed) Clostridial neurotoxins—the molecular pathogenesis of tetanus and botulism. Current topics in microbiology and immunology, vol 195. Springer, Berlin, pp 1–29

Raphael BH, Choudoir MJ, Luquez C, Fernandez R, Maslanka SE (2010) Sequence diversity of genes encoding botulinum neurotoxin type F. Appl Environ Microbiol 76(14):4805–4812

Sakaguchi Y, Hayashi T, Kurokawa K, Nakayama K, Oshima K, Fujinaga Y, Ohnishi M, Ohtsubo E, Hattori M, Oguma K (2005) The genome sequence of *Clostridium botulinum* type C neurotoxin-converting phage and the molecular mechanisms of unstable lysogeny. Proc Natl Acad Sci USA 102(48):17472–17477

Sakaguchi Y, Hayashi T, Yamamoto Y, Nakayama K, Zhang K, Ma S, Arimitsu H, Oguma K (2009) Molecular analysis of an extrachromosomal element containing the C2 toxin gene discovered in *Clostridium botulinum* type C. J Bacteriol 191(10):3282–3291

Sebaihia M, Peck MW, Minton NP, Thomson NR, Holden MT, Mitchell WJ, Carter AT, Bentley SD, Mason DR, Crossman L, Paul CJ, Ivens A, Wells-Bennik MH, Davis IJ, Cerdeno-Tarraga AM, Churcher C, Quail MA, Chillingworth T, Feltwell T, Fraser A, Goodhead I, Hance Z, Jagels K, Larke N, Maddison M, Moule S, Mungall K, Norbertczak H, Rabbinowitsch E, Sanders M, Simmonds M, White B, Whithead S, Parkhill J (2007) Genome sequence of a proteolytic (Group I) *Clostridium botulinum* strain Hall A and comparative analysis of the clostridial genomes. Genome Res 17(7):1082–1092

Seddon HR (1922) Bulbar paralysis in cattle due to the action of a toxicogenic bacillus, with a discussion on the relationship of the condition to the forage poisoning (botulism). Jour Compar Path Ther 35:147–190

Skarin H, Segerman B (2011) Horizontal gene transfer of toxin genes in *Clostridium botulinum*: Involvement of mobile elements and plasmids. Mob Genet Elem 1(3):213–215

Skarin H, Hafstrom T, Westerberg J, Segerman B (2011) *Clostridium botulinum* group III: a group with dual identity shaped by plasmids, phages and mobile elements. BMC Genomics 12:185

Smith LD, Sugiyama H (1988) Botulinum: the organism, its toxins, the disease. In:Smith LD, Sugiyama H (eds) Charles Thomas, Springfield, IL

Smith TJ, Lou J, Geren IN, Forsyth CM, Tsai R, Laporte SL, Tepp WH, Bradshaw M, Johnson EA, Smith LA, Marks JD (2005) Sequence variation within botulinum neurotoxin serotypes impacts antibody binding and neutralization. Infect Immun 73(9):5450–5457

Smith TJ, Hill KK, Foley BT, Detter JC, Munk AC, Bruce DC, Doggett NA, Smith LA, Marks JD, Xie G, Brettin TS (2007) Analysis of the neurotoxin complex genes in Clostridium botulinum A1–A4 and B1 strains: BoNT/A3,/Ba4 and/B1 clusters are located within plasmids. PLoS One 2(12):e1271

Sonnabend O, Sonnabend W, Heinzle R, Sigrist T, Dirnhofer R, Krech U (1981) Isolation of *Clostridium botulinum* type G and identification of type G botulinal toxin in humans: report of five sudden unexpected deaths. J Infect Dis 143(1):22–27

Sonnabend WF, Sonnabend UP, Krech T (1987) Isolation of *Clostridium botulinum* type G from Swiss soil specimens by using sequential steps in an identification scheme. Appl Environ Microbiol 53(8):1880–1884

van Ermengem E (1897) A new anaerobic bacillus and its relation to botulism. (Originally published as "Ueber einen neuen anaëroben Bacillus und seine Beziehungen zum Botulismus" in Zeitschrift für Hygiene und Infektionskrankheiten 26:1–56, 1897.). Rev Infect Dis 1(4):701–719

Zhou Y, Sugiyama H, Nakano H, Johnson EA (1995) The genes for the Clostridium botulinum type G toxin complex are on a plasmid. Infect Immun 63(5):2087–2091

Assembly and Function of the Botulinum Neurotoxin Progenitor Complex

Shenyan Gu and Rongsheng Jin

Abstract Botulinum neurotoxins (BoNTs) are among the most poisonous substances known to man, but paradoxically, BoNT-containing medicines and cosmetics have been used with great success in the clinic. Accidental BoNT poisoning mainly occurs through oral ingestion of food contaminated with *Clostridium botulinum*. BoNTs are naturally produced in the form of progenitor toxin complexes (PTCs), which are high molecular weight (up to ~900 kDa) multiprotein complexes composed of BoNT and several non-toxic neurotoxin-associated proteins (NAPs). NAPs protect the inherently fragile BoNTs against the hostile environment of the gastrointestinal (GI) tract and help BoNTs pass through the intestinal epithelial barrier before they are released into the general circulation. These events are essential for ingested BoNTs to gain access to motoneurons, where they inhibit neurotransmitter release and cause muscle paralysis. In this review, we discuss the structural basis for assembly of NAPs and BoNT into the PTC that protects BoNT and facilitate its delivery into the bloodstream.

Keywords M-progenitor toxin complex · Neurotoxin associated protein · Non-toxic non-hemagglutinin protein · Co-crystal structure · Botulinum neurotoxin

Abbreviations

PTC	Progenitor toxin complex
NTNHA	Non-toxic non-hemagglutinin protein
HA	Hemagglutinin
NAP	Neurotoxin-associated protein

S. Gu · R. Jin (✉)
Center for Neuroscience, Aging and Stem Cell Research, Sanford-Burnham Medical Research Institute, 10901 North Torrey Pines Road, La Jolla, CA 92037, USA
e-mail: rjin@sanfordburnham.org

A. Rummel and T. Binz (eds.), *Botulinum Neurotoxins*, Current Topics in Microbiology and Immunology 364, DOI: 10.1007/978-3-642-33570-9_2,

Contents

1 Introduction 22
2 BoNT and NAPs 23
 2.1 Molecular Architecture of BoNT 23
 2.2 NAPs: The Bodyguards of BoNTs 25
3 Structure of the M-PTC 26
 3.1 A Recombinant Expression System to Dissect the PTCs 26
 3.2 NTNHA-A Forms a Tight Protective Complex with BoNT/A 27
 3.3 Architecture of the M-PTC 27
 3.4 The Structure of NTNHA-A 29
 3.5 A Novel Conformation of BoNT/A in the M-PTC 31
4 Dynamic Association of the M-PTC 33
 4.1 BoNT/A and NTNHA-A Interactions 33
 4.2 pH-Dependent Association of the PTC 34
5 HA Proteins 35
 5.1 The Structure of the HA Proteins 35
 5.2 HAs Mediate Toxin–Host Interactions 37
6 Conclusions 38
References 39

1 Introduction

Botulinum neurotoxins (BoNTs) are produced by the anaerobic, Gram-positive, spore-forming bacteria *Clostridium botulinum*. Seven serotypes of BoNTs (termed A–G) have been identified, which cause botulism, a severe neurological disease associated with a life-threatening flaccid paralysis that affects both humans and animals (Binz et al. 1990; Hatheway 1990; Montecucco and Schiavo 1995; Sugiyama 1980).

Following toxin ingestion, there are two major points at which the interaction between BoNT and host cells is pivotal. To date, most attention has been paid to the BoNT-neuron interaction. Here, BoNTs act as sequence-specific endopeptidases that cleave soluble N-ethylmaleimide sensitive factor attachment protein receptors (SNAREs), blocking the release of acetylcholine at neuromuscular junctions (NMJs) and thus paralyzing the affected muscles (Blasi et al. 1993a; Schiavo et al. 1992). However, there is a large gap in our understanding of the critical events that precede the BoNT–neuron interaction, the most important of which is how the inherently fragile BoNTs survive the harsh environment (low pH, protease-rich) of the gastrointestinal (GI) tract and cross the intestinal epithelial barrier to enter the bloodstream.

The mechanism by which this is accomplished involves the secret weapons of BoNTs; the progenitor toxin complexes (PTCs). PTCs are high molecular weight (up to ~900 kDa) multiprotein complexes produced by *C. botulinum* and are composed of BoNT and several non-toxic neurotoxin-associated proteins (NAPs) (Collins and East 1998). The NAPs comprise four clostridial proteins; a non-toxic non-hemagglutinin (NTNHA) protein and three hemagglutinin proteins (HA-17, HA-33, and HA-70). The naturally occurring minimally functional PTC (M-PTC) is composed of BoNT (~ 150 kDa) and NTNHA (~ 140 kDa). Other PTCs, with molecular masses ranging from 500 to 900 kDa, are assembled by the addition of various combinations of HAs to the M-PTC through unknown mechanisms (Ito et al. 2011; Montecucco and Schiavo 1995).

The oral toxicity of BoNTs is markedly increased when coassembled with NAPs into PTCs (Ohishi et al. 1977). This is achieved through two complementary processes. NTNHA directly interacts with BoNT and plays a major role in shielding it in the hostile GI environment. In addition, HAs interact with intestinal epithelial cells and play an active role in BoNT transport (Fujinaga et al. 1997; Fujinaga et al. 2004; Niwa et al. 2007).

Clearly, a detailed knowledge of the molecular architecture of the PTC is required to understand the strategic interactions between BoNT and the host that take place in the GI tract. This review focuses on the structure–function relationship of BoNT PTCs, including the mechanism by which NAPs protect BoNTs, the regulatory mechanisms underlying PTC assembly, and the structural basis by which NAPs participate in BoNT–host interactions.

2 BoNT and NAPs

2.1 *Molecular Architecture of BoNT*

The seven BoNTs have a high degree of primary sequence conservation, although all are antigenically distinct (Lacy and Stevens 1999). BoNT is synthesized as a single polypeptide chain of ~ 150 kDa and post-translationally nicked by an unknown protease into a ~ 50 kDa light chain (LC) and a ~ 100 kDa heavy chain (HC) (Montecucco and Schiavo 1995). The LC and HC remain covalently linked through a disulfide bond until they encounter reducing conditions in the neuronal cytosol (Montecucco and Schiavo 1995). Crystal structures have been reported for full-length BoNT/A (PDB: 3BTA), BoNT/B (PDB: 1EPW), and BoNT/E (PDB: 3FFZ) (Kumaran et al. 2009; Lacy et al. 1998; Swaminathan and Eswaramoorthy 2000). All three structures are similar in that they exhibit a modular architecture comprising three domains. The LC is a protease. The HC is composed of two domains: the N-terminal domain (H_N, also known as the translocation domain) mediates translocation of LC across the endosomal membrane, whereas the C-terminal domain (H_C) is the cell surface receptor-binding domain.

The first point of attack at the NMJ involves a highly specific interaction between BoNT and the motoneuron. A well-accepted dual-receptor model proposes that BoNTs simultaneously bind to gangliosides, a class of glycosphingolipids enriched at nerve terminals, and to specific protein receptors (Binz and Rummel 2009; Chai et al. 2006; Jin et al. 2006; Montecucco et al. 2004; Stenmark

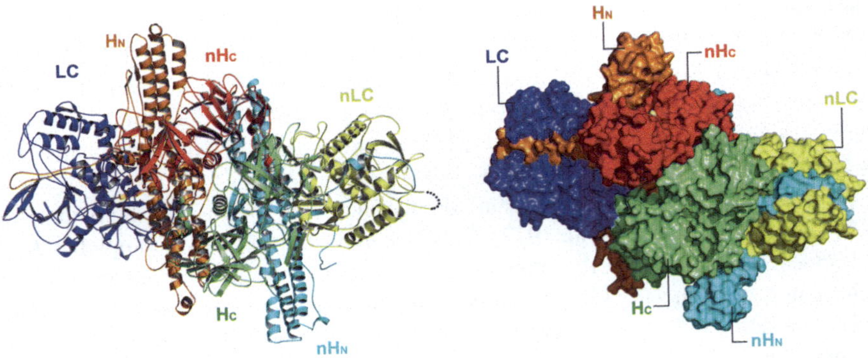

Fig. 1 The structure of the M-PTC of BoNT/A. BoNT/A domains are *blue* (LC), *orange* (H_N), and *green* (H_C). NTNHA-A domains are *yellow* (nLC), *cyan* (nH_N), and *red* (nH_C). The cartoon and surface representations of the M-PTC are shown in the *left* and *right* panels, respectively

et al. 2008; Rummel 2012). The BoNT–ganglioside interaction ensures a high toxin–cell encounter rate, while the protein receptors confer cell specificity. BoNT serotypes A, E, and F recognize synaptic vesicle protein 2 (SV2) as the protein receptor, whereas BoNT/B and G utilize synaptotagmin I and II (SytI and II) (Binz and Rummel 2009; Rummel et al. 2009).

The warhead of BoNTs is the LC, which is a Zn^{2+}-dependent endopeptidase that cleaves specific peptide bonds within the neuronal SNAREs (synaptobrevin, syntaxin, and SNAP-25) (Binz 2012). BoNT/A and E cleave SNAP-25 while serotypes B, D, F, and G cleave synaptobrevin. BoNT/C is unique in that it is able to hydrolyze both syntaxin and SNAP-25 (Blasi et al. 1993b; Foran et al. 1996; Schiavo et al. 1995; Vaidyanathan et al. 1999; Williamson et al. 1996).

Once the toxin is endocytosed, the next step is to deliver LC across the intracellular membrane to its targets in the cytosol (Schiavo et al. 2000). The H_N fragment has two long helices forming a coiled-coil (Fig. 1) and likely acts as both a channel and a transmembrane chaperone to ensure the successful transit of LC from the acidic endosomes into the cytosol (Fischer and Montal 2006, 2007a, b; Koriazova and Montal 2003). We speculate that H_C remains bound to receptors on the endosomal membrane where it brings the N-terminus of H_N in contact with the membrane prior to the H_N-facilitated translocation of LC. However, the details of the translocation process are largely unknown.

Before they can gain access to the motoneuron, BoNTs must traverse the GI tract and enter the bloodstream. Paradoxically, the highly sophisticated BoNTs are intrinsically vulnerable to the harsh physiological environment of the GI tract. For example, the free BoNT/A is readily degraded by digestive proteases such as trypsin and pepsin and is almost completely inactivated at pH 3 or less (Gu et al. 2012). In addition, in vitro assays have shown that trans-epithelial transcytosis of free BoNT/A is very inefficient compared with HA-containing PTCs (Fujinaga et al. 2009; Matsumura et al. 2008; Niwa et al. 2007). The responsibility for protecting BoNTs and facilitating their transit out of the GI tract falls to the NAPs.

2.2 NAPs: The Bodyguards of BoNTs

All serotypes of BoNT are produced by bacteria in their native form as stable and non-covalent complexes with NAPs to form PTCs. The PTC protects the toxin sufficiently to reduce the oral lethal dose by 10- to 100-fold compared with the free BoNT (Cheng et al. 2008; Ohishi et al. 1977; Sugii et al. 1977). In general, BoNT and the four NAPs (NTNHA and three HAs) are encoded in a gene cluster: the downstream segment containing *bont* and *ntnha* and the upstream segment containing the three HA genes, *ha-17, ha-33*, and *ha-70*. This gene arrangement is defined as the *ha70-ha17-ha33-botR-ntnha-bont* cluster, or the *ha* cluster (Hill et al. 2009; Hill and Smith 2012). Interestingly, the HA genes are absent from BoNT/E and BoNT/F strains, as well as some BoNT/A strains. Instead, they contain the *orfX*3-*orfX*2-*orfX*1-*(botR)-p*47-*ntnha-bont* complex genes (abbreviated as *orfX* cluster) (East et al. 1998; East and Collins 1994; Fujii et al. 1993; Hill et al. 2009; Hill and Smith 2012). The expression and function of the orfX proteins are unknown.

Many different forms of PTC have been identified, with molecular masses ranging from ~300 to ~900 kDa. It remains largely unknown how and why different PTCs exist. The M-PTC is ~300 kDa and is composed of BoNT and NTNHA, while other PTCs (e.g., 490, 610, 680, and 900 kDa) are assembled by the addition of various combinations of HAs to the M-PTC (Ito et al. 2011; Montecucco and Schiavo 1995). Historically, the three major forms of PTC are defined as M-PTC (~300 kDa), L-PTC (~500 kDa), and LL-PTC (~900 kDa), which are also termed the 12, 16, and 19S complex, respectively. The serotype A BoNT are complexed in all three forms and BoNT/B, C, D, and G in two forms, L-PTC and M-PTC. In contrast, bacterial serotypes E and F, which do not have the HA genes, may only produce BoNT complexed in the M-PTC form.

We now know that NTNHA directly interacts with BoNT and plays a major role in shielding it in the hostile GI environment. Furthermore, NTNHA binding and protection are exquisitely regulated to allow BoNT to be released upon entry into the circulation (Gu et al. 2012). The three HAs interact with intestinal epithelial cells and play an active role in BoNT transport (Fujinaga et al. 1997; Fujinaga et al. 2004; Niwa et al. 2007). Strong support for the functional role of HAs in this regard is provided by the finding that HAs directly bind to E-cadherin and disrupt the intercellular epithelial barrier (Ito et al. 2011; Matsumura et al. 2008; Sugawara et al. 2010; Fujinaga et al. 2012). These functions of NTNHA and the HAs will be discussed in detail later.

The NAP proteins clearly play crucial roles in the early stages of BoNT oral poisoning. Therefore, a better understanding of the structure and function of NAPs as well as their interplay with BoNT will help us find a novel approach to prevent BoNT entry into the general circulation.

3 Structure of the M-PTC

3.1 A Recombinant Expression System to Dissect the PTCs

Our limited knowledge of PTCs is in part due to the challenge associated with recombinant protein production. Most of the current research-grade BoNTs and NAPs as well as BoNT-containing medicines are partially purified from natural sources. Many studies on in vitro assembly of PTCs have employed native BoNT PTC, where individual NAPs had first to be stripped from the PTC using harsh conditions such as guanidine hydrochloride. A common problem with this approach is that the purified PTC components may be contaminated by other PTC components and/or other clostridial proteins, which can complicate the data interpretation and may even account for some of the discrepancies in experimental observations. Furthermore, as a Category A reagent, the extremely high toxicity of BoNTs prevents most laboratories from performing detailed biochemical and biophysical analyses, which require large quantities of toxins. Therefore, there is an urgent need to establish a safe and efficient recombinant expression system to enable comprehensive characterization of PTCs. To this end, several laboratories have developed novel approaches based on overexpression of recombinant BoNT and individual NAPs in *E. coli*, which allows their purification to high homogeneity (Gu et al. 2012; Miyata et al. 2009; Rummel et al. 2004). The following discussion will focus on BoNT/A1, which is one of the major causes of botulism in humans and the most commonly used medicine among the BoNTs.

In a recent paper, a genetically modified, catalytically inactive recombinant BoNT/A1 (termed BoNT/Ai), which carries three mutations in the catalytic site of LC; E224Q/R363A/Y366F, was produced in *E. coli* (Gu et al. 2012). It is known that either double mutation E224Q/Y366F or R363A/Y366F is sufficient to reduce the catalytic activity of BoNT/A-LC to an undetectable level (Binz et al. 2002; Breidenbach and Brunger 2004; Fu et al. 2006; Li et al. 2000). Consistent with this, BoNT/Ai has no detectable neurotoxicity in the mouse phrenic nerve toxicity assay when tested at 30,000-fold higher concentration than active BoNT/A (Binz et al. 2002; Gu et al. 2012; Li et al. 2000). The crystal structure of BoNT/Ai was determined at a resolution of 5 Å, which confirmed that the triple mutations do not alter the overall structure of BoNT/A (Gu et al. 2012).

In addition to the recombinant BoNT/Ai, the corresponding full-length NTNHA-A (NTNHA-A1, ~140 kDa) was also produced in *E. coli* (Gu et al. 2012). NTNHA-A shares low sequence identity (~20 %) with BoNT/A, and its structure was unknown. During long-term storage, NTNHA-A slowly self-processes into two polypeptide chains (~125 and 15 kDa) that remain non-covalently linked. N-terminal amino acid sequencing identified the cleavage site between Lys133 and Lys134. This is consistent with the known spontaneous nicking of native NTNHA in the homologous area, which occurs through an unknown mechanism (Fujita et al. 1995; Inoue et al. 1996; Miyata et al. 2009; Sagane et al. 2002). Recombinant forms of HA proteins for various BoNT serotypes have also been produced in several laboratories (Inoue et al. 2003; Nakamura et al. 2009; Nakamura et al. 2008).

3.2 NTNHA-A Forms a Tight Protective Complex with BoNT/A

The wild-type PTCs are stable at pH values of 6.25 or less but release of BoNTs occurs at pH values above 7.0 (Boroff et al. 1966; DasGupta and Boroff 1968; Eisele et al. 2011). The recombinant BoNT/Ai and NTNHA-A form a tight complex at acidic pH (pH 6.0) with dissociation constant (K_d) of ~30.8 nM and a stoichiometry of 1:1, as quantified by isothermal titration calorimetry and analytic ultracentrifugation (Gu et al. 2012). No interaction could be detected at pH 7.5. Thus, the recombinant M-PTC fully recapitulates the pH-dependent assembly of the wild-type PTC. This also suggests that BoNT and NTNHA, not the HA proteins, are mainly responsible for the pH-dependent assembly of PTC.

It is well known that the oral toxicity of free BoNTs is significantly less than that of PTCs (Cheng et al. 2008; Ohishi et al. 1977; Sugii et al. 1977). In vitro studies also showed that free BoNTs are much more easily detoxified than those in PTC by pepsin and pancreatin, and by gastric and intestinal juices (Ohishi et al. 1977; Sugii et al. 1977; Sugiyama 1980). To reproduce the physiological conditions faced by BoNT in the GI tract, the response of free BoNT/A and the M-PTC to low pH and digestive proteases was examined. It is found that free BoNT/A was almost completely inactivated by incubation at pH 3 or less, but remained fully active at pH 4–8. Similarly, the free BoNT/A and NTNHA-A were easily degraded into short peptides by digestive proteases, as is observed with most other proteins (Gu et al. 2012). In contrast, the in vitro-reconstituted M-PTC fully protected BoNT/A from protease- and low pH-induced inactivation, mimicking the behavior of the wild-type PTC. No such mutual protection was observed at neutral pH, which induced dissociation of the M-PTC. Thus, NTNHA plays a major role in protecting and stabilizing BoNTs under the low pH and protease-rich conditions prevailing in the GI tract, whereas HA proteins likely play a minor role in this process.

3.3 Architecture of the M-PTC

To understand the molecular mechanisms underlying BoNT/A protection and pH-dependent assembly of the M-PTC, we set out to determine the crystal structure of the M-PTC. Prior to this work, the best structural information available on a PTC was a projection map of the BoNT/A LL-PTC at 30 Å resolution (Burkard et al. 1997) and a negative stain transmission microscopic image of BoNT/D L-PTC (Hasegawa et al. 2007).

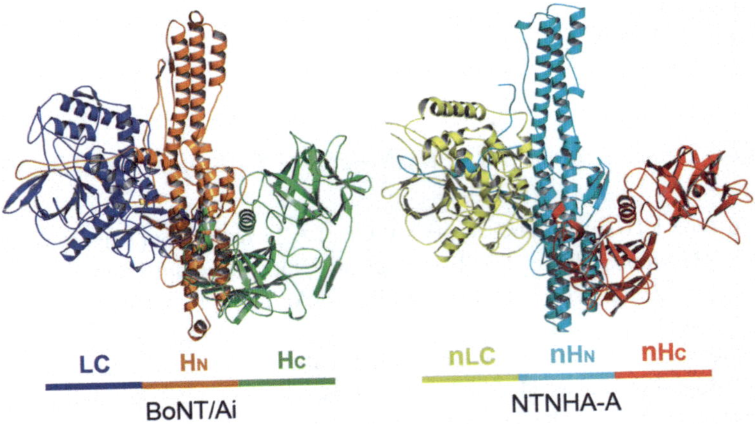

Fig. 2 Structures of BoNT/A and NTNHA-A in the M-PTC. The domain colors are as described for Fig. 1

Not surprisingly, large molecular mass multiprotein complexes are notoriously difficult to crystallize, and if they do, their diffraction is often limited to low resolution. After years of effort and numerous crystals screened at synchrotrons, the structure of the M-PTC composed of the full-length BoNT/Ai and NTNHA-A could be determined at 3.9 Å resolution. Despite the high quality of the structure model, which was built based on the unbiased experimental phase, it could not provide the detailed information needed to interrogate the extensive protein interactions between BoNT/Ai and NTNHA-A.

The breakthrough came only with the successful nanobody-facilitated crystallization of the M-PTC. Nanobodies are the smallest antigen-binding fragments (~12–15 kDa) of naturally occurring heavy-chain-only antibodies (VHH) present in camelids (Hamers-Casterman et al. 1993; Muyldermans et al. 2001) and they have been exploited as inhibitors for BoNTs (Conway et al. 2010; Dong et al. 2010; Mukherjee et al. 2012; Tremblay et al. 2010). VHH provided additional sites to facilitate a more compact crystal packing (Korotkov et al. 2009; Lam et al. 2009), which significantly improved the diffraction quality of crystals to 2.7 Å resolution. Although the free and the VHH-bound M-PTC were crystallized in different crystal forms under different crystallization conditions, the structures of the M-PTC are indistinguishable, demonstrating that this structure is a faithful representation of the physiological conformation of the M-PTC, independent of VHH binding and crystal packing.

The M-PTC structure has revealed a wealth of information and shows many unanticipated features (Fig. 1). First, BoNT/Ai adopts a distinct quaternary arrangement in the M-PTC. Pair-wise structural comparisons with the free BoNT/A (Lacy et al. 1998) yield root-mean-square deviations (r.m.s.d) of 0.6, 1.1, and 1.3 Å for LC, H_N, and H_C, respectively, suggesting the existence of conserved structures within each domain. Unexpectedly, H_C in the M-PTC differs from H_C in free BoNT/A in that it rotates about 140° though a linker between H_N and H_C.

Second, NTNHA-A is remarkably similar to BoNT/A. The three domains of NTNHA-A, when pair-wise compared with LC, H_N, or H_C, yield r.m.s.d. of 2.2, 2.3, and 1.9 Å, respectively, and thus were named nLC, nH_N, and nH_C (Figs. 1, 2). BoNT/Ai and NTNHA-A form an interlocked compact complex, reminiscent of a handshake. In the center of this molecular complex is the H_C fragment that is embraced by all three domains of NTNHA-A (Fig. 1). Interestingly, LC does not directly contact NTNHA-A.

3.4 The Structure of NTNHA-A

Despite an overall similarity in domain organization and individual domain structure to BoNT/A, NTNHA has several unique features and has clearly lost most of BoNTs features necessary to fulfill the neuron-specific attack (Fig. 2).

The protease activity of LC is absolutely dependent on a bound Zn^{2+} whereas nLC does not bind zinc. This has been carefully verified by inspecting an anomalous difference electron density map and is consistent with the fact that no protease activity has been observed for nLC. Primary sequence analysis of LC revealed an "HELIH+E" signature motif in all BoNTs (Ile is highly conserved except in BoNT/C and D where it is replaced by Asn and Thr, respectively). The His and the second Glu residues coordinate a Zn^{2+}, while the first Glu coordinates a water molecule for hydrolysis. This motif is found in a variety of Zn^{2+}-dependent metalloproteases such as thermolysin, which suggests that LCs may utilize a similar enzymatic mechanism (Binz et al. 2002; Kurazono et al. 1992). However, such a motif is not found in NTNHA-A or any other serotype. Interestingly, a structure-based sequence alignment shows that this motif is replaced by a highly conserved "KCLIK" motif in NTNHAs (Gu et al. 2012).

A second difference between the two molecules is that a long loop connecting LC and HC in BoNT/A is missing in NTNHA. This loop is cleaved post-translationally to activate BoNT/A, while LC and HC remain linked via an essential disulfide bond (Cys430–Cys454 in BoNT/A) that is conserved in all BoNTs (de Paiva et al. 1993). In all available crystal structures of the full-length BoNTs, most of this loop (Ile435–Ala449 in BoNT/A) has no visible electron density, indicating a highly flexible structure in this area. It is suggested that this loop keeps LC attached to HC via the conserved disulfide bond during transportation of LC, and then releases LC upon reduction in the cytosol. This disulfide bond and the homologous loop in BoNT/A are not found in any serotype of NTNHA.

The structure of NTNHA also suggests that nH_C is unlikely to have neuron-binding capacity due to the many deletions of surface exposed loops in H_C. Notably, none of the GT1b-binding residues of BoNT/A-H_C are conserved in NTNHA-A, including the highly conserved ganglioside-binding motif present in many BoNT serotypes (Rummel et al. 2009; Rummel et al. 2004; Stenmark et al.

2008). Accordingly, NTNHA-A did not bind to immobilized ganglioside mixtures (Rummel unpublished results).

Despite these differences, NTNHA does retain an intriguing structural feature of BoNTs; the so-called translocation domain belt. The belt region (Asn493–Asp546 in BoNT/A) is a mostly extended loop that wraps around LC and is suggested to be a pseudosubstrate inhibitor of the LC protease, and/or a chaperone during translocation of LC into the cytosol (Brunger et al. 2007). NTNHA-A maintains the homologous belt region (Asp451–Asn496, referred to as the nBelt) as part of nH_N, which wraps around nLC in an extended conformation. The function of the nBelt remains to be discovered.

It is worth noting that NTNHA has a structural feature that is not present in the equivalent part of BoNT/A, which is a large insert in nLC. Sequence analysis shows that a 33-residue fragment in this area (Gly116–Ala148 in NTNHA-A, termed the nLoop), is conserved in NTNHA serotypes A1, B, C, D, and G, but is missing in serotypes A2, E, and F. Many of the known spontaneous nicking sites in NTNHA are located in the nLoop (Fujita et al. 1995; Inoue et al. 1996; Miyata et al. 2009; Sagane et al. 2002). The whole nLoop has no visible electron density in the M-PTC structure, indicating high flexibility in this area. It has been reported that the nicking sites in NTNHA are masked in the HA-bound PTC, and that the M-PTC that contains the nicked NTNHA can no longer assemble with HAs (Kouguchi et al. 2002; Sagane et al. 2002). These data suggest that the nLoop of NTNHA is likely part of the binding site for HAs, and thus participates in assembly of the larger sized PTCs (Gu et al. 2012). Consistent with this hypothesis, NTNHA-A2, E, and F, which do not have the nLoop, also lack the accompanying HA proteins and only form the HA-negative M-PTC (East and Collins 1994; Fujii et al. 1993; Lin et al. 2010).

Collectively, these findings corroborate the hypothesis that BoNT/A and NTNHA-A might have evolved from a common ancestor by gene duplication, which has resulted in a similar general architecture but distinct biological functions. On the basis of amino acid sequence alignment, there is high sequence identity among the different serotypes of NTNHA. For example, NTNHA-A shows 82.9, 65.8, 65.9, 65.8, 74.9, and 72.2 % identity with NTNHA of serotypes B, C, D, E, F, and G, respectively (Krebs and Lebeda 2008). This high sequence identity suggests that the overall structural architecture of NTNHA should be conserved across the different serotypes. On the other hand, the sequence identity is much lower among the different serotypes of BoNTs: BoNT/A shares only 39.4, 32.5, 33.1, 40.4, 39.0, and 39.6 % identity with BoNT/B, C, D, E, F, and G, respectively (Krebs and Lebeda 2008; Hill and Smith 2012). This raises several interesting questions for future studies. Why are the amino acid sequences of NTNHAs relatively static compared with those of the BoNTs? Are interactions between BoNT and NTNHA serotype-specific? Does NTNHA have functions other than protecting BoNT?

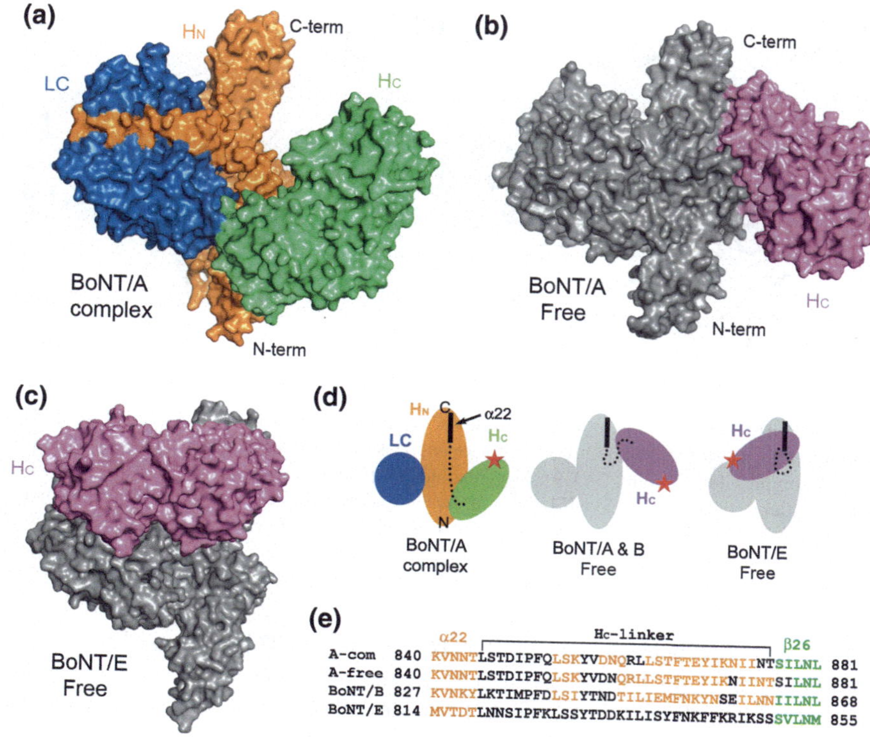

Fig. 3 The H_C of BoNT/A can adopt multiple conformations. Surface representations of the structures of BoNT/A in the M-PTC (**a**), free forms of BoNT/A (BoNT/B adopts an identical conformation) (**b**), and free BoNT/E (**c**). **d** Cartoon model showing the three conformations of the H_C fragment depicted in (**a**)–(**c**). The receptor-binding site in H_C is indicated by a *red star*. **e** Sequence alignment of the H_N–H_C linker. Residues that adopt an α-helix or β-sheet are in *orange* or *green*, respectively

3.5 A Novel Conformation of BoNT/A in the M-PTC

BoNT/Ai adopts a novel conformation in the M-PTC. Specifically, H_C rotates about 140° through a linker between H_N and H_C, which brings the receptor-binding site on H_C into the vicinity of the C-terminus of H_N (Fig. 3). In contrast, the LC-H_N moiety of complexed BoNT/Ai maintains the same structure as that observed in the free BoNT/A, BoNT/B, and a truncated BoNT/A composed of LC-H_N (Masuyer et al. 2009). The H_C reorientation of BoNT/Ai is likely induced by NTNHA-A rather than pH because the same conformation is adopted by all structures of free BoNT/A or BoNT/B crystallized at pHs ranging from 5.0 to 7.0 (Chai et al. 2006; Eswaramoorthy et al. 2004; Lacy et al. 1998; Swaminathan and Eswaramoorthy 2000).

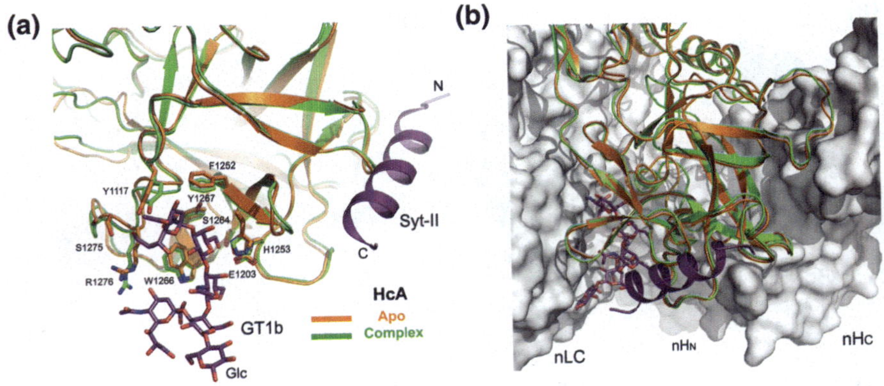

Fig. 4 NTNHA binding does not directly block the receptor-binding sites in BoNT/A. **a** The structure of H_CA in the M-PTC (complex, *green*) is superimposed with an H_CA–GT1b (apo, *orange*) complex (PDB code: 2VU9). Key residues that interact with GT1b (*purple sticks*) are shown in stick representation. Syt-II (ribbon in *purple*) is modeled based on H_CB–Syt-II complex (PDB code: 2NM1). **b** The structures in panel **a** are shown in the context of the M-PTC with NTNHA-A in a surface representation (*gray*)

The conformational changes of BoNT/A in the M-PTC are caused mostly by rotation of H_C, with the LC-H_N moiety as the base, via a linker composed of residues Leu845–Thr876 in BoNT/A. In the free BoNT/A, this linker contains an α-helix (Gln860–Lys871) and two 3_{10}-helices (Leu853–Lys855 and Ile873–Thr876) and is structurally conserved in the free BoNT/B. However, the linker adopts a different structure in the complex form, comprising an α-helix (Leu863–Ile874) and two 3_{10}-helices (Leu853–Lys855 and Asp858–Gln860) (Fig. 3e). Interestingly, the equivalent linker in BoNT/E (Leu819–Ser850) has no rigid secondary structure, which could explain the flexible domain organization of BoNT/E revealed by the electron microcopy and crystal structures (Fischer et al. 2008; Kumaran et al. 2009).

Mutagenesis studies suggested that the correct reorientation of H_C is crucial for effective H_N-mediated translocation (Gu et al. 2012). It was speculated that the flexible H_N–H_C linker may play a role in coordinating H_C-mediated receptor binding and H_N-mediated translocation, given that the membrane-anchored receptors impose some geometric restrictions on the position of H_C with respect to the membrane surface (Jin et al. 2006), and the long helical H_N needs to strategically orientate on the membrane to achieve efficient translocation of LC to the cytosol (Baldwin et al. 2007; Kumaran et al. 2009; Montal 2010). This is consistent with the hypothesis that the more rapid translocation of LC to the cytosol observed with BoNT/E than with BoNT/A is due to the more flexible BoNT/E-like linker that allows an "easier" reorientation of H_C to achieve the optimal conformation for translocation (Fischer et al. 2008; Keller et al. 2004; Kumaran et al. 2009; Wang et al. 2008).

Mechanistically, H_C-mediated receptor binding is the earliest step during neuron invasion, and likely one of the most crucial, because damage to H_C will jeopardize the enrichment of BoNT/A on the neuron membrane (Montecucco 1986). Structural modeling shows that the ganglioside-binding pocket in H_C is mostly unchanged upon binding to NTNHA (Fig. 4) (Stenmark et al. 2008). The potential binding site for the protein receptor, predicted to be homologous to the Syt-binding site in BoNT/B (Jin et al. 2006), is also exposed on the molecular surface of the M-PTC (Fig. 4). It is thus reasonable to predict that BoNT/A will still be capable of binding its dual receptors in the context of the M-PTC.

4 Dynamic Association of the M-PTC

4.1 BoNT/A and NTNHA-A Interactions

The molecular handshake of BoNT/Ai and NTNHA-A buries an unusually large solvent-accessible area (~3,200 Å^2 per molecule), resulting in a tightly bound complex at pH 6.0. The binding is predominantly driven by enthalpy (K_d = 30.8 nM, $\Delta H = -17.6$ kcal mol^{-1}, and $\Delta S = -24.8$ cal mol^{-1} K^{-1}). It is interesting to note that most of the intra-complex interactions are mediated by H_C, which forms extensive interfaces with all three domains of NTNHA-A. The unique M-PTC architecture is consistent with a comprehensive antibody mapping study, which showed that all epitopes on BoNT/A that are masked in the PTC are mapped to H_C, whereas epitopes in LC are all exposed in the PTC (Chen et al. 1997).

To investigate the interactions of BoNT/A within the M-PTC, we took advantage of the recombinant expression system to perform a systematic domain truncation study. We found that the binding behavior of the full-length BoNT/A could largely be replicated by the isolated H_C, which binds to NTNHA-A with a high affinity at pH 6.0 (K_d = 48.3 nM, $\Delta H = -12.3$ kcal mol^{-1}, and $\Delta S = -8.0$ cal mol^{-1} K^{-1}) but not at pH 7.5. Consistent with this, a truncated BoNT/A lacking H_C no longer binds NTNHA-A. Somewhat surprisingly, contributions to the association of the complex are not distributed evenly across the large interacting interface. A single point mutation (e.g. H_C-K1000A) could significantly decrease the binding affinity of BoNT/A for NTNHA-A by up to 7-fold and H_C-K1000A/K1039A/K1121A virtually disrupted the M-PTC (Gu et al. 2012). Thus, there may be "hot spots" within the intra-complex interactions that are crucial to maintain a stable M-PTC.

The clamp-like binding of H_C by NTNHA-A is reminiscent of a diverse array of protein chaperones (Stirling et al. 2006). However, NTNHA-A differs from the conventional role of chaperones, which facilitate protein folding, by providing large and multivalent binding surfaces for BoNT/A to shield it from the hostile gut environment. Interestingly, it has been reported that H_C of the free BoNT is more susceptible to proteolytic cleavage than are the LC and H_N domains, suggesting that protection of H_C is likely critical (Chen et al. 1997; Shone et al. 1985). Future studies should be guided by the structure of the M-PTC to determine how, or if, the apparent "biased" protection toward H_C has evolved as an optimal strategy to protect BoNT/A.

4.2 pH-Dependent Association of the PTC

One of the most fascinating features of the M-PTC is that BoNT/A can be released from the fully armored complex by a simple pH change. It was reasoned that the pH-dependent assembly of the M-PTC may be regulated by specific residues (referred to as pH sensors) that adopt pH-dependent conformational changes. Attempting to identify pH sensors in a 300 kDa protein complex is a daunting task. Fortunately, the structure of the M-PTC has prompted to focus on H_C, which is responsible for most of the intra-complex interactions and maintains the pH-sensing feature observed in the full-length BoNT/A.

Which residues could serve as pH sensors? Titratable residues in the acidic environment are histidine (His), glutamate (Glu), and aspartate (Asp), which have typical side chain pK_a values (where K_a is the acid dissociation constant) around 6.1, 4.3, and 3.9, respectively. Protonation of these residues has been reported to be involved in low pH-sensing in the pH-gated urea channel, chloride channel, histidine kinases, and viral membrane fusion proteins (Mueller et al. 2008; Perez and Groisman 2007; Stroffekova et al. 1998; Weeks and Sachs 2001).

Using structure-based mutagenesis and in vitro binding assays, Glu982 and Asp1037 of BoNT/A were identified as two key residues that mediate the pH-dependent binding between BoNT/A and NTNHA-A. A surface electrostatic potential analysis showed that Glu982 and Asp1037 are located in a generally positively charged surface in H_C, whereas the opposing NTNHA surface is negatively charged. Thus, the negative charge of the deprotonated Glu982 and Asp1037 would potentially weaken the local complementary electrostatic interactions in a pH-dependent manner.

The pK_a describes the tendency of a titratable residue to give up a proton. For internal ionizable groups in proteins in different microenvironments, their pKa values and charged states could be substantially different from their intrinsic values (Pace et al. 2009). Based on the crystal structures, the pK_a of Glu982 and Asp1037 were calculated to increase to neutral or mild alkaline pH when in the complex form compared to the free BoNT/A (Li et al. 2005). Clearly, the assembly of the M-PTC is a finely tuned dynamic process. It can be hypothesized that when the BoNT/A and NTNHA-A approach each other to form the transient complex, the pK_a of Glu982 and Asp1037 of BoNT/A will gradually increase, partly because of the desolvation effect when they are buried in the complex. In an acidic environment, pH sensors, including Glu982 and Asp1037 of BoNT/A, will be protonated and subsequently stabilize complex assembly. In contrast, the pH sensors will be deprotonated in neutral or alkaline environments, resulting in repulsive charge interactions that destabilize the transient complex.

It is worth noting that we are just beginning to understand the pH-sensing mechanism underlying M-PTC assembly. Other pH-sensing components in addition to Glu982 and Asp1037 must exist in BoNT/A. Strong evidence for this has come from the thermodynamic studies on the mutated H_C-E982A, which is able to bind NTNHA-A at pH 7.5. Nevertheless, a significant loss of enthalpy ($\Delta H \sim$ 6.3 kcal mol^{-1}) was observed for binding at pH 7.5 compared to that at pH 6.0. This suggests that the network of protein–protein interactions between H_C and NTNHA-A and solvation of the H_C surface, contributed by the combined surface pH-sensing elements, are different at pH 7.5 and pH 6.0. To add another layer of complexity, the key residues in BoNT/A that mediate BoNT–NTNHA interactions, including Glu982 and Asp1037, are not conserved in other BoNT serotypes. The structure of BoNT/A M-PTC should provide a framework for future studies to fully dissect the comprehensive pH-sensing mechanism across different serotypes.

5 HA Proteins

5.1 The Structure of the HA Proteins

The HA proteins account for up to 60 % of the molecular mass of PTC, but we have a poor understanding of their role in the pathogenesis of BoNTs. The crystal structures of the isolated HAs have been reported for selected serotypes, including HA-17 (BoNT/D), HA-33 (BoNT/A, C, and D), and HA-70 (BoNT/C) (Arndt et al. 2005; Hasegawa et al. 2007; Inoue et al. 2003; Nakamura et al. 2009; Nakamura et al. 2008). However, it is largely unknown how the HAs assemble with each other and subsequently with BoNT and NTNHA.

HA-70 of BoNT/C consists of two domains, HA-70a and HA-70b, which are proteolytically cleaved from the full-length HA-70. The HA-70a is a single domain fragment with three α-helices and eight β-strands. The HA-70b subcomponent consists of three subdomains (I, II, and III). Domain I of HA-70b is very similar to HA-70a. Domains II and III adopt a similar jelly roll-like β-sandwich structure (Nakamura et al. 2009). Intriguingly, HA-70 forms a three-bladed propeller-like trimer in crystals, with a pore located at the center of the trimer (Fig. 5). The pore, estimated to be 33 Å in length and 21 Å in diameter, is a compact cylindrical β-barrel composed of 12 β-strands from HA-70a and the domain I of HA-70b (Nakamura et al. 2009).

HA-33 has a dumbbell-like shape, composed of two β-trefoil domains linked by an α-helix (Fig. 5). The β-trefoil fold is a common structural module found in many proteins, including many lectins, where it acts as an oligosaccharide-binding unit. The β-trefoil domains of HA-33 from BoNT/A, C, and D are highly similar. However, the domains are capable of twisting against each other via the helical linker, suggesting significant inter-domain conformational plasticity (Arndt et al.

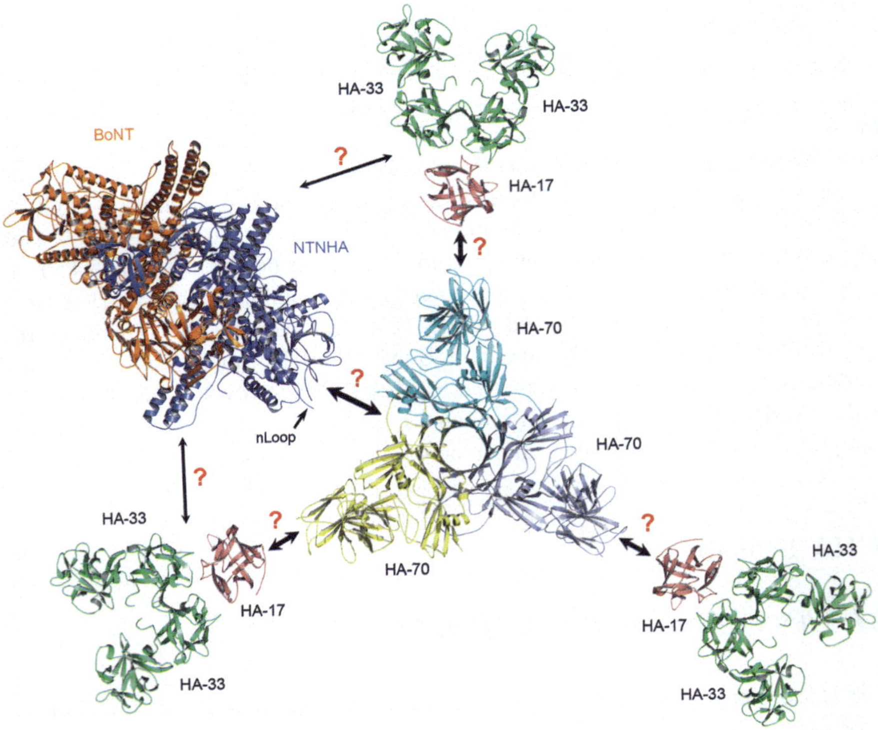

Fig. 5 A model of the assembly of the PTC. The structures are shown in a cartoon representation: HA-17 (salmon, BoNT/D, PDB code 2E4M), HA-33 (*green*, BoNT/D, PDB code 2E4M), HA-70 (*cyan/yellow/gray*, BoNT/C, PDB code 2ZS6), BoNT (*gold*, BoNT/A1, PDB code 3V0A), and NTNHA (*blue*, BoNT/A1, PDB code 3V0A)

2005; Hasegawa et al. 2007; Inoue et al. 2003). HA-17 seems to tightly associate with HA-33, forming a complex that remains bound even in the presence of high concentration of guanidine hydrochloride (Kouguchi et al. 2002). The crystal structure shows that HA-17 of BoNT/D also adopts a β-trefoil fold, with each HA-17 associating with two HA-33 molecules by binding to their N-terminal β-trefoil domains (Hasegawa et al. 2007).

Several different structural models of the fully assembled PTC have been proposed (Hasegawa et al. 2007; Inoue et al. 1996; Kouguchi et al. 2002; Lietzow et al. 2008; Mutoh et al. 2003; Suzuki et al. 2005). The discrepancies among the models are largely due to the lack of systematic biochemical and biophysical characterization of HAs in solution. For example, we do not yet know if the trimeric HA-70 or the HA-17/33 complex that were observed in crystal structures represent the physiologically relevant conformations in solution. Furthermore, future studies should address whether and how HA-17/33 interacts with HA-70, and how BoNT and/or NTNHA may interact with the HA proteins (Fig. 5).

5.2 HAs Mediate Toxin–Host Interactions

HA proteins appear to play an important role in absorption of PTC by interacting with oligosaccharides on intestinal epithelial cells (Fujinaga et al. 1997; Fujinaga et al. 2004; Inoue et al. 2001; Kojima et al. 2005; Nakamura et al. 2007) (also see chapter Uptake of Botulinum Neurotoxin in the Intestine). One N-acetylneuraminic acid (Neu5Ac) binding site has been reported in HA-70b of BoNT/C, which is located in a cleft formed by domains II and III (Nakamura et al. 2009). The HA-70–Neu5Ac interaction was not well-defined in the reported crystal structure, likely due to the low binding affinity of Neu5Ac to HA-70. Improved crystal structures are reported recently, which show that HA-70 of BoNT/C can recognize both α2,3- and α2,6-sialylated oligosaccharides, and that most of the HA-sugar interactions are contributed by Neu5Ac (Yamashita et al. 2012). A binding assay of HA-70 of BoNT/C against α2,3-sialyllactosamine- or α2,6-sialyllactosamine-conjugated BSA suggests that it might have a higher affinity for α2,3-sialylated oligosaccharide (Yamashita et al. 2012). However, the crystal structures show that α2,6-sialyllactosamine forms more hydrogen bonds with HA-70 compared with α2,3-sialyllactosamine.

HA-33 of BoNT/C has been reported to have three binding sites that could interact with several sugars, including Neu5Ac, galactose (Gal), and N-acetylgalactosamine (GalNAc) (Nakamura et al. 2008). This finding is largely derived from crystal-soaking experiments, where HA-33 crystals were soaked in high-concentration sugar solutions prior to collection of diffraction data for structure determination. It is worth noting that all three sugars examined in this study bind differently to the two identical HA-33 molecules in an asymmetric unit in the crystal, suggesting there may be crystallographic artifacts. Furthermore, inconsistent results have been reported regarding the GalNAc-binding site(s) in HA-33 of BoNT/C (Nakamura et al. 2008; Nakamura et al. 2011). Thus, the physiological role of HA-33–sugar interactions awaits clarification in future studies.

One of the most exciting recent breakthroughs is the discovery that E-cadherin (E-cad), an epithelial cell surface adhesion molecule, is a host receptor for HA proteins. It was suggested that HA proteins directly bind to and disrupt E-cad-mediated cell-to-cell adhesion, thus facilitating the absorption of BoNT through the intestinal epithelium via a paracellular route (Ito et al. 2011; Matsumura et al. 2008; Sugawara et al. 2010; Fujinaga et al. 2012). Consistent with this hypothesis, the interaction is highly specific: HAs bind only to E-cad and not to N-cad or VE-cad (Sugawara et al. 2010). Furthermore, the HA–E-cad interaction is serotype- and species-specific. For example, the HAs of BoNT/A or B efficiently bind to human, bovine, or mouse E-cad, but not to chicken E-cad; whereas the HA of BoNT/C does not recognize human E-cad (Sugawara et al. 2010). These observations correlate well with the epidemiology of botulism. For instance, birds are more resistant to oral intoxication of BoNT/A (Notermans et al. 1980) and there are few reports of human botulism caused by BoNT/C (Coffield et al. 1997).

Collectively, these observations suggest that the host susceptibility to botulism is likely determined in part by the interplay between HA proteins and host receptors in the GI tract. Future studies should focus on dissecting the poorly understood HA-mediated interplay between toxin and host, including the identity of the host carbohydrate and protein receptors. A better understanding of the molecular mechanisms controlling these steps will facilitate the development of novel strategies to block the early stages of oral intoxication by preventing BoNT entry into the general circulation.

6 Conclusions

We have observed an unprecedented growth in our knowledge of the structure and function of botulinum neurotoxin progenitor complexes, due in large part to worldwide collaborations among scientists from different fields with complementary expertise. Among the crucial findings, the first crystal structure of the minimal progenitor toxin complex of BoNT/A has been reported recently (Gu et al. 2012), which has rationalized a large set of genetic and biochemical observations.

BoNT/A and NTNHA-A form an interlocked handshake-like complex, which lends both proteins extraordinary stability against low pH and digestive proteases. Interestingly, the NTNHA-A has low sequence identity to BoNT/A but adopts a similar architecture. Despite this, NTNHA-A is devoid of the characteristic structural features of BoNT/A that are crucial to its biological functions (e.g., endopeptidase activity, host receptor binding). The M-PTC structure also helps to pinpoint several pH-sensing residues that are key players in balancing the seemingly contradictory needs of BoNT/A and NTNHA-A: strong binding for protection in the gut and timely release upon gaining entry to the general circulation.

The fascinating mode of BoNT/A protection and assembly should also facilitate mechanistic studies on other BoNT serotypes and other bacterial toxins. However, many questions remain to be addressed. How does the M-PTC assemble with the HA proteins? Does the ~500 kDa HA complex further protect BoNT/A against low pH and proteases? How does the HA complex interact with host receptors and facilitate translocation across the epithelium? How does the PTC disassemble upon reaching the relative safety of the bloodstream? The answers to these questions await rigorous structural and functional analyses in the future.

On another front, a better understanding of the protection and transportation functions of the PTC may help to identify the Achilles' heel of this complex and expedite the design of novel approaches to counteract BoNT intoxication. This could be accomplished, for example, by small molecules that promote premature disassembly and destruction of the PTC in the GI tract, or which disrupt HA-mediated toxin–host recognition in the small intestine to prevent BoNT entry into the circulation. Such preventive countermeasures for oral BoNT intoxication will be crucial in situations such as during an outbreak of botulism, and could be a strong deterrent against the use of BoNT as a bioterrorism weapon.

Finally, the unique features of the complex involved in BoNT protection, absorption, and dynamic assembly suggest that PTC-based vehicles could be engineered to shield proteinaceous drugs from GI destruction and thus allow their oral administration. Furthermore, retargeted BoNT with altered cell tropism could be constructed that would allow tissue- and organ-specific drug delivery. If successful, this would be an enormous step forward in the development of therapeutic biologics.

Acknowledgments This work was partially supported by NIH grant 5R01AI091823 to R.J.

References

Arndt JW, Gu J, Jaroszewski L, Schwarzenbacher R, Hanson MA, Lebeda FJ, Stevens RC (2005) The structure of the neurotoxin-associated protein HA33/A from *Clostridium botulinum* suggests a reoccurring beta-trefoil fold in the progenitor toxin complex. J Mol Biol 346:1083–1093

Baldwin MR, Kim JJ, Barbieri JT (2007) Botulinum neurotoxin B-host receptor recognition: it takes two receptors to tango. Nat Struct Mol Biol 14:9–10

Binz T, Bade S, Rummel A, Kollewe A, Alves J (2002) Arg(362) and Tyr(365) of the botulinum neurotoxin type a light chain are involved in transition state stabilization. Biochemistry 41:1717–1723

Binz T, Kurazono H, Wille M, Frevert J, Wernars K, Niemann H (1990) The complete sequence of botulinum neurotoxin type A and comparison with other clostridial neurotoxins. J Biol Chem 265:9153–9158

Binz T, Rummel A (2009) Cell entry strategy of clostridial neurotoxins. J Neurochem 109:1584–1595

Binz T (2012) Clostridial Neurotoxin Light Chains: Devices for SNARE Cleavage Mediated Blockade of Neurotransmission 10.1007/978-3-642-33570-9_7

Blasi J, Chapman ER, Link E, Binz T, Yamasaki S, De Camilli P, Sudhof TC, Niemann H, Jahn R (1993a) Botulinum neurotoxin A selectively cleaves the synaptic protein SNAP-25. Nature 365:160–163

Blasi J, Chapman ER, Yamasaki S, Binz T, Niemann H, Jahn R (1993b) Botulinum neurotoxin C1 blocks neurotransmitter release by means of cleaving HPC-1/syntaxin. EMBO J 12:4821–4828

Boroff DA, Townend R, Fleck U, DasGupta BR (1966) Ultracentrifugal analysis of the crystalline toxin and isolated fractions of *Clostridium botulinum* type A. J Biol Chem 241:5165–5167

Breidenbach MA, Brunger AT (2004) Substrate recognition strategy for botulinum neurotoxin serotype A. Nature 432:925–929

Brunger AT, Breidenbach MA, Jin R, Fischer A, Santos JS, Montal M (2007) Botulinum neurotoxin heavy chain belt as an intramolecular chaperone for the light chain. PLoS Pathog 3:e113

Burkard F, Chen F, Kuziemko GM, Stevens RC (1997) Electron density projection map of the botulinum neurotoxin 900-kilodalton complex by electron crystallography. J Struct Biol 120:78–84

Chai Q, Arndt JW, Dong M, Tepp WH, Johnson EA, Chapman ER, Stevens RC (2006) Structural basis of cell surface receptor recognition by botulinum neurotoxin B. Nature 444:1096–1100

Chen F, Kuziemko GM, Amersdorfer P, Wong C, Marks JD, Stevens RC (1997) Antibody mapping to domains of botulinum neurotoxin serotype A in the complexed and uncomplexed forms. Infect Immun 65:1626–1630

Cheng LW, Onisko B, Johnson EA, Reader JR, Griffey SM, Larson AE, Tepp WH, Stanker LH, Brandon DL, Carter JM (2008) Effects of purification on the bioavailability of botulinum neurotoxin type A. Toxicology 249:123–129

Coffield JA, Bakry N, Zhang RD, Carlson J, Gomella LG, Simpson LL (1997) In vitro characterization of botulinum toxin types A, C and D action on human tissues: combined electrophysiologic, pharmacologic and molecular biologic approaches. J Pharmacol Exp Ther 280:1489–1498

Collins MD, East AK (1998) Phylogeny and taxonomy of the food-borne pathogen *Clostridium botulinum* and its neurotoxins. J Appl Microbiol 84:5–17

Conway JO, Sherwood LJ, Collazo MT, Garza JA, Hayhurst A (2010) Llama single domain antibodies specific for the 7 botulinum neurotoxin serotypes as heptaplex immunoreagents. PLoS ONE 5:e8818

DasGupta BR, Boroff DA (1968) Separation of toxin and hemagglutinin from crystalline toxin of *Clostridium botulinum* type A by anion exchange chromatography and determination of their dimensions by gel filtration. J Biol Chem 243:1065–1072

de Paiva A, Poulain B, Lawrence GW, Shone CC, Tauc L, Dolly JO (1993) A role for the interchain disulfide or its participating thiols in the internalization of botulinum neurotoxin A revealed by a toxin derivative that binds to ecto-acceptors and inhibits transmitter release intracellularly. J Biol Chem 268:20838–20844

Dong J, Thompson AA, Fan Y, Lou J, Conrad F, Ho M, Pires-Alves M, Wilson BA, Stevens RC, Marks JD (2010) A single-domain llama antibody potently inhibits the enzymatic activity of botulinum neurotoxin by binding to the non-catalytic alpha-exosite binding region. J Mol Biol 397:1106–1118

East AK, Bhandari M, Hielm S, Collins MD (1998) Analysis of the botulinum neurotoxin type F gene clusters in proteolytic and nonproteolytic *Clostridium botulinum* and *Clostridium barati*. Curr Microbiol 37:262–268

East AK, Collins MD (1994) Conserved structure of genes encoding components of botulinum neurotoxin complex M and the sequence of the gene coding for the nontoxic component in nonproteolytic *Clostridium botulinum* type F. Curr Microbiol 29:69–77

Eisele KH, Fink K, Vey M, Taylor HV (2011) Studies on the dissociation of botulinum neurotoxin type A complexes. Toxicon 57:555–565

Eswaramoorthy S, Kumaran D, Keller J, Swaminathan S (2004) Role of metals in the biological activity of *Clostridium botulinum* neurotoxins. Biochemistry 43:2209–2216

Fischer A, Garcia-Rodriguez C, Geren I, Lou J, Marks JD, Nakagawa T, Montal M (2008) Molecular architecture of botulinum neurotoxin E revealed by single particle electron microscopy. J Biol Chem 283:3997–4003

Fischer A, Montal M (2006) Characterization of *Clostridial botulinum* neurotoxin channels in neuroblastoma cells. Neurotox Res 9:93–100

Fischer A, Montal M (2007a) Crucial role of the disulfide bridge between botulinum neurotoxin light and heavy chains in protease translocation across membranes. J Biol Chem 282:29604–29611

Fischer A, Montal M (2007b) Single molecule detection of intermediates during botulinum neurotoxin translocation across membranes. Proc Natl Acad Sci U S A 104:10447–10452

Foran P, Lawrence GW, Shone CC, Foster KA, Dolly JO (1996) Botulinum neurotoxin C1 cleaves both syntaxin and SNAP-25 in intact and permeabilized chromaffin cells: correlation with its blockade of catecholamine release. Biochemistry 35:2630–2636

Fu Z, Chen S, Baldwin MR, Boldt GE, Crawford A, Janda KD, Barbieri JT, Kim JJ (2006) Light chain of botulinum neurotoxin serotype A: structural resolution of a catalytic intermediate. Biochemistry 45:8903–8911

Fujii N, Kimura K, Yokosawa N, Yashiki T, Tsuzuki K, Oguma K (1993) The complete nucleotide sequence of the gene encoding the nontoxic component of *Clostridium botulinum* type E progenitor toxin. J Gen Microbiol 139:79–86

Fujinaga Y, Inoue K, Watanabe S, Yokota K, Hirai Y, Nagamachi E, Oguma K (1997) The haemagglutinin of *Clostridium botulinum* type C progenitor toxin plays an essential role in binding of toxin to the epithelial cells of guinea pig small intestine, leading to the efficient absorption of the toxin. Microbiology 143(Pt 12):3841–3847

Fujinaga Y, Inoue K, Watarai S, Sakaguchi Y, Arimitsu H, Lee JC, Jin Y, Matsumura T, Kabumoto Y, Watanabe T, Ohyama T, Nishikawa A, Oguma K (2004) Molecular characterization of binding subcomponents of *Clostridium botulinum* type C progenitor toxin for intestinal epithelial cells and erythrocytes. Microbiology 150:1529–1538

Fujinaga Y, Matsumura T, Jin Y, Takegahara Y, Sugawara Y (2009) A novel function of botulinum toxin-associated proteins: HA proteins disrupt intestinal epithelial barrier to increase toxin absorption. Toxicon 54:583–586

Fujinaga Y, Matsumura T, Sugawara Y (2012) Uptake of botulinum neurotoxin in the intestine. 10.1007/978-3-642-33570-9_3

Fujita R, Fujinaga Y, Inoue K, Nakajima H, Kumon H, Oguma K (1995) Molecular characterization of two forms of nontoxic-nonhemagglutinin components of *Clostridium botulinum* type A progenitor toxins. FEBS Lett 376:41–44

Gu S, Rumpel S, Zhou J, Strotmeier J, Bigalke H, Perry K, Shoemaker CB, Rummel A, Jin R (2012) Botulinum neurotoxin is shielded by NTNHA in an interlocked complex. Science 335:977–981

Hamers-Casterman C, Atarhouch T, Muyldermans S, Robinson G, Hamers C, Songa EB, Bendahman N, Hamers R (1993) Naturally occurring antibodies devoid of light chains. Nature 363:446–448

Hasegawa K, Watanabe T, Suzuki T, Yamano A, Oikawa T, Sato Y, Kouguchi H, Yoneyama T, Niwa K, Ikeda T, Ohyama T (2007) A novel subunit structure of *clostridium botulinum* serotype D toxin complex with three extended arms. J Biol Chem 282:24777–24783

Hatheway CL (1990) Toxigenic clostridia. Clin Microbiol Rev 3:66–98

Hill KK, Xie G, Foley BT, Smith TJ, Munk AC, Bruce D, Smith LA, Brettin TS, Detter JC (2009) Recombination and insertion events involving the botulinum neurotoxin complex genes in *Clostridium botulinum* types A, B, E and F and *Clostridium butyricum* type E strains. BMC Biol 7:66

Hill K & Smith TJ (2012) Genetic Diversity Within Clostridium Botulinum Serotypes, Botulinum Neurotoxin Gene Clusters and Toxin Subtypes 10.1007/978-3-642-33570-9_1

Inoue K, Fujinaga Y, Honke K, Arimitsu H, Mahmut N, Sakaguchi Y, Ohyama T, Watanabe T, Inoue K, Oguma K (2001) *Clostridium botulinum* type A haemagglutinin-positive progenitor toxin (HA(+)-PTX) binds to oligosaccharides containing Gal beta1-4GlcNAc through one subcomponent of haemagglutinin (HA1). Microbiology 147:811–819

Inoue K, Fujinaga Y, Watanabe T, Ohyama T, Takeshi K, Moriishi K, Nakajima H, Inoue K, Oguma K (1996) Molecular composition of *Clostridium botulinum* type A progenitor toxins. Infect Immun 64:1589–1594

Inoue K, Sobhany M, Transue TR, Oguma K, Pedersen LC, Negishi M (2003) Structural analysis by X-ray crystallography and calorimetry of a haemagglutinin component (HA1) of the progenitor toxin from *Clostridium botulinum*. Microbiology 149:3361–3370

Ito H, Sagane Y, Miyata K, Inui K, Matsuo T, Horiuchi R, Ikeda T, Suzuki T, Hasegawa K, Kouguchi H, Oguma K, Niwa K, Ohyama T, Watanabe T (2011) HA-33 facilitates transport of the serotype D botulinum toxin across a rat intestinal epithelial cell monolayer. FEMS Immunol Med Microbiol: 1–9

Jin R, Rummel A, Binz T, Brunger AT (2006) Botulinum neurotoxin B recognizes its protein receptor with high affinity and specificity. Nature 444:1092–1095

Keller JE, Cai F, Neale EA (2004) Uptake of botulinum neurotoxin into cultured neurons. Biochemistry 43:526–532

Kojima S, Eguchi H, Ookawara T, Fujiwara N, Yasuda J, Nakagawa K, Yamamura T, Suzuki K (2005) Clostridium botulinum type A progenitor toxin binds to Intestine-407 cells via N-acetyllactosamine moiety. Biochem Biophys Res Commun 331:571–576

Koriazova LK, Montal M (2003) Translocation of botulinum neurotoxin light chain protease through the heavy chain channel. Nat Struct Biol 10:13–18

Korotkov KV, Pardon E, Steyaert J, Hol WG (2009) Crystal structure of the N-terminal domain of the secretin GspD from ETEC determined with the assistance of a nanobody. Structure 17:255–265

Kouguchi H, Watanabe T, Sagane Y, Sunagawa H, Ohyama T (2002) In vitro reconstitution of the *Clostridium botulinum* type D progenitor toxin. J Biol Chem 277:2650–2656

Krebs KM, Lebeda FJ (2008) Comparison of the structural features of botulinum neurotoxin and NTNH, a non-toxic accessory protein of the progenitor complex. Botulinum J 1:116–134

Kumaran D, Eswaramoorthy S, Furey W, Navaza J, Sax M, Swaminathan S (2009) Domain organization in *Clostridium botulinum* neurotoxin type E is unique: its implication in faster translocation. J Mol Biol 386:233–245

Kurazono H, Mochida S, Binz T, Eisel U, Quanz M, Grebenstein O, Wernars K, Poulain B, Tauc L, Niemann H (1992) Minimal essential domains specifying toxicity of the light chains of tetanus toxin and botulinum neurotoxin type A. J Biol Chem 267:14721–14729

Lacy DB, Stevens RC (1999) Sequence homology and structural analysis of the clostridial neurotoxins. J Mol Biol 291:1091–1104

Lacy DB, Tepp W, Cohen AC, DasGupta BR, Stevens RC (1998) Crystal structure of botulinum neurotoxin type A and implications for toxicity. Nat Struct Biol 5:898–902

Lam AY, Pardon E, Korotkov KV, Hol WG, Steyaert J (2009) Nanobody-aided structure determination of the EpsI:EpsJ pseudopilin heterodimer from *Vibrio vulnificus*. J Struct Biol 166:8–15

Li H, Robertson AD, Jensen JH (2005) Very fast empirical prediction and rationalization of protein pKa values. Proteins 61:704–721

Li L, Binz T, Niemann H, Singh BR (2000) Probing the mechanistic role of glutamate residue in the zinc-binding motif of type A botulinum neurotoxin light chain. Biochemistry 39:2399–2405

Lietzow MA, Gielow ET, Le D, Zhang J, Verhagen MF (2008) Subunit stoichiometry of the *Clostridium botulinum* type A neurotoxin complex determined using denaturing capillary electrophoresis. Protein J 27:420–425

Lin G, Tepp WH, Pier CL, Jacobson MJ, Johnson EA (2010) Expression of the *Clostridium botulinum* A2 neurotoxin gene cluster proteins and characterization of the A2 complex. Appl Environ Microbiol 76:40–47

Masuyer G, Thiyagarajan N, James PL, Marks PM, Chaddock JA, Acharya KR (2009) Crystal structure of a catalytically active, non-toxic endopeptidase derivative of *Clostridium botulinum* toxin A. Biochem Biophys Res Commun 381:50–53

Matsumura T, Jin Y, Kabumoto Y, Takegahara Y, Oguma K, Lencer WI, Fujinaga Y (2008) The HA proteins of botulinum toxin disrupt intestinal epithelial intercellular junctions to increase toxin absorption. Cell Microbiol 10:355–364

Miyata K, Yoneyama T, Suzuki T, Kouguchi H, Inui K, Niwa K, Watanabe T, Ohyama T (2009) Expression and stability of the nontoxic component of the botulinum toxin complex. Biochem Biophys Res Commun 384:126–130

Montal M (2010) Botulinum neurotoxin: a marvel of protein design. Annu Rev Biochem 79:591–617

Montecucco C (1986) How do tetanus and botulinum neurotoxins bind to neuronal membranes? Trends Biochem Sci 11:314–317

Montecucco C, Rossetto O, Schiavo G (2004) Presynaptic receptor arrays for clostridial neurotoxins. Trends Microbiol 12:442–446

Montecucco C, Schiavo G (1995) Structure and function of tetanus and botulinum neurotoxins. Q Rev Biophys 28:423–472

Mueller DS, Kampmann T, Yennamalli R, Young PR, Kobe B, Mark AE (2008) Histidine protonation and the activation of viral fusion proteins. Biochem Soc Trans 36:43–45

Mukherjee J, Tremblay JM, Leysath CE, Ofori K, Baldwin K, Feng X, Bedenice D, Webb RP, Wright PM, Smith LA, Tzipori S, Shoemaker CB (2012) A novel strategy for development of recombinant antitoxin therapeutics tested in a mouse botulism model. PLoS ONE 7:e29941

Mutoh S, Kouguchi H, Sagane Y, Suzuki T, Hasegawa K, Watanabe T, Ohyama T (2003) Complete subunit structure of the *Clostridium botulinum* type D toxin complex via intermediate assembly with nontoxic components. Biochemistry 42:10991–10997

Muyldermans S, Cambillau C, Wyns L (2001) Recognition of antigens by single-domain antibody fragments: the superfluous luxury of paired domains. Trends Biochem Sci 26:230–235

Nakamura T, Kotani M, Tonozuka T, Ide A, Oguma K, Nishikawa A (2009) Crystal structure of the HA3 subcomponent of *Clostridium botulinum* type C progenitor toxin. J Mol Biol 385:1193–1206

Nakamura T, Takada N, Tonozuka T, Sakano Y, Oguma K, Nishikawa A (2007) Binding properties of *Clostridium botulinum* type C progenitor toxin to mucins. Biochim Biophys Acta 1770:551–555

Nakamura T, Tonozuka T, Ide A, Yuzawa T, Oguma K, Nishikawa A (2008) Sugar-binding sites of the HA1 subcomponent of *Clostridium botulinum* type C progenitor toxin. J Mol Biol 376:854–867

Nakamura T, Tonozuka T, Ito S, Takeda Y, Sato R, Matsuo I, Ito Y, Oguma K, Nishikawa A (2011) Molecular diversity of the two sugar-binding sites of the beta-trefoil lectin HA33/C (HA1) from *Clostridium botulinum* type C neurotoxin. Arch Biochem Biophys 512:69–77

Niwa K, Koyama K, Inoue S, Suzuki T, Hasegawa K, Watanabe T, Ikeda T, Ohyama T (2007) Role of nontoxic components of serotype D botulinum toxin complex in permeation through a Caco-2 cell monolayer, a model for intestinal epithelium. FEMS Immunol Med Microbiol 49:346–352

Notermans S, Dufrenne J, Kozaki S (1980) Experimental botulism in Pekin ducks. Avian Dis 24:658–664

Ohishi I, Sugii S, Sakaguchi G (1977) Oral toxicities of *Clostridium botulinum* toxins in response to molecular size. Infect Immun 16:107–109

Pace CN, Grimsley GR, Scholtz JM (2009) Protein ionizable groups: pK values and their contribution to protein stability and solubility. J Biol Chem 284:13285–13289

Perez JC, Groisman EA (2007) Acid pH activation of the PmrA/PmrB two-component regulatory system of *Salmonella enterica*. Mol Microbiol 63:283–293

Rummel A, Hafner K, Mahrhold S, Darashchonak N, Holt M, Jahn R, Beermann S, Karnath T, Bigalke H, Binz T (2009) Botulinum neurotoxins C, E and F bind gangliosides via a conserved binding site prior to stimulation-dependent uptake with botulinum neurotoxin F utilising the three isoforms of SV2 as second receptor. J Neurochem 110:1942–1954

Rummel A, Mahrhold S, Bigalke H, Binz T (2004) The HCC-domain of botulinum neurotoxins A and B exhibits a singular ganglioside binding site displaying serotype specific carbohydrate interaction. Mol Microbiol 51:631–643

Rummel A (2012) Double Receptor Anchorage of Botulinum Neurotoxins Accounts for Their Exquisite Neurospecificity 10.1007/978-3-642-33570-9_4

Sagane Y, Watanabe T, Kouguchi H, Sunagawa H, Obata S, Oguma K, Ohyama T (2002) Spontaneous nicking in the nontoxic-nonhemagglutinin component of the *Clostridium botulinum* toxin complex. Biochem Biophys Res Commun 292:434–440

Schiavo G, Benfenati F, Poulain B, Rossetto O, Polverino de Laureto P, DasGupta BR, Montecucco C (1992) Tetanus and botulinum-B neurotoxins block neurotransmitter release by proteolytic cleavage of synaptobrevin. Nature 359:832–835

Schiavo G, Matteoli M, Montecucco C (2000) Neurotoxins affecting neuroexocytosis. Physiol Rev 80:717–766

Schiavo G, Shone CC, Bennett MK, Scheller RH, Montecucco C (1995) Botulinum neurotoxin type C cleaves a single Lys-Ala bond within the carboxyl-terminal region of syntaxins. J Biol Chem 270:10566–10570

Shone CC, Hambleton P, Melling J (1985) Inactivation of *Clostridium botulinum* type A neurotoxin by trypsin and purification of two tryptic fragments. Proteolytic action near the COOH-terminus of the heavy subunit destroys toxin-binding activity. Eur J Biochem 151:75–82

Stenmark P, Dupuy J, Imamura A, Kiso M, Stevens RC (2008) Crystal structure of botulinum neurotoxin type A in complex with the cell surface co-receptor GT1b-insight into the toxin-neuron interaction. PLoS Pathog 4:e1000129

Stirling PC, Bakhoum SF, Feigl AB, Leroux MR (2006) Convergent evolution of clamp-like binding sites in diverse chaperones. Nat Struct Mol Biol 13:865–870

Stroffekova K, Kupert EY, Malinowska DH, Cuppoletti J (1998) Identification of the pH sensor and activation by chemical modification of the ClC-2G Cl- channel. Am J Physiol 275:C1113–C1123

Sugawara Y, Matsumura T, Takegahara Y, Jin Y, Tsukasaki Y, Takeichi M, Fujinaga Y (2010) Botulinum hemagglutinin disrupts the intercellular epithelial barrier by directly binding E-cadherin. J Cell Biol 189:691–700

Sugii S, Ohishi I, Sakaguchi G (1977) Correlation between oral toxicity and in vitro stability of Clostridium botulinum type A and B toxins of different molecular sizes. Infect Immun 16:910–914

Sugiyama H (1980) Clostridium botulinum neurotoxin. Microbiol Rev 44:419–448

Suzuki T, Watanabe T, Mutoh S, Hasegawa K, Kouguchi H, Sagane Y, Fujinaga Y, Oguma K, Ohyama T (2005) Characterization of the interaction between subunits of the botulinum toxin complex produced by serotype D through tryptic susceptibility of the isolated components and complex forms. Microbiology 151:1475–1483

Swaminathan S, Eswaramoorthy S (2000) Structural analysis of the catalytic and binding sites of *Clostridium botulinum* neurotoxin B. Nat Struct Biol 7:693–699

Tremblay JM, Kuo CL, Abeijon C, Sepulveda J, Oyler G, Hu X, Jin MM, Shoemaker CB (2010) Camelid single domain antibodies (VHHs) as neuronal cell intrabody binding agents and inhibitors of Clostridium botulinum neurotoxin (BoNT) proteases. Toxicon 56:990–998

Vaidyanathan VV, Yoshino K, Jahnz M, Dorries C, Bade S, Nauenburg S, Niemann H, Binz T (1999) Proteolysis of SNAP-25 isoforms by botulinum neurotoxin types A, C, and E: domains and amino acid residues controlling the formation of enzyme-substrate complexes and cleavage. J Neurochem 72:327–337

Wang J, Meng J, Lawrence GW, Zurawski TH, Sasse A, Bodeker MO, Gilmore MA, Fernandez-Salas E, Francis J, Steward LE, Aoki KR, Dolly JO (2008) Novel chimeras of botulinum neurotoxins A and E unveil contributions from the binding, translocation, and protease domains to their functional characteristics. J Biol Chem 283:16993–17002

Weeks DL, Sachs G (2001) Sites of pH regulation of the urea channel of *Helicobacter pylori*. Mol Microbiol 40:1249–1259

Williamson LC, Halpern JL, Montecucco C, Brown JE, Neale EA (1996) Clostridial neurotoxins and substrate proteolysis in intact neurons: botulinum neurotoxin C acts on synaptosomal-associated protein of 25 kDa. J Biol Chem 271:7694–7699

Yamashita S, Yoshida H, Uchiyama N, Nakakita Y, Nakakita SI, Tonozuka T, Oguma K, Nishikawa A, Kamitori S (2012) Carbohydrate recognition mechanism of HA70 from *Clostridium botulinum* deduced from X-ray structures in complexes with sialylated oligosaccharides. FEBS Lett

Uptake of Botulinum Neurotoxin in the Intestine

Yukako Fujinaga, Yo Sugawara and Takuhiro Matsumura

Abstract Foodborne and intestinal botulism are the most common forms of human botulism; both result from the absorption of botulinum neurotoxin (BoNT) from the digestive tract into the circulation. BoNT is a large protein toxin (approximately 150 kDa), but it is able to pass through the epithelial barrier in the digestive tract. Recent cellular and molecular biology studies have begun to unravel the mechanisms by which this large protein toxin crosses the intestinal epithelial barrier. This review provides an overview of current knowledge relating to the absorption of botulinum toxins (BoNT and BoNT complex) from the gastrointestinal tract, with particular emphasis on the interaction of these toxins with the intestinal epithelial barrier.

Keywords Hemagglutinin · Transcytosis · Epithelial monolayer · E-cadherin · Botulinum neurotoxin complex

Abbreviations

BoNT	Botulinum neurotoxin
HA	Hemagglutinin
HC	Heavy chain, H chain
H_C	50 kDa carboxyl-terminal half of heavy chain
LC	Light chain, L chain
NAP	Non-toxic neurotoxin–associated proteins
NTNHA	Non-toxic non-hemagglutinin protein
PTC	Progenitor toxin complex

Y. Fujinaga (✉) · Y. Sugawara · T. Matsumura
Laboratory for Infection Cell Biology, International Research Center for Infectious Diseases, Research Institute for Microbial Diseases, Osaka University, Yamada-oka 3-1, Suita, Osaka 565-0871, Japan
e-mail: yukafuji@biken.osaka-u.ac.jp

A. Rummel and T. Binz (eds.), *Botulinum Neurotoxins*, Current Topics in Microbiology and Immunology 364, DOI: 10.1007/978-3-642-33570-9_3,

Contents

1 Introduction 46
2 Site of Uptake of BoNT and BoNT Complex in the Digestive Tract 48
3 Non-toxic Proteins Protect BoNT from Gastrointestinal Degradation 49
4 Mechanisms Involved in BoNT Invasion Across the Intestinal Epithelial Barrier 49
4.1 H_C-Mediated Transcytosis of BoNT (Route i) 51
4.2 HA-Mediated Binding and Transcytosis of BoNT (Route ii) 51
4.3 HA-Mediated Disruption of Epithelial Barrier (Route iii) 52
5 Conclusion 55
References 56

1 Introduction

Botulinum neurotoxin (BoNT) (serotypes A–G, BoNT/A–G, MW ~150 kDa) is produced by several strains of the anaerobic spore-forming bacteria *Clostridium botulinum*, *C. butyricum*, and *C. baratii*. The toxin causes botulism, a severe neurological disease characterized by flaccid paralysis (reviewed in Schiavo et al. 2000). Among the seven BoNT serotypes, types A, B, E, and F are associated with botulism in both humans and animals, but BoNT/C and D primarily cause disease in domestic animals. Type G-producing organisms have been isolated from soil, and were later reported to be present in several necropsy specimens of humans (Sonnabend et al. 1981, 1987; reviewed in Poulain et al. 2008). Foodborne and intestinal botulism are the primary forms of botulism affecting humans (Fox et al. 2005). In the former disease, ingestion of BoNT-contaminated food causes botulism, whereas in the latter, BoNT-producing clostridia colonize the intestinal lumen of infants (infant botulism) or rarely adults, leading to paralysis. In these conditions, intoxication becomes apparent only if BoNT is absorbed from the digestive tract, enters the systemic circulation, and reaches peripheral nerves (Schiavo et al. 2000).

BoNT-producing clostridia always produce BoNT as a complex consisting of BoNT and non-toxic proteins (non-toxic neurotoxin-associated proteins, NAP) (Sakaguchi 1982; Minton 1995; Collins and East 1998; Oguma et al. 1999; Poulain et al. 2008) (Fig. 1). NAPs include a non-toxic non-hemagglutinin protein (NTNHA) and a hemagglutinin (HA) component. The HA component is composed of three different proteins; HA1 (33 kDa), HA2 (17 kDa), and HA3 (70 kDa). Depending on the strain, the bacterium produces one, two or all of the three main forms of the BoNT complex: 12S toxin (M toxin/M-PTC), 16S toxin (L toxin/L-PTC), and 19S toxin (LL toxin/LL-PTC) (Sakaguchi 1982; Oguma et al. 1999). 12S toxin is composed of a BoNT and an NTNHA. 16S toxin is composed of a BoNT, an NTNHA and an HA component. 19S toxin has the same components as 16S toxin, and is presumed to be a dimer of two 16S toxins linked by one of the HA proteins, HA1. BoNT dissociates from NAP under slightly alkaline conditions (pH > ~ 7–8, depending on the toxin serotype) (reviewed in Sakaguchi 1982). In

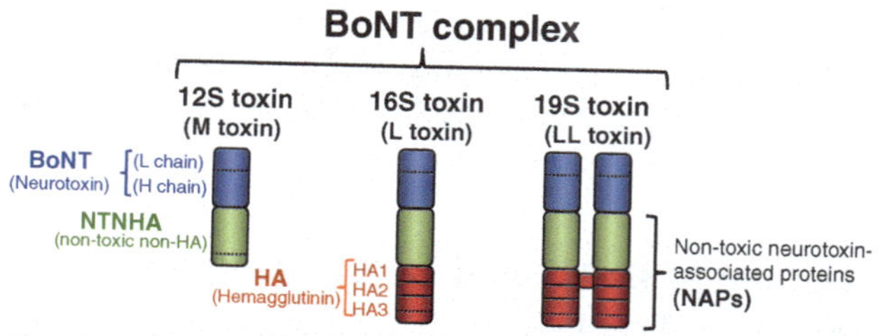

Toxin type	12S toxin	16S toxin	19S toxin
A (A1)	+	+	+
B, C, D	+	+	-
E, F, A2, A3	+	-	-
G	-	+	-

Fig. 1 Schematic structure of botulinum neurotoxin (BoNT) complexes. *Upper panel* Three major forms of BoNT complexes, 12S toxin (M toxin), 16S toxin (L toxin), and 19S toxin (LL toxin), are presented (reviewed in Sakaguchi 1982; Oguma et al. 1999; Poulain et al. 2008). 12S toxin is composed of a BoNT and a non-toxic non-HA (NTNHA). 16S toxin is composed of a BoNT, an NTNHA and a HA component. 19S toxin has the same components as 16S toxin, and is presumed to be a dimer of two 16S toxins linked by one of the HA proteins, HA1. The HA component has three different proteins; HA1, HA2, and HA3. In some strains, HA3 is cleaved into HA3a and 3b. Dashed line denotes the proteolytic cleavage site. *Lower panel* BoNT/A1 is produced by *C. botulinum* in three forms: 12, 16, and 19S toxins. BoNT/B, C, and D are produced in two forms: 12S and 16S. BoNT/A2, A3, E, and F are produced as 12S. BoNT/G complex is produced as 16S, but there is a lack of HA1 gene in the BoNT/G-cluster of genes (Bhandari et al. 1997). + presence of the complex forms; − not present

this review, we cannot precisely refer to BoNT or the BoNT complex (i.e., BoNT plus NAP), we will generally refer to the toxin as "botulinum toxin".

When BoNT is incorporated into complexes, it exhibits greater oral toxicity than uncomplexed BoNT (reviewed in Sakaguchi 1982). This implies that the BoNT complex has the ability to evade the mucosal barrier system in the digestive tract. On the other hand, the gastrointestinal tract has evolved multiple barriers that protect the host from harmful external elements (reviewed in Turner 2009). The gastrointestinal tract is lined with a continuous monolayer of epithelial cells that restrict the passage of potentially harmful molecules from the luminal surface into the surrounding tissues. On the luminal side of enterocytes, a layer of membrane-anchored glycoproteins (filamentous brush border glycocalyx, FBBG), in conjunction with an exterior hydrated gel layer formed by mucins, limit the access of macromolecules to the epithelial surface. In addition, the lumen of the small intestine contains membrane-bound proteases (brush border proteases) and many proteolytic enzymes that are secreted by the pancreas and enterocytes (reviewed in Woodley 1994). BoNT must overcome these barriers in order to access the systemic circulation. Although the molecular mechanisms by which BoNT crosses these barriers are not completely understood, recent studies have significantly advanced our understanding of the strategies exploited by BoNT and BoNT complex in order to surmount these barriers. Here, we describe recent advances in our understanding of the transport pathway used by this toxin to exit the gut lumen, cross the intestinal epithelial barrier, and enter the systemic circulation.

2 Site of Uptake of BoNT and BoNT Complex in the Digestive Tract

BoNT can be absorbed from a mucosal surface or a skin lesion, but not from the intact skin (Arnon et al. 2001). In patients with foodborne or intestinal botulism, toxins are absorbed across the digestive tract. The site of toxin absorption in the digestive tract has been studied previously, using gastrointestinal tract isolated by ligature, in various species of animals (mainly laboratory animals such as rabbits, guinea pigs, and rats). Although the toxin can be absorbed across the epithelial barrier in all areas of the alimentary canal, including the buccal cavity, stomach, and colon, the upper small intestine is considered to be the most important site of absorption (reviewed in Bonventre 1979; Sakaguchi 1982). In the cases of rabbits exposed to type A botulinum toxin, and rats exposed to type B botulinum toxin, absorbed toxin directly enters the lymphatics (reviewed in Bonventre 1979; Sakaguchi 1982). Consequently, orally ingested toxin reaches the circulatory system via the thoracic lymph duct, rather than the hepatic portal system. BoNT/B complexes do not dissociate in the digestive tract (Sugii et al. 1977a); instead, in a study using the rat ligated duodenum loop assay, toxin complexes were absorbed intact from the intestine into the lymphatics (Sugii et al. 1977b). In this system, dissociation occurs immediately after BoNT complexes are absorbed into the lymphatics (Sugii et al. 1977a) (reviewed in Sakaguchi 1982).

In infant botulism, the large intestine (cecum, transverse colon, and recto-sigmoid colon), where anaerobic conditions allow vegetation of *C. botulinum* spores, is the site of *C. botulinum* colonization and toxin production (Mills and Arnon 1987). Furthermore, infant botulism differs from food-borne botulism by its slow and chronic development (Fenicia and Anniballi 2009). This protracted form of the disease may develop when small, sub-lethal doses of BoNT are absorbed over several days. Nevertheless, in some cases, patients' feces contain high levels of BoNT (Paton et al. 1982, 1983). These data, which were obtained from invaluable clinical cases, can be interpreted to mean that the large intestine, where *C. botulinum* spores germinate and toxin is produced, is less efficient as a toxin absorption site than the small intestine.

There are significant differences in the abilities of toxins of various serotypes to pass through the digestive tract in various host species, as indicated by oral/i.v. LD_{50} ratios (Smith 1988). Further accumulation of in vivo experiments using a combination of highly susceptible hosts and multiple toxin serotypes will provide more precise information regarding the absorption sites of the botulinum toxin and transport mechanism into the circulation.

3 Non-toxic Proteins Protect BoNT from Gastrointestinal Degradation

When BoNT is experimentally isolated from BoNT complexes and orally administered to mice, it only induces slight toxicity. However, this oral toxicity increases incrementally as BoNT associates with NAP (reviewed in Sakaguchi 1982). In mouse, among the six toxin serotypes (A–F), the oral toxicity of type B toxin exhibits the most pronounced changes in association with NAP; type B 16S toxin is approximately 700 times more potent than 12S toxin, and 12S toxin is approximately 20 times more potent than BoNT/B (Ohishi et al. 1977; Sakaguchi 1982). Although the mechanism(s) that contribute to the increased oral toxicity of the BoNT complexes is not fully understood, it is known that both NTNHA and HA protect BoNT against the low pH and proteases present in the digestive tract (Sakaguchi 1982). Based on a recently reported X-ray crystal structure of type A 12S toxin, remarkable progress has been made in our understanding of the molecular basis by which NTNHA confers resistance to low pH and proteolysis within the gastrointestinal tract (Gu et al. 2012); (Hill and Smith 2012). In that report, Gu et al. showed that NTNHA has a structure highly similar to that of BoNT, and that NTNHA in 12S toxin provides large and multivalent binding interfaces that shield the sensitive 50 kDa H_C-fragment of BoNT from proteolysis. They also identified a putative HA binding site, "nloop", which is present in nLC of NTNHA remote from the BoNT–NTNHA interface. A structure model for the 16S toxin of BoNT/D, obtained by the combination of transmission electron microscopy and X-ray crystallography, also indicates that HA is localized on the opposite side of the BoNT–NTNHA interface, and does not completely insulate the BoNT (Hasegawa et al. 2007). Thus, the protective function of HA may not be sufficient to explain the mechanisms by which HA reduces the oral median lethal dose (LD_{50}) of 16S BoNT/B by ~700-fold relative to 12S toxin. This phenomenon could be explained, at least partially, by HA-mediated binding and transcytosis of BoNT (Sect. 4.2) and/or HA-mediated disruption of the epithelial barrier (Sect. 4.3).

4 Mechanisms Involved in BoNT Invasion Across the Intestinal Epithelial Barrier

In recent years, several laboratories have described the interaction of BoNT and BoNT complex with the epithelial barrier using in vitro model systems to simulate the intestinal epithelial barrier. A large body of research has shown that BoNT/A and BoNT/B bind to human intestinal epithelial monolayers (Caco-2 and T84) and undergo transcytosis from the apical to the basolateral side (Maksymowych and Simpson 1998, 2004; Ahsan et al. 2005; Couesnon et al. 2008, 2009, 2012; Matsumura et al. 2008). It has also become clear that HA proteins tether BoNT to the apical surface of the intestinal epithelial monolayer via their lectin activities,

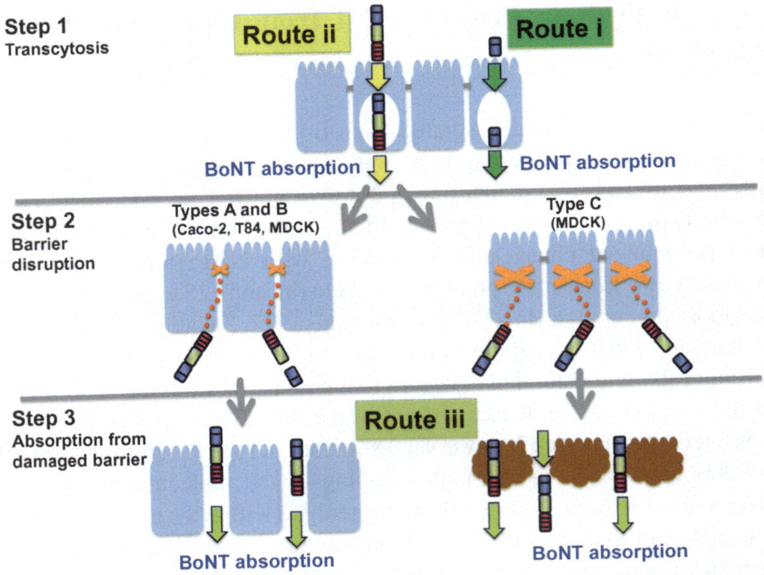

Fig. 2 Proposed model for the penetration of BoNT complexes across the intestinal epithelial barrier. [Step 1] The HC/H_C-fragment of BoNT mediates its binding and transcytosis (*Arrows in green*, "Route i") (Maksymowych and Simpson 1998, 2004; Ahsan et al. 2005; Couesnon et al. 2008, 2009, 2012; Matsumura et al. 2008). Within the BoNT complex, HA also mediates the binding and transcytosis of luminal BoNT complexes across the epithelium, without disrupting the epithelial barrier (*arrows in yellow*, "Route ii") (Matsumura et al. 2008). [Step 2] After translocation to the basolateral surface, HA disrupts the epithelial barrier (Matsumura et al. 2008; Jin et al. 2009). Type A and B HA proteins bind E-cadherin and disrupt the intercellular barrier without causing cytotoxic effects in epithelial cells of susceptible hosts (Sugawara et al. 2010). Type C HA may disrupt barrier function by exerting cytotoxic effects on the epithelial cells of susceptible animals (Jin et al. 2009). [Step 3] Large quantities of BoNT complex pass across the damaged epithelial intercellular barrier, and BoNT accumulates on the serosal side (*arrows in light green*, "Route iii") (Matsumura et al. 2008). Adapted from (Fujinaga 2010)

which presumably leads to transcytosis of BoNT across the monolayer (Fujinaga et al. 1997; Niwa et al. 2007; Arimitsu et al. 2008; Matsumura et al. 2008). Furthermore, HA proteins disrupt the epithelial barrier function (Matsumura et al. 2008; Jin et al. 2009); E-cadherin is the target molecule of type B HA (a complex of HA1/HA2/HA3, HA component of type B 16S toxin) (Sugawara et al. 2010; Sugawara and Fujinaga 2011). In light of these studies, we are currently in a position to speculate that any or all of the identified routes for passing through the epithelial barrier in in vitro models is involved in the mechanisms by which the BoNT crosses the gut epithelium in vivo on the road to pathogenesis. Based on these findings, we propose that the botulinum toxin traverses the intestinal epithelial barrier in three steps (Fig. 2). In Step 1, a small amount of luminal BoNT, either alone (Route i) (Maksymowych and Simpson 1998; Maksymowych and Simpson 2004; Ahsan et al. 2005; Couesnon et al. 2008, 2009, 2012; Matsumura et al. 2008) or in BoNT complexes (Route ii) (Matsumura et al. 2008), undergoes transcytosis across the epithelium without disrupting the epithelial intercellular barrier. Once on the basolateral surface, the HA moiety of the 16S toxin mediates disruption of the intercellular barrier (Step 2) (Matsumura et al. 2008; Jin et al. 2009; Sugawara et al. 2010). Finally, in Step 3, the absence of an intact intercellular barrier allows paracellular passage of large amounts of BoNT complexes and/or BoNT alone (Route iii) (Matsumura et al. 2008). The following sections describe each of these three routes in detail.

4.1 H_C-Mediated Transcytosis of BoNT (Route i)

In the late 1990s, the Simpson lab performed pioneering investigations of the interaction between BoNT and the intestinal epithelial barrier. Those authors showed that BoNT/A and B bind to polarized human intestinal epithelial cell lines (Caco-2 and T84) and undergo transcytosis from the apical to the basolateral side (Maksymowych and Simpson 1998). However, the estimated rate of transport of BoNT/A (specifically, iodinated BoNT/A) through the epithelial barrier is at the low end of the range for other intact proteins (Maksymowych and Simpson 1998). Subsequently, it was shown that the heavy chain (HC) and the 50 kDa H_C-fragment of BoNT/A mediates toxin binding to intestinal cell lines (Maksymowych and Simpson 2004; Couesnon et al. 2008; Couesnon et al. 2009), and that ganglioside (GD1b and GT1b) and SV2C (or SV2C-related protein) expressed on the membranes of these cells are involved in the process of transcytosis (Couesnon et al. 2008). Moreover, the Popoff group demonstrated that a mouse intestinal crypt-like cell line (m-ICc12), which expresses higher levels of SV2C (or SV2C-related protein) than Caco-2, exhibits greater binding and 10-fold higher transcytosis rate of BoNT/A (Couesnon et al. 2008). The transcytosis rate for biologically active BoNT/A in m-ICc12 was estimated to be $\sim 1\ \%$ within 2 h (Couesnon et al. 2008). Although the entire transcytosis pathway followed by BoNT/A has not yet been clearly defined, it was recently reported that BoNT/A H_C is taken up by Caco-2 and m-ICc12 cells via a Cdc42-dependent and clathrin-independent pathway, and reaches an early endosomal compartment (Couesnon et al. 2009). More recently, the Popoff group showed that BoNT/A H_C is preferentially taken up by a subset of neuroendocrine intestinal crypt cells; those authors proposed that these neuroendocrine intestinal crypt cells are the entry site of the BoNT/A across the intestinal barrier (Couesnon et al. 2012).

4.2 HA-Mediated Binding and Transcytosis of BoNT (Route ii)

The carbohydrate-binding activity of HA mediates the agglutination of red blood cells. Although the BoNT complex has been known for 60 years to possess hemagglutination activity (Lamanna 1948), the significance of this activity in the pathogenesis of botulism remained unknown until more recently. In the late 1990s, Fujinaga et al. reported that type C 16S toxin, but neither 12S toxin nor BoNT alone, binds to carbohydrates expressed by guinea pig intestinal epithelium (Fujinaga et al. 1997), suggesting that HA facilitates adhesion of 16S toxin to the intestinal epithelium and may, following uptake of BoNT, contribute to the high oral toxicity of the 16S toxin. Subsequently, HAs from other serotypes of the botulinum toxin complex were shown to bind to intestinal epithelial cells [type A (Fujinaga et al. 2000; Kojima et al. 2005), type D (Niwa et al. 2007), type B (Arimitsu et al. 2008)]. Type C and type D HA in the BoNT complex facilitate the passage of the BoNT across the intestinal epithelial barrier models using human or rat derived cell lines (Niwa et al. 2007, 2010; Inui et al. 2010; Ito et al. 2011). Furthermore, type C 16S toxin, but not the 12S toxin or BoNT alone, induces internalization of the toxin complex and transport of BoNT into the Golgi in the human colon carcinoma cell line HT-29 (Nishikawa et al. 2004; Uotsu et al. 2006). Additionally, HA as constituent of the BoNT/B complex, as well as the recombinant HA (a complex of HA1/HA2/HA3), is transported from the apical side to the basolateral side via transcytosis (Matsumura et al. 2008; Sugawara et al. 2010). These data lead us to surmise that HA binding to cell-expressed carbohydrates induces the transport of BoNT across the intestinal epithelial barrier via transcytosis.

4.3 HA-Mediated Disruption of Epithelial Barrier (Route iii)

Recently, we found that type B HA disrupts the intestinal epithelial barrier by acting on the basolateral membrane (Matsumura et al. 2008). Type B HA disrupts the localization of tight and adherens junction proteins (e.g., occludin, ZO-1, β-catenin, and E-cadherin) at cell-to-cell boundaries, without affecting the viability of cultured epithelial cells. We also showed that $\sim$10-fold more BoNT/B crosses a Caco-2 monolayer when the cells are treated with 16S toxin, compared with cells treated with purified BoNT or 12S toxin over a 24 h time course (Matsumura et al. 2008). Furthermore, we showed that HA facilitates the paracellular transport of macromolecules, including 12S toxin, in an in vivo loop assay in mouse. Both type A and type B HA proteins exhibit similar barrier-disrupting activity on epithelial monolayers derived from several species, including human, but type C HA lacks this activity with respect to human cell lines such as Caco-2 and T84 (Table 1) (Jin et al. 2009). However, type C HA disrupts the epithelial barrier in cultured cells derived from other species, including rat, but these effects are the result of HA-mediated cytotoxicity, and thus fundamentally differ from the mechanism of epithelial disruption caused by type A and B HAs (Jin et al. 2009) (Sugawara and Fujinaga, "unpublished data"). At this time, the effects of type C HA-induced epithelial barrier disruption have only been examined using cell lines in vitro; in vivo studies using susceptible animals are required in order to validate its pathological role.

Table 1 Effects of type A, B, and C 16S toxin complexes on epithelial monolayers and cells

	Caco-2		T84		MDCKI		ACL-15	RCN-9
	apical	basal	apical	basal	apical	basal		
Barrier disruption								
type A*	+	++	±	+	+	++	ND	ND
type B	±	++	–	+	–	++	ND	ND
type C	–	–	–	–	–	+	ND	ND
Cytotoxicity								
type A*	–		–		–		+	+
type B	–		–		–		–	–
type C	±		–		+		+	+

Barrier disruption
++ 50 % decrease in TER after 12 h incubation was induced by <5 nM of toxin complex
+ 50 % decrease in TER after 12 h incubation was induced by <300 nM of toxin complex
± 20 % decrease in TER after 24 h incubation was induced by <300 nM of toxin complex
– decrease in TER was not induced within 24 h by 300 nM of toxin complex
Cytotoxicity (non-polarised cells)
+ WST-8 reduction was decreased to <50 % by 300 nM of toxin complex
± WST-8 reduction was decreased to <80 % by 300 nM of toxin complex
– no decrease in WST-8 reduction was induced by 300 nM of toxin complex
ND not determined
* type A: A mixture of 16S toxin and 19S toxin
Reprinted with permission from Jin et al. (2009)

More recently, we determined that epithelial cadherin (E-cadherin) is a target of type B HA (Fig. 3) (Sugawara et al. 2010; Sugawara and Fujinaga 2011). E-cadherin is a canonical member of the cadherin superfamily of adhesion molecules, and mediates calcium-dependent cell-to-cell adhesion (Takeichi 1991). In epithelial cells, E-cadherin is present at adherens junctions, and is essential for the function of tight junctions (Citi 1993). Type B HA directly binds to E-cadherin, and disrupts E-cadherin-mediated cell-to-cell adhesion (Sugawara et al. 2010). All three HA subunits are required for full binding to E-cadherin, although a complex consisting of only HA2 and HA3 exhibits weak binding activity. The carbohydrate-binding activity of type B HA is not required for this interaction. Type B HA binds to human, bovine, and mouse E-cadherins, but not the rat or chicken homologs. Additionally, type B HA does not interact with other members of the classic cadherin family, such as N- and VE-cadherin. Residue 20 of E-cadherin is critically involved in its interaction with type B HA (Fig. 4a); this residue is located in the distal extracellular cadherin domain (EC1) near the cadherin trans-dimer interface (Boggon et al. 2002). Therefore, it is conceivable that type B HA binding to E-cadherin sterically hinders E-cadherin trans-dimer formation. Anatomically, this presents a conundrum because E-cadherin is found exclusively on the basolateral surface of epithelial cells, whereas orally ingested HA in the BoNT complex is initially localized on the apical surface. Consistent with this, we showed in an in vitro epithelial model that the HA target site is on the

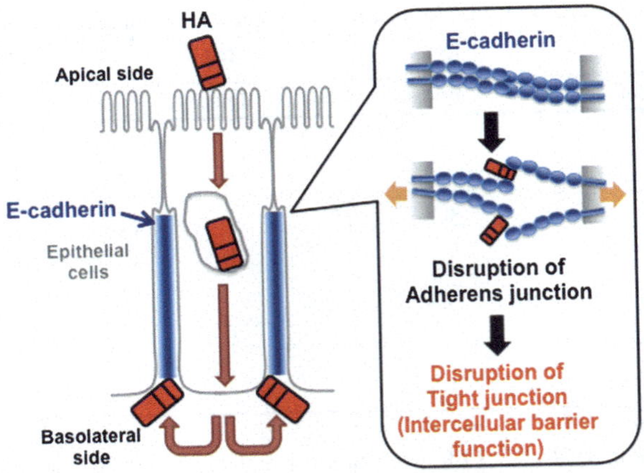

Fig. 3 Schematic representation of the mechanism of action of type B HA (Sugawara et al. 2010). HA (HA1/2/3 complex) bounds to the apical surface of epithelial cells gains access to the basolateral compartment by transcytosis. Translocated HA interacts with the extracellular domain of E-cadherin, leading to the disruption of adherens and tight junctions, which play a vital role in intercellular barrier function

basolateral surface, requiring HA to be transported from the apical to the basolateral surface in order to co-localize with E-cadherin. Finally, we showed that transcytosis mediates the transport of HA, at least in vitro [Fig. 3,(Matsumura et al. 2008; Sugawara et al. 2010), also see Sect. 4.2].

E-cadherin binding by HA is species specific, and the observed species specificity correlates with the epidemiology of botulism (Sugawara et al. 2010). For example, type B HA does not interact with chicken E-cadherin, consistent with the rarity of avian botulism caused by BoNT/B. Furthermore, in an experimental setting, birds are resistant to the BoNT/B complex, especially when administered orally and also upon intravenous administration (Gross and Smith 1971; Notermans et al. 1980). Additionally, the ability of BoNT/A and B complexes to interact with human E-cadherin, and the inability of BoNT/C complex to do so, correlates with the frequency of botulism caused by these types in humans (i.e., types A and B are much more common than type C) (Sugawara et al. 2010). These differences are not solely due to the differential ability of HA to bind E-cadherin. Chickens and rats are not typically susceptible to type B botulinum toxin, and their orthologs of VAMP-1 (also known as synaptobrevin-1), one of the intracellular substrates for BoNT/B, are resistant to BoNT-mediated cleavage due to a single amino acid change (Q78V) at the position P1 (Fig. 4b, Patarnello et al. 1993). Patarnello et al. proposed that this amino acid change at the site of BoNT's catalytic action in these different species is an example of convergent molecular evolution driven by BoNT. Additionally, both of these species have an identical amino acid substitution at residue 20 of E-cadherin (Sugawara et al. 2010). Thus, rats and chickens have evolved two distinct mechanisms to evade intoxication by type B botulinum toxin. Taken together, these findings suggest that the HA-E-cadherin interaction is

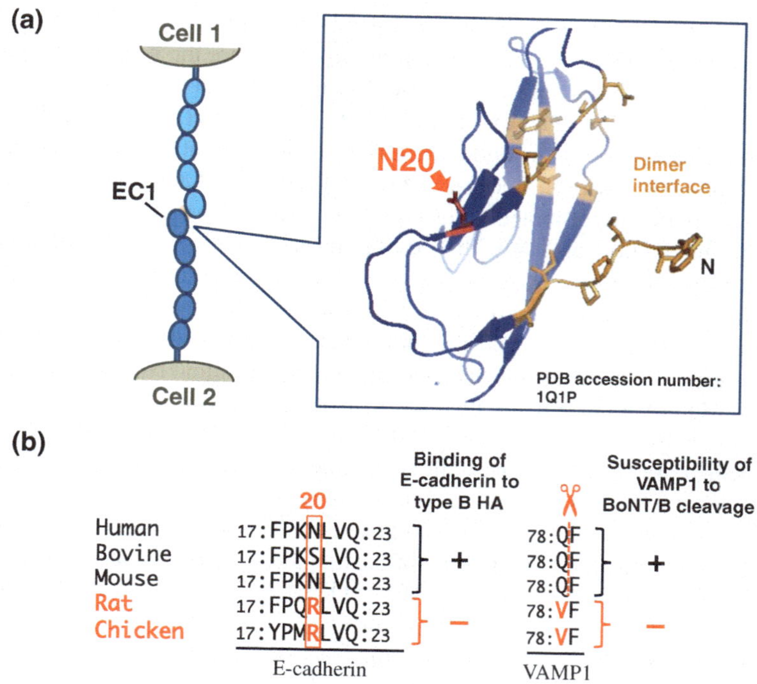

Fig. 4 E-cadherin binding of type B HA. a Topology of E-cadherin in trans (juxtaposed cell) interaction and the binding site for type B HA is shown. *Left panel* Schematic representation of E-cadherin in a trans-dimer interaction, deduced from X-ray crystal structure studies (Boggon et al. 2002; Haussinger et al. 2004). *Right panel* 3D structure of the EC1 domain of E-cadherin (from 1Q1P.PDB). Residue 20 of E-cadherin, which is critically important to its interaction with HA, is shown in *red* (Sugawara et al. 2010). Residues that make direct contact with E-cadherin in the context of homodimer formation are shown in *orange* (Boggon et al. 2002). **b** Amino acid sequences of the E-cadherin region that contains the critical site for type B HA binding, and of the neuronal VAMP-1 region that contains the site of proteolytic cleavage of BoNT/B. Sequences derived from human, bovine, mouse, rat, and chicken are shown (Patarnello et al. 1993; Sugawara et al. 2010)

an important factor in the pathogenesis of food-borne botulism. Additional studies using genetically modified animals expressing HA-resistant isoforms of E-cadherin are required in order to test these hypotheses.

5 Conclusion

In the past few years, we have made significant progress in understanding the interaction of botulinum toxin with the intestinal epithelial barrier. Accumulating evidence suggests that BoNT can be itself undergo transcytosis, and that the interaction of BoNT/A H_C with cellular ligands including gangliosides (GD1b and GT1b series) and SV2 (or SV2-related protein), which are the neuronal ligands for the BoNT/A, facilitates this process (Maksymowych and Simpson 1998, 2004; Maksymowych et al. 1999; Ahsan et al. 2005; Couesnon et al. 2008, 2009, 2012; Matsumura et al. 2008). In addition, HA proteins present in BoNT complexes disrupt epithelial barrier function (Matsumura et al. 2008; Jin et al. 2009); E-cadherin is the target molecule of type B HA (Sugawara et al. 2010). Furthermore, the interaction of HA with E-cadherin varies with HA serotype and host species. Similarly, different serotypes of BoNT recognize different molecules on the surface of neuronal cells (reviewed in Binz and Rummel 2009). Taken together, these findings suggest that the individual botulinum toxin serotypes exploit diverse molecular mechanisms in order to cross the intestinal barrier of susceptible animals. Moreover, it remains possible that other pathways are involved in botulinum toxin absorption. For example, 12S toxins (M-PTCs) of BoNT/A2, A3, E, and F, which are not assembled with HAs, may use novel mechanisms in order to enhance their intestinal absorption. Interestingly, *C. botulinum* strains A2 ~ A4, A(B), E, and F possess the *orf X* gene-cluster, which contains open reading frames (ORFs) of unknown function, instead of the *ha* gene-cluster (Jacobson et al. 2008). The role of *orf X* cluster proteins (ORFX1, ORFX2, ORFX3, and P47) in the oral toxicity of the BoNTs should be examined.

Different host species are exposed to different BoNT-producing clostridia based on their environmental niches and geographical distribution; both hosts and toxins have evolved to reflect these differences. Further in vivo and in vitro studies using appropriate host species are required in order to fully understand the strategies that have evolved to promote (and prevent) toxin absorption. These studies will provide important insights into the molecular mechanisms underlying food borne and intestinal botulism, and they may promote the development of novel methods for therapeutically modulating the integrity of the intestinal epithelial barrier.

Acknowledgments I thank Drs. Andreas Rummel and Thomas Binz for critically reading the manuscript. This work was granted by the Japan Society for the Promotion of Science (JSPS) through the "Funding Program for Next Generation World-Leading Researchers (NEXT Program)," initiated by the Council for Science and Technology Policy (CSTP). The author has no conflicting financial interest.

References

Ahsan CR, Hajnoczky G, Maksymowych AB, Simpson LL (2005) Visualization of binding and transcytosis of botulinum toxin by human intestinal epithelial cells. J Pharmacol Exp Ther 315:1028–1035

Arimitsu H, Sakaguchi Y, Lee JC, Ochi S, Tsukamoto K, Yamamoto Y, Ma S, Tsuji T, Oguma K (2008) Molecular properties of each subcomponent in *Clostridium botulinum* type B haemagglutonin complex. Microb Pathog 45:142–149

Arnon SS, Schechter R, Inglesby TV, Henderson DA, Bartlett JG, Ascher MS, Eitzen E, Fine AD, Hauer J, Layton M, Lillibridge S, Osterholm MT, O'Toole T, Parker G, Perl TM, Russell PK,

Swerdlow DL, Tonat K (2001) Botulinum toxin as a biological weapon: medical and public health management. JAMA 285:1059–1070

Bhandari M, Campbell KD, Collins MD, East AK (1997) Molecular characterization of the clusters of genes encoding the botulinum neurotoxin complex in *Clostridium botulinum* (Clostridium argentinense) type G and nonproteolytic *Clostridium botulinum* type B. Curr Microbiol 35:207–214

Binz T, Rummel A (2009) Cell entry strategy of clostridial neurotoxins. J Neurochem 109:1584–1595

Boggon TJ, Murray J, Chappuis-Flament S, Wong E, Gumbiner BM, Shapiro L (2002) C-cadherin ectodomain structure and implications for cell adhesion mechanisms. Science 296:1308–1313

Bonventre PF (1979) Absorption of botulinal toxin from the gastrointestinal tract. Rev Infect Dis 1:663–667

Citi S (1993) The molecular organization of tight junctions. J Cell Biol 121:485–489

Collins MD, East AK (1998) Phylogeny and taxonomy of the food-borne pathogen *Clostridium botulinum* and its neurotoxins. J Appl Microbiol 84:5–17

Couesnon A, Molgo J, Connan C, Popoff MR (2012) Preferential entry of botulinum neurotoxin A Hc domain through intestinal crypt cells and targeting to cholinergic neurons of the mouse intestine. PLoS Pathog 8:e1002583

Couesnon A, Pereira Y, Popoff MR (2008) Receptor-mediated transcytosis of botulinum neurotoxin A through intestinal cell monolayers. Cell Microbiol 10:375–387

Couesnon A, Shimizu T, Popoff MR (2009) Differential entry of botulinum neurotoxin A into neuronal and intestinal cells. Cell Microbiol 11:289–308

Fenicia L, Anniballi F (2009) Infant botulism. Ann Ist Super Sanita 45:134–146

Fox CK, Keet CA, Strober JB (2005) Recent advances in infant botulism. Pediatr Neurol 32:149–154

Fujinaga Y (2010) Interaction of botulinum toxin with the epithelial barrier. J Biomed Biotechnol 2010:974943

Fujinaga Y, Inoue K, Nomura T, Sasaki J, Marvaud JC, Popoff MR, Kozaki S, Oguma K (2000) Identification and characterization of functional subunits of *Clostridium botulinum* type A progenitor toxin involved in binding to intestinal microvilli and erythrocytes. FEBS Lett 467:179–183

Fujinaga Y, Inoue K, Watanabe S, Yokota K, Hirai Y, Nagamachi E, Oguma K (1997) The haemagglutinin of *Clostridium botulinum* type C progenitor toxin plays an essential role in binding of toxin to the epithelial cells of guinea pig small intestine, leading to the efficient absorption of the toxin. Microbiology 143(Pt 12):3841–3847

Gross WB, Smith LD (1971) Experimental botulism in gallinaceous birds. Avian Dis 15:716–722

Gu S, Rumpel S, Zhou J, Strotmeier J, Bigalke H, Perry K, Shoemaker CB, Rummel A, Jin R (2012) Botulinum neurotoxin is shielded by NTNHA in an interlocked complex. Science 335:977–981

Hasegawa K, Watanabe T, Suzuki T, Yamano A, Oikawa T, Sato Y, Kouguchi H, Yoneyama T, Niwa K, Ikeda T, Ohyama T (2007) A novel subunit structure of *Clostridium botulinum* serotype D toxin complex with three extended arms. J Biol Chem 282:24777–24783

Haussinger D, Ahrens T, Aberle T, Engel J, Stetefeld J, Grzesiek S (2004) Proteolytic E-cadherin activation followed by solution NMR and X-ray crystallography. EMBO J 23:1699–1708

Hill KK, Smith TJ (2012) Genetic diversity within Clostridium botulinum serotypes, botulinum neurotoxin Gene. Cluster and toxin subtypes. doi: 10.1007/978-3-642-33570-9-1

Inui K, Ito H, Miyata K, Matsuo T, Horiuchi R, Ikeda T, Watanabe T, Ohyama T, Niwa K (2010) Involvement of sialic acid in transport of serotype C1 botulinum toxins through rat intestinal epithelial cells. J Vet Med Sci 72:1251–1255

Ito H, Sagane Y, Miyata K, Inui K, Matsuo T, Horiuchi R, Ikeda T, Suzuki T, Hasegawa K, Kouguchi H, Oguma K, Niwa K, Ohyama T, Watanabe T (2011) HA-33 facilitates transport of the serotype D botulinum toxin across a rat intestinal epithelial cell monolayer. FEMS Immunol Med Microbiol 61:323–331

Jacobson MJ, Lin G, Raphael B, Andreadis J, Johnson EA (2008) Analysis of neurotoxin cluster genes in *Clostridium botulinum* strains producing botulinum neurotoxin serotype A subtypes. Appl Environ Microbiol 74:2778–2786

Jin Y, Takegahara Y, Sugawara Y, Matsumura T, Fujinaga Y (2009) Disruption of the epithelial barrier by botulinum haemagglutinin (HA) proteins–differences in cell tropism and the mechanism of action between HA proteins of types A or B, and HA proteins of type C. Microbiology 155:35–45

Kojima S, Eguchi H, Ookawara T, Fujiwara N, Yasuda J, Nakagawa K, Yamamura T, Suzuki K (2005) *Clostridium botulinum* type A progenitor toxin binds to Intestine-407 cells via N-acetyllactosamine moiety. Biochem Biophys Res Commun 331:571–576

Lamanna C (1948) Hemagglutination by botulinal toxin. Proc Soc Exp Biol Med 69:332–336

Maksymowych AB, Reinhard M, Malizio CJ, Goodnough MC, Johnson EA, Simpson LL (1999) Pure botulinum neurotoxin is absorbed from the stomach and small intestine and produces peripheral neuromuscular blockade. Infect Immun 67:4708–4712

Maksymowych AB, Simpson LL (1998) Binding and transcytosis of botulinum neurotoxin by polarized human colon carcinoma cells. J Biol Chem 273:21950–21957

Maksymowych AB, Simpson LL (2004) Structural features of the botulinum neurotoxin molecule that govern binding and transcytosis across polarized human intestinal epithelial cells. J Pharmacol Exp Ther 310:633–641

Matsumura T, Jin Y, Kabumoto Y, Takegahara Y, Oguma K, Lencer WI, Fujinaga Y (2008) The HA proteins of botulinum toxin disrupt intestinal epithelial intercellular junctions to increase toxin absorption. Cell Microbiol 10:355–364

Mills DC, Arnon SS (1987) The large intestine as the site of *Clostridium botulinum* colonization in human infant botulism. J Infect Dis 156:997–998

Minton NP (1995) Molecular genetics of clostridial neurotoxins. Curr Top Microbiol Immunol 195:161–194

Nishikawa A, Uotsu N, Arimitsu H, Lee JC, Miura Y, Fujinaga Y, Nakada H, Watanabe T, Ohyama T, Sakano Y, Oguma K (2004) The receptor and transporter for internalization of *Clostridium botulinum* type C progenitor toxin into HT-29 cells. Biochem Biophys Res Commun 319:327–333

Niwa K, Koyama K, Inoue S, Suzuki T, Hasegawa K, Watanabe T, Ikeda T, Ohyama T (2007) Role of nontoxic components of serotype D botulinum toxin complex in permeation through a Caco-2 cell monolayer, a model for intestinal epithelium. FEMS Immunol Med Microbiol 49:346–352

Niwa K, Yoneyama T, Ito H, Taira M, Chikai T, Kouguchi H, Suzuki T, Hasegawa K, Miyata K, Inui K, Ikeda T, Watanabe T, Ohyama T (2010) Sialic acid-dependent binding and transcytosis of serotype D botulinum neurotoxin and toxin complex in rat intestinal epithelial cells. Vet Microbiol 141:312–320

Notermans S, Dufrenne J, Kozaki S (1980) Experimental botulism in Pekin ducks. Avian Dis 24:658–664

Oguma K, Inoue K, Fujinaga Y, Yokota K, Watanabe T, Ohyama T, Takeshi K, Inoue K (1999) Structure and function of *Clostridium botulinum* progenitor toxin. J Toxicol-Toxin Rev 18:17–34

Ohishi I, Sugii S, Sakaguchi G (1977) Oral toxicities of *Clostridium botulinum* toxins in response to molecular size. Infect Immun 16:107–109

Patarnello T, Bargelloni L, Rossetto O, Schiavo G, Montecucco C (1993) Neurotransmission and secretion. Nature 364:581–582

Paton JC, Lawrence AJ, Manson JI (1982) Quantitation of *Clostridium botulinum* organisms and toxin in the feces of an infant with botulism. J Clin Microbiol 15:1–4

Paton JC, Lawrence AJ, Steven IM (1983) Quantities of *Clostridium botulinum* organisms and toxin in feces and presence of *Clostridium botulinum* toxin in the serum of an infant with botulism. J Clin Microbiol 17:13–15

Poulain B, Popoff MR, Molgo J (2008) How do the Botulinum Neurotoxins block neurotransmitter release: from botulinum to the molecular mechanism of action. Botulinum J 1:14–87

Sakaguchi G (1982) *Clostridium botulinum* toxins. Pharmacol Ther 19:165–194
Schiavo G, Matteoli M, Montecucco C (2000) Neurotoxins affecting neuroexocytosis. Physiol Rev 80:717–766
Smith LDS (1988) Botulism: the organism, its toxins, the disease. Springfield, Illinois, USA
Sonnabend O, Sonnabend W, Heinzle R, Sigrist T, Dirnhofer R, Krech U (1981) Isolation of *Clostridium botulinum* type G and identification of type G botulinal toxin in humans: report of five sudden unexpected deaths. J Infect Dis 143:22–27
Sonnabend WF, Sonnabend UP, Krech T (1987) Isolation of *Clostridium botulinum* type G from Swiss soil specimens by using sequential steps in an identification scheme. Appl Environ Microbiol 53:1880–1884
Sugawara Y, Fujinaga Y (2011) The botulinum toxin complex meets E-cadherin on the way to its destination. Cell Adh Migr 5:34–36
Sugawara Y, Matsumura T, Takegahara Y, Jin Y, Tsukasaki Y, Takeichi M, Fujinaga Y (2010) Botulinum hemagglutinin disrupts the intercellular epithelial barrier by directly binding E-cadherin. J Cell Biol 189:691–700
Sugii S, Ohishi I, Sakaguchi G (1977a) Correlation between oral toxicity and in vitro stability of *Clostridium botulinum* type A and B toxins of different molecular sizes. Infect Immun 16:910–914
Sugii S, Ohishi I, Sakaguchi G (1977b) Intestinal absorption of botulinum toxins of different molecular sizes in rats. Infect Immun 17:491–496
Takeichi M (1991) Cadherin cell adhesion receptors as a morphogenetic regulator. Science 251:1451–1455
Turner JR (2009) Intestinal mucosal barrier function in health and disease. Nat Rev Immunol 9:799–809
Uotsu N, Nishikawa A, Watanabe T, Ohyama T, Tonozuka T, Sakano Y, Oguma K (2006) Cell internalization and traffic pathway of *Clostridium botulinum* type C neurotoxin in HT-29 cells. Biochim Biophys Acta 1763:120–128
Woodley JF (1994) Enzymatic barriers for GI peptide and protein delivery. Crit Rev Ther Drug Carrier Syst 11:61–95

Double Receptor Anchorage of Botulinum Neurotoxins Accounts for their Exquisite Neurospecificity

Andreas Rummel

Abstract The high potency of the botulinum neurotoxins (BoNT) and tetanus neurotoxin (TeNT) is mainly due to their neurospecific binding which is mediated by the interaction with two receptor components. TeNT and all BoNT bind first to complex polysialo-gangliosides abundantly present on the outer leaflet of neuronal membranes. The ganglioside binding occurs in BoNT/A, B, E, F and G via a conserved ganglioside binding pocket within the most carboxyl-terminal 25 kDa domain H_{CC} whereas TeNT, BoNT/C and D display two different ganglioside binding sites within their H_{CC}-domain. Subsequently, upon exocytosis the intraluminal domains of synaptic vesicle proteins are exposed and can be accessed by the surface accumulated neurotoxins. BoNT/B and G bind with their H_{CC}-domain to a 20-mer membrane juxtaposed segment of the intraluminal domain of synaptotagmin-I and -II, respectively. BoNT/A and E employ the intraluminal domain 4 of the synaptic vesicle glycoprotein 2 (SV2) as protein receptor. Whereas the 50 kDa cell binding domain H_C of BoNT/A interacts with all three SV2 isoforms, BoNT/E H_C only binds SV2A and SV2B. Also, BoNT/D, F, and TeNT employ SV2 for binding and uptake. Thereafter, the synaptic vesicle is recycled and the anchored neurotoxin is endocytosed. Acidification of the vesicle lumen triggers membrane insertion of the translocation domain followed by pore formation and finally translocation of the enzymatically active light chain to its site of action leading to block of neurotransmitter release.

Keywords Botulinum neurotoxin · Ganglioside binding site · Protein receptor binding site · Synaptic vesicle protein

A. Rummel (✉)
Institut für Toxikologie,
Medizinische Hochschule Hannover, 30623 Hannover, Germany
e-mail: rummel.andreas@mh-hannover.de

A. Rummel and T. Binz (eds.), *Botulinum Neurotoxins*, Current Topics in Microbiology and Immunology 364, DOI: 10.1007/978-3-642-33570-9_4,

Abbreviations

BoNT	botulinum neurotoxin
BoNT/X	serotype X of BoNT, X = A–G
BoNT/XY	subtype Y of serotype BoNT/X, Y = 1–8
Cer	ceramide
CNT	clostridial neurotoxins
NMJ	neuromuscular junction
Gal	β-galactose
GBS	ganglioside binding site
Glc	β-glucose
GPI	glycosylphosphatidylinositol
GT1b, GD1b, GD1a, GM1, GM3, GT2	sialo ganglioside
HC/X	heavy chain of BoNT serotype X/TeNT
H_CX	50 kDa cell binding fragment of BoNT/TeNT
H_{CC}	25 kDa C-terminal domain of the H_C
H_{CN}	25 kDa N-terminal domain of the H_C
H_NX	50 kDa translocation domain of BoNT/TeNT
Lac	lactose
LacCer	lactose-ceramide
LC/X	light chain, catalytic domain of BoNT serotype X/TeNT
MS	mass spectrometry
NAcGal	N-acetyl-β-galactosamine
NAcNeu	N-acetylneuraminic acid, sialic acid, Sia
SNAP-25	synaptosomal associated protein of 25 kDa
SNARE	soluble N-ethylmaleimide-sensitive factor attachment protein receptor
SV2	synaptic vesicle glycoprotein 2
Syt-X	Isoform X of synaptotagmin, X = I-XV
syntaxin-X	Isoform X of syntaxin, X = 1–19
TeNT	tetanus neurotoxin
VAMP-2	synaptobrevin-2

Contents

1 Introduction 63
2 Complex Polysialo Gangliosides Accumulate BoNT on the Neuronal Membrane 63
3 The Four Domain Structure of TeNT and BoNT 66

4 The Mode of Molecular CNT-Ganglioside Interaction 68
5 Synaptic Vesicle Proteins are Receptors of Botulinum Neurotoxins 75
6 Mode of the BoNT-Protein Receptor Interaction 80
7 The BoNT Membrane Approach 82
8 Outlook 83
References 84

1 Introduction

Tetanus neurotoxin (TeNT) and the seven botulinum neurotoxin serotypes (BoNT/A-G) exert an extreme toxicity (i.p. <1 ng/kg bodyweight) which is caused by their high neuronal specificity, an efficient uptake in clathrin-coated vesicles (Bercsenyi et el. 2012) and subsequent delivery of a Zn^{2+}-endoprotease into the neuronal cytosol (Fischer 2012), where the metalloprotease specifically cleaves one member of the SNARE proteins (Binz 2012; Ahnert-Hilger et al. 2012) resulting in blockade of neurotransmitter release for a long duration (Shoemaker and Oyler 2012).

BoNT bind highly specific to non-myelinated areas of cholinergic motor nerve terminals (Dolly et al. 1984). Like for cholera toxin, gangliosides, complex polysialic acid containing glycolipids, play an important role in binding of TeNT and BoNT. However, a protease sensitive interaction of TeNT and BoNT with neuronal membranes led to the dual receptor hypothesis which postulates an interaction with gangliosides and a proteinaceous receptor (Montecucco 1986). Subsequently, the individual type of gangliosides and the mode of interaction as well as the nature of the proteinaceous receptors have been characterised extensively.

2 Complex Polysialo Gangliosides Accumulate BoNT on the Neuronal Membrane

TeNT, sharing a 35 % amino acid identity with BoNT (Niemann et al. 1994), was first identified to bind gangliosides (ganglion + glycoside), glycosphingolipids that are found particularly in the outer leaflet of neuronal cell membranes (van Heyningen 1958, 1959; van Heyningen and Miller 1961).

Polysialo-gangliosides consist of a ceramide part (Cer) which is incorporated in the membrane and a carbohydrate portion (Fig. 1a, b). The Cer is built up by sphingosine connecting a fatty acid chain and an alkyl chain to the carbohydrate part. The first carbohydrate is a β-glucose (Glc) linked 1,4 to β-galactose (Gal). This lactose-ceramide (LacCer) is the starting material for synthesis of complex polysialo-gangliosides like GT1b (nomenclature according to (Svennerholm

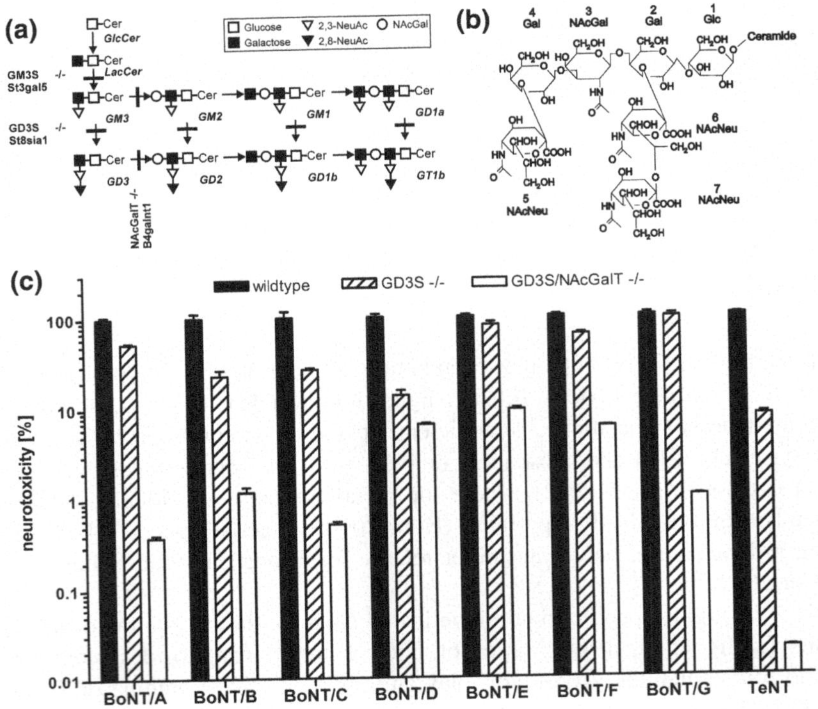

Fig. 1 **a** Schematic representation of the carbohydrate building blocks and the biosynthetic pathway of complex polysialo-gangliosides. The enzymes GM3-synthetase (GM3S), GD3-synthetase (GD3S) and N-acetylgalactosamine-transferase (NAcGalT) are involved in ganglioside biosynthesis. Their encoding genes are deleted in the respective knock-out mice leading to an altered ganglioside expression pattern. **b** Chemical structure of polysialo-ganglioside GT1b. **c** TeNT and BoNT show reduced neurotoxicity when applied to phrenic nerve hemidiaphragm preparations obtained from GD3S or combined GD3S/NAcGalT knock-out mice. These mice only express GM3, GM2, GM1 and GD1a, or GM3-only, respectively (n = 3–7, mean ± SD) (Rummel et al. 2007; Rummel et al. 2009; Strotmeier et al. 2010)

1994), Fig. 1a). GM3-synthetase attaches the first sialic acid (α-*N*-acetylneuraminic acid, NAcNeu) 2,3 to the Gal yielding GM3. The GD3-synthetase (GD3S) elongates the NAcNeu-branch by fusing a second α-NAcNeu 2,8 to the first NAcNeu resulting in the ganglioside GD3. Furthermore, the NAcGal-transferase (NAcGalT) extends the LacCer by connecting first β-*N*-acetylgalactosamine (NAcGAl) 1,4 and then adding Gal 1,3 leading to GM2 and GM1, respectively, the latter being the receptor for cholera toxin (Holmgren et al. 1980).

The knowledge of ganglioside biosynthesis allowed the genetic approach to create mice lacking complex polysialo-gangliosides. This is achieved by deleting the genes St8sia1 and B4galnt1 encoding NAcGalT and GD3S, respectively (Fig. 1a). NAcGalT-deficient mice only express LacCer, GM3 and GD3 and displayed resistance against treatment with TeNT as well as BoNT/A, B, D and G in time-to-death experiments (Peng et al. 2011; Kitamura et al. 1999; Dong et al.

2007). Furthermore, nerve stimulation-evoked endplate potentials at isolated neuromuscular junctions derived from NAcGalT-deficient mice remained unaltered upon incubation with BoNT/A (Bullens et al. 2002). In addition, binding and entry of the seven BoNT serotypes was reduced in cultured hippocampal neurons of NAcGalT-deficient mice, but could be rescued by adding exogenous bovine brain ganglioside mix (Peng et al. 2011; Dong et al. 2007, 2008). On the other hand, GD3S knock-out mice expressing only LacCer and the a-series gangliosides GM3, GM2, GM1 and GD1a were resistant to TeNT in the time-to-death assay, but kept their sensitivity towards BoNT/A, B and E (Kitamura et al. 2005) which indicates that the disialyl moiety of GT1b plays a minor role in binding of BoNT/A, B and E. Isolating the phrenic nerve hemidiaphragm of such knock-out mice for an ex vivo assay (mouse phrenic nerve (MPN) assay) allows the precise quantitation of the individual BoNT-ganglioside dependence at their site of action, the motor nerve terminals. Hereto the hemidiaphragms of mice lacking NAcGalT and/or GD3S were intoxicated with TeNT and BoNT/A-G [Fig. 1c; (Rummel et al. 2007, 2009; Strotmeier et al. 2010)]. The lack of the disialyl group in GD3S-/- tissue hardly affected the action of BoNT/A, E, F and G, whereas the potency of BoNT/B, C, D and especially TeNT was interfered. A conclusive combination of both gene knock-outs resulted in GM3-only mice which displayed high resistance towards all seven BoNT serotypes and especially TeNT indicating the importance of the terminal NAcGal-Gal-NAcNeu moiety being present in GD1a and GT1b as well as the disialyl group of GD1b and GT1b.

Biochemical approaches at the cellular level like removal of sialic acid residues by neuraminidase treatment of cultured cells isolated from spinal cord (Bigalke et al. 1986) and adrenergic chromaffin cells (Marxen et al. 1989) demonstrated a reduced potency of BoNT/A and TeNT (Critchley et al. 1986). Furthermore, binding of BoNT/C to neuroblastoma cell lines as well as rat brain synaptosomes was diminished upon neuraminidase treatment (Tsukamoto et al. 2005; Yokosawa et al. 1989), all indicating interactions between sialic acid moieties and BoNT/A, C and TeNT. Conversely, bovine chromaffin cells lacking complex polysialo-gangliosides were rendered sensitive to TeNT and BoNT/A by pretreatment with gangliosides (Marxen and Bigalke 1989; Marxen et al. 1991). In addition, a monoclonal antibody to GT1b antagonised the action of BoNT/A on rat superior cervical ganglions (Kozaki et al. 1998). The inhibition of ganglioside biosynthesis with fumonisin in primary spinal cord neurons or with D,L-threo-l-phenyl-2-hexadecanoylamino-3-morpholino-propanol in the mouse neuroblastoma cell line Neuro-2a resulted in insensitivity towards TeNT and BoNT/A, respectively (Williamson et al. 1999; Yowler et al. 2002).

After it became clear that gangliosides act as acceptor it was demonstrated that preincubation of BoNT/A, B, D or E with gangliosides, especially GT1b resulted in their detoxification (Simpson and Rapport 1971a, b). Extensive studies characterised the ganglioside specificity of BoNT and TeNT first in overlay binding assays employing ganglioside mixtures separated by thin layer chromatography. It was demonstrated that BoNT/A, B, C, E and F bind to GT1b, GD1b and GD1a with varying affinities (Tsukamoto et al. 2005; Kozaki et al. 1987; Takamizawa et al. 1986; Kamata et al. 1986; Ochanda et al. 1986; Kitamura et al. 1980). BoNT/A, B, and E adhered to GT1b better than to GD1a and much less to GM1 at low ionic strength, and as the ionic strength increased to physiological conditions, less binding was observed (Schengrund et al. 1991). However, employing surface plasmon resonance, BoNT/A bound to GT1b when the ionic strength was increased from 0.06 to 0.16 M with a similar K_D ($\sim 10^{-7}$ M) for each ionic strength (Yowler and Schengrund 2004). Furthermore, MALDI-TOF mass spectroscopy demonstrated binding of isolated GT1b to the cell binding domain of BoNT/A, B, D and TeNT (Strotmeier et al. 2010; Rummel et al. 2003, 2004a). Use of immobilised, isolated, individual gangliosides on polystyrene surfaces of microtiter plates complemented the understanding of ganglioside preference. TeNT prefers the b-series gangliosides GT1b, GD1b and GQ1b (Rummel et al. 2003; Chen et al. 2008; Angstrom et al. 1994). Isolated GT1b also binds BoNT/A, B and with higher affinity BoNT/G (Rummel et al. 2004a; Schmitt et al. 2010), whereas BoNT/F interacts predominantly with GD1a and GT1b but hardly with GD1b or GM1 (Fu et al. 2009). GD1a is bound best by BoNT/F, followed by BoNT/E and A (Benson et al. 2011) thereby supporting the GD3S-knock-out mice data (Fig. 1c). In contrast, BoNT/C is efficiently immobilised by GD1b and to a lesser extent by GT1b and GD1a while the closely related mosaic serotype BoNT/DC binds preferentially GM1 and much weaker GD1a but hardly GT1b and GD1b (Karalewitz et al. 2010). BoNT/D displays a ganglioside preference similar to TeNT by interacting with GT1b, GD1b and GD2 pinpointing the absolute requirement for a disialyl moiety (Kroken et al. 2011).

Combining the data derived from ganglioside-deficient mice and biochemical assays, BoNT/A, E, F and G display a preference for the terminal NAcGal-Gal-NAcNeu moiety being present in GD1a and GT1b, whereas BoNT/B, C, D and TeNT require the disialyl motif found in GD1b, GT1b and GQ1b. In conclusion, abundant complex polysialo-gangliosides such as GD1a, GD1b and GT1b are essential to specifically accumulate all BoNT serotypes and TeNT on the surface of neuronal cells as the first step of intoxication.

3 The Four Domain Structure of TeNT and BoNT

TeNT and BoNT are initially synthesised as ~a 150 kDa single chain protein, which is subsequently cleaved by specific bacterial or host proteases. The resulting ~50 kDa light chain (LC) and ~100 kDa heavy chain (HC) remain attached via a single disulphide bond and non-covalent interactions mediated by an HC-derived peptide loop wrapping around the LC within the substrate cleft. The LC represent the active components which operate as zinc endoproteases with strict substrate specificities. Their apo structures have all been determined [reviewed in (Brunger and Rummel 2009)]. The structural differences among the LC are mostly limited to solvent-exposed loops and potential substrate interaction sites. The LC are non-toxic molecules but become highly poisonous agents when linked to the HC which

(a)

aa identity %	H_{CC}										pdb codes of crystal structures covering H_C	
H_C	A1	B1	C	CD	DC	D	E1	F1	G	T	apo structures	complex structures
BoNT/A1	**100**	24.5	18.7	15.8	19.3	16.3	31.9	37.3	19.1	22.1	3BTA, 2VUA, 3FUO	2VU9
BoNT/B1	32.6	**100**	19.2	17.9	19.3	18.4	24.3	22.8	42.1	25.2	1EPW, 1S0B-G , 1Z0H	1F31, 1I1E , 2NM1, 2NP0
BoNT/C	22.8	21.5	**100**	24.0	57.4	23.0	17.8	17.1	19.2	15.7	3N7K, 3R4U	3R4S
BoNT/CD	24.2	25.8	33.6	**100**	23.0	93.4	18.4	19.2	19.9	16.3	3PME,	4F83
BoNT/DC	22.6	22.0	74.9	33.6	**100**	21.9	18.4	13.5	19.8	17.3	3N7L, 3N7M, 3AZV, 3AZW	
BoNT/D	24.6	26.1	32.9	94.5	33.4	**100**	18.9	19.2	19.9	16.3	3N7J, 3OBR, 3OGG, 3RMX, 3RMY	3OBT
BoNT/E1	37.7	24.0	21.5	21.3	20.5	20.8	**100**	45.9	18.4	17.3	3FFZ	
BoNT/F1	42.6	25.4	21.9	22.0	17.6	21.6	54.8	**100**	21.2	14.5	3FUQ,	3RSJ
BoNT/G	32.6	42.4	23.3	26.1	23.7	26.8	26.7	29.6	**100**	22.3	2VXR, 3MPP	
TeNT	24.2	23.4	18.0	18.7	18.1	18.7	21.8	18.8	25.1	**100**	1A8D, 1AF9	1DLL, 1DFQ, 1FV2, 1FV3, 1YXW, 1YYN, 3HMY, 3HN1...

(b)

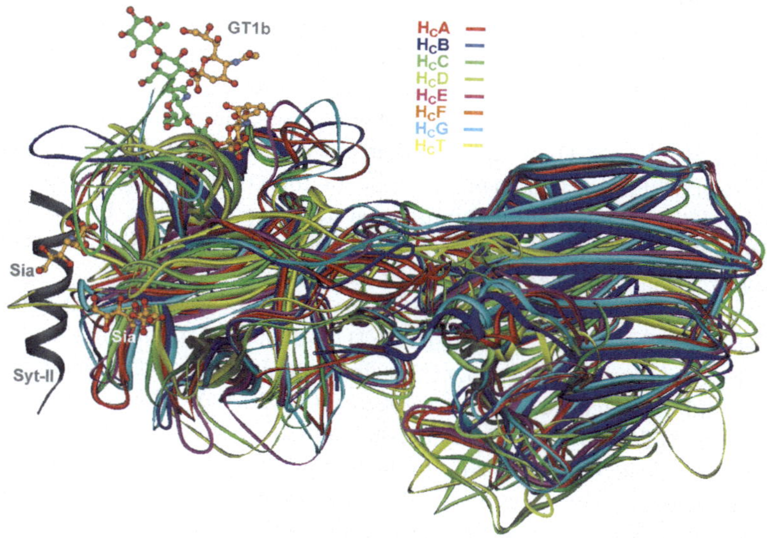

Fig. 2 **a** Amino acid sequence identities among BoNT and TeNT H_C-fragments (*lower left*) and H_{CC}-domains (*upper right*). Identity values of aligned sequences were calculated using ClustalW. To date, structural information available covering the H_C-fragment are shown in the right column. **b** Superimposition of the H_C-fragment crystal structures of TeNT and of all seven BoNT serotypes. Syt-II (*grey ribbon*) bound to H_CB (*dark blue ribbon*, 2NM1.pdb) was superimposed with H_CA (*red ribbon*) in complex with GT1b (*ball & stick*, 2UV9.pdb), H_CC (*dark green*) in complex with sialic acid (*ball & stick*, 3R4S.pdb), H_CD (*light green*) in complex with sialic acid (*ball & stick*, 3OBT.pdb), H_CE (pink, 3FFZ.pdb), H_CF (*orange*, 3FQU.pdb), H_CG (*light blue*, 2VXR.pdb) and H_CT in complex with GT2 (*yellow ribbon*, 3HMY.pdb)

ensures that the catalytic LC reach their site of action, the cytosol of neuronal target cells. Hereto, the HC comprise two functional subunits, a ~50 kDa largely α–helical domain at the N-terminus, called H_N, and at the C-terminus a ~50 kDa fragment, H_C, in which the two ~25 kDa domains H_{CN} and H_{CC} can be confined (Fig. 2b). Comparison of amino acid sequences of the H_{CC}-domains of the seven serotypes and the two mosaic BoNT/CD and DC, which were evolved by mutual exchange of their cell binding domain H_C of the closely related serotypes BoNT/C and D (Moriishi et al. 1996) (Hill and Smith 2012; Fig. 2a), revealed only 24 % identity on average, whereas the H_{CN}-domains displayed 34 % identity which corresponds to the overall identity of the full-length BoNT and TeNT. Highest sequence identities within H_{CC} of different serotypes are found for BoNT/E and F with 46 % and BoNT/B and G with 42 %, respectively, whereas the H_{CC} of the BoNT/C & D group share the lowest identity with the other serotypes (<20 %; Fig. 2a). Nevertheless, the β–trefoil architecture of the H_{CC} domain consisting of 12 conserved 5-mer clusters allows a high sequence divergence while maintaining similar core structures (Ginalski et al. 2000). Structural comparison among the H_C-fragments of BoNT/A [2VU9.pdb (Stenmark et al. 2008)], B [2NM1.pdb (Jin et al. 2006)], C [3R4S.pdb (Strotmeier et al. 2011)], D [3OBT.pdb (Strotmeier et al. 2010)], E [3FFZ.pdb (Kumaran et al. 2009)], F [3FUQ.pdb (Fu et al. 2009)], G [2VXR.pdb (Stenmark et al. 2010)] and TeNT [3HMY.pdb (Chen et al. 2009); Fig. 2a] showed that there is a varying twist between H_{CN} and H_{CC} culminating in H_CC by about 17.2° (Fig. 2b). Nevertheless, separate pair-wise structure comparisons of all eight H_{CN}- and H_{CC}-domains demonstrated that the structures are conserved within each domain. In spite of the conserved core structure, large structural differences are found in many surface-exposed loops. Five of such areas reside in H_{CN} and nine loops in the H_{CC}-domain (Fig. 2b). These variations are likely to account for the differences in receptor recognition.

The function of the lectin-like jelly role H_{CN}–domain of H_C that connects H_N and H_{CC} is still unresolved. Whereas the H_{CN}–domain of TeNT does not bind to rat primary dorsal root ganglia cells (Figueiredo et al. 1995) and nerve growth factor (NGF)-differentiated PC12 cells (Herreros et al. 2000a), a low affinity binding of H_{CN} of BoNT/A to phosphatidylinositol monophosphate incorporated in sphingomyelin-enriched microdomains of the immortalised motor neuron cell line NSC-34 was reported (Muraro et al. 2009). An involvement of H_{CN} of BoNT/A in the pH-dependent triggering of the translocation step is hypothesised (Fischer et al. 2008). On the other hand, the β–trefoil domain H_{CC} harbours the main features required for target cell recognition and internalisation, as demonstrated for TeNT and several BoNTs (see below).

4 The Mode of Molecular CNT-Ganglioside Interaction

It is many years since it was shown that the specific binding to peripheral nerve endings at the neuromuscular junction involves just the H_C-fragment (Simpson 1984a, b; Simpson 1985; Evinger and Erichsen 1986; Fishman and Carrigan 1987; Lalli et al. 1999) and initially complex polysialo-gangliosides. In one of the early approaches to identify the ganglioside binding site a cross-linking experiment employing ^{125}I-azido-GD1b and TeNT H_C-fragment led to radiolabelling of H1293 in the proximity of a large cavity within the H_{CC}-domain (Shapiro et al. 1997). The neighbourhood of H1293 to the ganglioside binding pocket was confirmed in a mutagenesis study showing reduced in vitro binding of the TeNT H_C mutant H1293A to isolated ganglioside GT1b (Sinha et al. 2000). The mutation of the TeNT residue Y1290, forming the bottom of this cavity, to phenylalanine, serine and alanine, disturbed the integrity of the local structure and therefore also reduced the affinity to GT1b as well as binding to synaptosomal membranes (Sutton et al. 2001). Independently, the anticancer drug doxorubicin, an anthracycline antibiotic, could be docked in silico into this pocket and was able to inhibit the binding of the TeNT H_C-fragment to liposome-integrated GT1b (Lightstone et al. 2000). First structural insight was revealed by co-crystallisation of the TeNT H_C-fragment and four carbohydrate subunits of GT1b which exhibited four distinct binding sites, including the one in the proximity of H1293 (Emsley et al. 2000). Here, a lactose molecule interacts with the residues D1222, T1270, S1287, W1289, Y1290 and G1300 (Fig. 3f; 1DLL.pdb). A separate site comprising R1226 as the key residue coordinated either a molecule of NAcGal (1D0H.pdb) or sialic acid (1DFQ.pdb). Two additional sites were identified in complexes of TeNT H_C with Gal or NAcGal (Emsley et al. 2000). However, the latter two sites were considered not to function as binding pockets for polysialo-gangliosides, due to insufficient space (Gal) for a complete ganglioside to bind or a too flexible peptide backbone (NAcGal). Isaacs et al. refined their co-crystallisation approach by using a synthetic GT1b-β oligosaccharide. Indeed, the terminal disaccharide NAcGal-Gal bound to the lactose binding site next to H1293, while the disialic acid branch of another GT1b-β molecule interacted with the sialic acid binding site comprising R1226 (Fotinou et al. 2001) (Fig. 3f; 1FV3.pdb). Independent co-crystallisation of TeNT H_C–fragment with disialyllactose (1YYN.pdb) as well as the carbohydrate portion of GT2 (3HMY.pdb) confirmed the binding of the disialic acid branch to the sialic acid binding site (Chen et al. 2009; Jayaraman et al. 2005). Mutation of residues D1222, H1271 and W1289 in the lactose binding site led to reduced binding of TeNT H_C-fragment to GT1b in surface plasmon resonance experiments and NGF-differentiated PC12 cells (Louch et al. 2002). The importance of the lactose binding site for binding and entry of TeNT was conclusively demonstrated by the application of corresponding, recombinant full-length TeNT mutants using the MPN assay leading to a 350-fold reduction in neurotoxicity in case of the single amino acid mutation W1289L (Rummel et al. 2003). Furthermore, these experiments demonstrated that also the sialic acid binding site is essential for TeNT action, since, e.g. the TeNT mutants R1226F/L possess a 70-fold reduced activity in the MPN assay. Mass spectroscopy experiments indicated simultaneous binding of two molecules GT1b to the TeNT H_C-fragment, but no ganglioside mediated cross-linking was observed in size exclusion chromatography experiments (Rummel et al. 2003). Based on these results, the ganglioside specificity of the lactose and sialic acid binding sites could be biochemically refined to GM1/GD1a and GD1b/GT1b, respectively (Chen et al. 2008), thereby confirming the above-mentioned crystallographic studies by Isaac et al. and Swaminathan et al. Although binding of a ganglioside to the sialic acid binding pocket was shown and it was also reported that TeNT high-affinity binding to neurons is mediated solely by gangliosides (Chen et al. 2009), it is conceivable that subsequently either a GPI-anchored glycoprotein (Herreros et al. 2000b, 2001; Munro et al. 2001) or sialic acids containing synaptic vesicle glycoprotein 2 (SV2) (Yeh et al. 2010) substitute the gangliosides because no ganglioside could be detected in TeNT containing vesicles of motoneurons (Deinhardt et al. 2006). Also, it is still unclear whether the

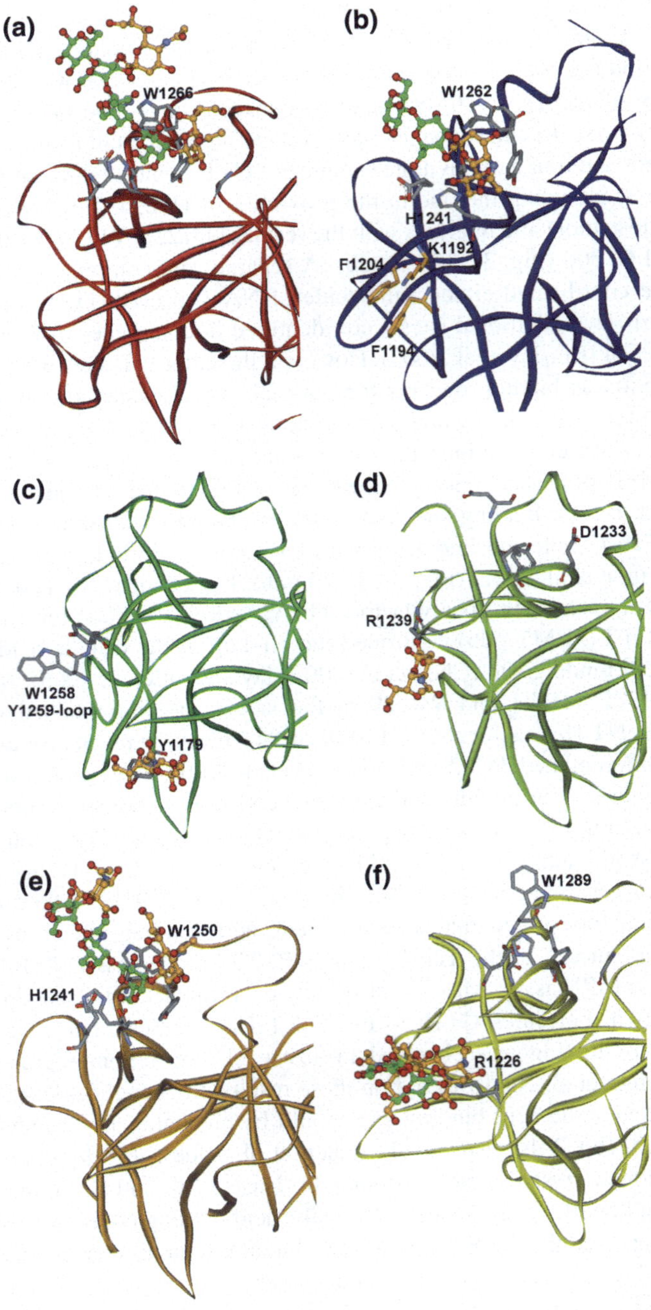
(a)
W1266
(b)
W1262
H1241
K1192
F1204
F1194
(c)
W1258
Y1259-loop
Y1179
(d)
D1233
R1239
(e)
W1250
H1241
(f)
W1289
R1226

◀**Fig. 3** Carbohydrate binding sites in the H_{CC}-domain of clostridial neurotoxins. Neurotoxin residues of binding sites are displayed as stick presentations whereas the bound carbohydrates are displayed in ball and stick presentation. **a** Synthetic GT1b binds via its terminal NAcGalβ3-1Galβ moiety (Gal carbon scaffold in *green*, NAcNeu carbon scaffold in *orange*) to the conserved ganglioside binding site within the BoNT/A H_{CC}-domain (2VU9.pdb). The hydrophobic side of the Gal ring packs parallel against the key residue W1266 (*grey stick* representation) whereas the terminal 3,2-NAcNeu sticks out of the pocket and displaces Y1117. The 8,2-NAcNeuα could not be resolved by X-ray crystallography. **b** Binding of sialyllactose in the conserved ganglioside binding site within the H_{CC}-domain of BoNT/B (Gal carbon scaffold in green, NAcNeu carbon scaffold in orange) (1F31.pdb). The terminal NAcNeu packs parallel against W1262 which is the key residue of the conserved GBS motif (*grey stick* representation). K1192, F1194 and F1204 are key residues for interaction with Syt-II (*orange stick* representation). **c** BoNT/C H_{CC} lacks the conserved ganglioside binding site but instead displays the WY-loop (W1258, Y1259) and a unique sialic acid binding site. The coordination of NAcNeu (carbon scaffold in *orange*) by Y1179 is of hydrophobic nature and an arginine like in BoNT/D and TeNT is missing (3R4S.pdb). **d** BoNT/D H_{CC} possesses a ganglioside binding site at the homologous position but with different amino acid configuration (D1233.Y1235…V1251.N1253; grey stick representation; 3OBT.pdb). In addition, H_{CC}D displays a sialic acid binding site like in TeNT. R1239 (*grey stick* representation) coordinates the carboxyl group of the NAcNeu (carbon scaffold in orange). **e** Synthetic GD1a binds via its terminal NAcGalβ3-1Galβ moiety (Gal carbon scaffold in *green*, NAcNeu carbon scaffold in *orange*) to the conserved ganglioside binding site within the BoNT/F H_{CC}-domain (3RSJ.pdb). As for BoNT/A, the terminal 3,2-NAcNeu points away from the pocket and only interacts hydrophilically with R1111 and R1256. The hydrophobic side of the Gal ring packs parallel against the key residue W1250 (*grey stick* representation). The Glc could not be resolved by X-ray crystallography. **f** Binding of disialyllactose via its NAcNeuα8-2NAcNeuα element (sialic acid carbon scaffold in *orange*) to the sialic acid binding site in TeNT H_{CC}-domain (1YYN.pdb). R1226 (*grey stick* representation) mediates the main salt bridge to the carboxyl group of the terminal NAcNeu. W1289 is the key residue of the conserved ganglioside binding site (*grey stick* representation)

two ganglioside binding sites of TeNT relate to its retrograde intra-axonal transport (read more in Bercsenyi et al. 2012).

The lactose binding site in TeNT is built by the peptide motif D…H…SXWY…G (Fig. 3f; Table 1), which is conserved among BoNT/A, B, E, F and G. This cavity displays the typical features necessary for carbohydrate interaction found also in other protein toxins such as ricin and cholera toxin. An aromatic residue, preferable tryptophan or tyrosine, supplies the surface for the hydrophobic face of the sugar ring. Polar residues like aspartate or glutamate and serine are located opposite to interact with the sugar hydroxyl groups. These polar interactions are supported by a histidine or in BoNT/E by a lysine.

Co-crystallisation studies with BoNT/B and sialyllactose (1F31.pdb) or doxorubicin (1I1E.pdb) suggested that the lactose binding site of TeNT is the ganglioside binding site (GBS) in BoNTs (Swaminathan and Eswaramoorthy 2000; Eswaramoorthy et al. 2001). Detailed mutational analyses defined the contribution of various residues within the homologous GBS of BoNT/A and B (Fig. 3a, b). Again, the mutations of the conserved aromatic key residues, W1266 and W1262 in BoNT/A and B, respectively, to leucine led to dramatic reductions of neurotoxicity using the MPN assay (Rummel et al. 2004a). In contrast to TeNT, mass spectroscopy data revealed the binding of only a single GT1b molecule to the H_C-

Table 1 *Identified neurotoxin receptors and corresponding binding sites*

	AA motif in the conserved ganglioside binding site	Key AA of the sialic acid binding site	SV protein receptor[#]	Key AA of SV protein receptor binding site
BoNT/A	E…H…SXWY…G		SV2C/A/B[*]	?
BoNT/B	E…H…SXWY…G		Syt-II/Syt-I	K1192, F1194, F1204
BoNT/C	W1258, Y1259; WY-loop[§]	Y1179	?	?
BoNT/DC	W1252, F1253; WF-loop[§]	Y1175	Syt-II/Syt-I	M1179, N1185, V1191, L1235, I1264
BoNT/D	DXY…VXN	R1239	SV2B/C/A	?
BoNT/E	E…K…SXWY…G		SV2A/B	?
BoNT/F	E…H…SXWY…G		SV2A/C/B	?
BoNT/G	Q…G…SXWY…G		Syt-I/Syt-II	Q1200, F1202, F1212
TeNT	D…H…SXWY…G	R1226	SV2	?

§ no conserved ganglioside binding site present
# order expresses decreasing neurotoxin affinity
* N-glycosylation of LD4 is not obligatory

fragment of BoNT/A and B (Rummel et al. 2004a). Recently, these physiological and biochemical data were confirmed by a crystal structure of a synthetic GT1b-oligosaccharide bound to the sole GBS of BoNT/A H_C (Stenmark et al. 2008) (Fig. 3a; 2UV9.pdb). Interestingly, whereas the mutational data of the conserved GBS suggests a shared ganglioside binding mode of BoNT/A and B differing from that of TeNT, the crystallographic results indicate that BoNT/A like TeNT interacts predominantly with the terminal NAcGal-Gal moiety and ignores the NAcNeu, whereas BoNT/B predominantly interacts with the NAcNeu of sialyllactose. It cannot be ruled out that different crystallisation conditions and the use of sialyllactose instead of GT1b causes these discrepancies. However, the different sensitivity of GD3S-/- mice towards BoNT/A and B supports this observation (Fig. 1c).

Structural data of BoNT/E, BoNT/F H_C and BoNT/G H_C displayed the presence of a conserved GBS like in BoNT/A and B (Fu et al. 2009; Kumaran et al. 2009; Stenmark et al. 2010), but only the mutation of W1268 in BoNT/G, W1224 in BoNT/E and W1250 in BoNT/F, which are all part of the conserved GBS motif E(D)...H(K)...SXWY...G, to leucine demonstrated their key role in ganglioside interaction and biological activity (Rummel et al. 2007, 2009) (Table 1). Just recently, a complex structure of BoNT/F H_C and GD1a-oligosaccharide (3RSJ.pdb) along with mutational analysis of the conserved GBS in BoNT/F and E (Benson et al. 2011) confirmed previous data (Rummel et al. 2009; Fu et al. 2009). Here, the terminal NAcGal-Gal moiety of GD1a interacts with W1250 analogously to that of GT1b with W1266 in BoNT/A (Stenmark et al. 2008). Additionally, two arginine residues (R1111, R1256) of BoNT/F facilitate ganglioside binding by interacting with the terminal NAcNeu. Remarkably, replacing the imidazole ring of H1241 (Fig. 3e), which maintains the hydrophilic interactions with the hydroxyl groups of the Gal, by lysine improved the binding of BoNT/F H_C to GD1a and GM1 ~65-fold (Benson et al. 2011).

Although mutation of W1258 in BoNT/C, which aligns in the similar motif GXWY, clearly reduced binding to gangliosides in synaptosomal membranes as well as biological activity (Rummel et al. 2009; Tsukamoto et al. 2008), the conserved GBS is absent in the BoNT/C H_C structure (Karalewitz et al. 2010; Strotmeier et al. 2011) (Figs. 2b, 3c; 3N7K.pdb, 3R4U.pdb). W1258 is part of a long loop extending out of the core structure of H_{CC}C. Molecular dynamics simulations indicate that this WY-loop is not a crystallographic artefact (Strotmeier et al. 2011). However, the mode of ganglioside coordination at this peptide remote from the H_{CC}C core structure remains unclear. Furthermore, a second GBS within H_{CC}C was exhibited by co-crystallisation of sialic acid with H_CC (Figs. 2b, 3c; 3R4S.pdb). This pocket called Sia-1 site locates at the tip of the H_{CC}-domain in the neighbourhood of the WY-loop but constitutes an autonomous ganglioside binding pocket and displays a mode of sialic acid binding completely different from the one observed in the sialic acid binding site of TeNT (Strotmeier et al. 2011). The closely related H_C-fragment of BoNT/DC, displaying 75 % amino acid sequence identity to H_CC (Fig. 2a), also lacks the ganglioside binding motif but presents an analogous WF-loop in the crystal structure (3N7L.pdb).

Consequently, H_CDC mutant W1252A completely lost binding to GM1, the preferred ganglioside of BoNT/DC, although the WF-loop backbone of the mutant (3N7M.pdb) nicely superimposes with the H_CDC wild-type structure. (Karalewitz et al. 2010). Nevertheless, partial electron density of sialyllactose was detected at a position homologous to the conserved GBS (3AZW.pdb). A subsequent mutational analysis identified residues important for binding of BoNT/DC H_C to GM1 and P19 cells, neuronally differentiated embryonal carcinoma cells, within the conserved GBS as well as the WF-loop (Nuemket et al. 2011).

Due to high amino acid sequence divergence and the lack of crystal structures, no GBS has been identified in BoNT/D until recently. Co-crystallisation of sialic acid with BoNT/D H_C (3OBT.pdb) revealed a sialic acid binding site in a position similar to the sialic acid binding site in TeNT (Strotmeier et al. 2010) (Figs. 2b, 3d). In addition, mutagenesis studies identified R1239 within the sialic acid binding site coordinating sialic acid like R1226 in TeNT. Alanine mutation of the neighbouring W1238 and F1240 also had a drastic impact on ganglioside binding, whereas the H_CD F1240W mutant behaved like wild-type (Strotmeier et al. 2010). Structural data exhibited that mutation W1238A disorders the backbone thereby also influencing R1239 (3RMY.pdb), whereas F1240 seems to be directly involved in ganglioside binding (3RMX.pdb) (Kroken et al. 2011). Furthermore, a second carbohydrate binding site at the location of the conserved GBS in BoNT/A, B, E, F, G and TeNT, but completely different amino acid configuration, and a loop which is structurally related to the WY-loop of BoNT/C and putatively interacts with the membrane were identified. But the absence of the GBS motif E(D)…H(K)…SXWY…G might explain the low affinity of BoNT/D to gangliosides (Strotmeier et al. 2010). Previously, it was reported that H_C of BoNT/D and the 94 % identical H_C of the mosaic BoNT/CD bind to phosphatidylethanolamine (PE) in overlay assays (Tsukamoto et al. 2005). Later, Kozaki's finding was confirmed by precipitation of BoNT/D H_C and BoNT/CD H_C through liposome integrated PE (Zhang et al. 2010, 2011). Interestingly, Cinnamycin, a tetracyclic peptide antibiotic associating with PE, inhibited the binding of H_CD and H_CCD but not H_CC to synaptosomes in a dose-dependent manner (Tsukamoto et al. 2008). Superior binding to PE and synaptosomes of H_CCD versus H_CD was ascribed to K1118 and K1136 which are replaced by E1114 and G1132, respectively, in BoNT/D (Tsukamoto et al. 2008; Zhang et al. 2011). Structural analysis of H_CCD (3PME.pdb) exhibits a sialic acid binding site consisting of W1242, R1243 and F1244 homologous to that of BoNT/D.

In conclusion, BoNT/A, B, E, F and G harbour a single GBS made up of the conserved amino acid motif E(Q)…H(K)…SXWY…G at a homologous location within the H_{CC}-domain (Table 1). The lactose binding site of TeNT is highly similar to this conserved GBS, whereas the second site complexing sialic acid is closely related to the one in BoNT/D and BoNT/CD. The other carbohydrate binding site in BoNT/D and CD is unique with respect to its amino acid configuration but locates similar to the conserved GBS. BoNT/C and BoNT/DC display a unique sialic acid site and a WY/F-loop outside their core structure; thus neither their amino acid configuration nor their positions are related to any other known carbohydrate binding pocket of CNT.

5 Synaptic Vesicle Proteins are Receptors of Botulinum Neurotoxins

As mentioned above, many evidence argued against gangliosides as the sole receptors of clostridial neurotoxins at nerve terminals like a much higher affinity of BoNT in vivo compared to that observed in binary interaction studies in vitro or the marked reduction of TeNT binding to rat brain membranes upon protease pretreatment (Lazarovici and Yavin 1986; Pierce et al. 1986). To account for these findings, a dual-receptor-model involving two sequential binding steps was proposed (Montecucco 1986) which suggests an initial low affinity accumulation of neurotoxin by the abundant polysialo-gangliosides and a subsequent binding to the sparsely distributed protein receptor(s). Simultaneous interaction with ganglioside and protein receptor results in a high affinity binding and is a prerequisite for the subsequent specific endocytotic step (Bercsenyi et al. 2012).

Nevertheless, the nature of protein receptors for BoNT remained obscure for a long time although several studies demonstrated an accelerated uptake of BoNT/A (Black and Dolly 1986; Hughes and Whaler 1962; Simpson 1980), BoNT/C (Rummel et al. 2009; Simpson 1982), BoNT/D (Rummel et al. 2009), BoNT/E (Rummel et al. 2009; Lawrence et al. 2007) and BoNT/F (Rummel et al. 2009) into the phrenic nerve prepared together with the adjacent diaphragm upon electrical stimulation which resulted in an earlier onset of neurotransmitter blockade. Others reported that high K^+ concentrations stimulating neurons lead to an accelerated uptake of BoNT/A, B, D, E, F, G and TeNT into spinal cord and hippocampal preparations (Peng et al. 2011; Dong et al. 2007, 2008; Fu et al. 2009; Kroken et al. 2011; Keller et al. 2004; Blum et al. 2012; Matteoli et al. 1996). As neuronal stimulation causes increased rates of exo- and endocytosis of synaptic vesicles, the intravesicular domains of synaptic vesicle membrane proteins are frequently extracellularly exposed and thus accessible for the membrane-associated BoNT. However, Schiavo et al. reported that uptake of TeNT into the motor nerve terminals and cultured motoneurons is largely independent (Deinhardt et al. 2006; Schmitt et al. 1981) or only partially modulated (Simpson 1985) by synaptic activity. This remarkable difference suggests that TeNT may use a different endocytic mechanism in motoneurons from other neuronal types such as inhibitory interneurons and hippocampal neurons which lead to a different sorting (Bercsenyi et al. 2012).

Congruously, classical biochemical separation techniques of total rat brain protein, non-selective searches among synaptic vesicle proteins, identified the synaptic vesicle protein synaptotagmin I (Syt-I) as the protein receptor for BoNT/B (Nishiki et al. 1994). The two homologous synaptotagmin isoforms I and II of this type I transmembrane protein (Perin et al. 1990; Geppert et al. 1991) are

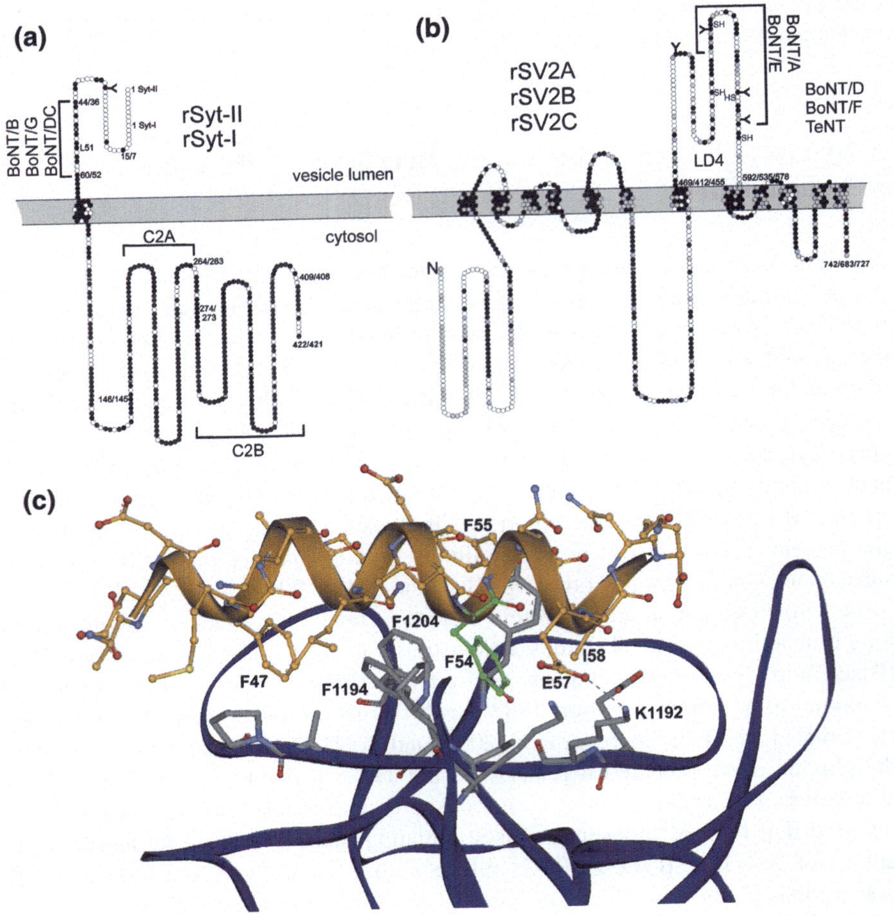

Fig. 4 Membrane topology of botulinum neurotoxin protein receptors and mode of BoNT/B-Syt-II interaction. **a** Rat Syt-II/Syt-I are type I membrane proteins consisting of 422/421 amino acids. The 65/57 N-terminal amino acids are localised in the lumen of synaptic vesicles. The cytosolic part forms two C2-domains coordinating two Ca^{2+} (indicted by brackets; C2A; C2B). Residues being conserved in Syt-I are shown in *black*, and the segment mediating binding of BoNT/B, DC and G is indicated by a *bracket*. "Y" represents an *N*-glycosylation site, which apparently does not contribute to binding of BoNT. **b** Rat SV2A/B/C represent membrane glycoproteins of 742/683/727 amino acids with 12 putative transmembrane domains. Strictly conserved residues are shown in *black*, those conserved in SV2C and either in SV2A or in SV2B in *grey*. The C-terminal part (indicated by a *bracket*) of the large luminal domain (LD4) which comprises 125 amino acids and up to four N-glycosylation sites (indicated by "Y", of which sites 1–3 are conserved in SV2A and SB2B) mediates the interaction with BoNT/A and E. SH denotes cysteine residues that may form disulphide bridges. Note that the binding of BoNT/E to SV2A and B strictly depends on glycosylation of the third N-glycosylation site N573. **c** Binding of the transmembrane juxtaposed 17-mer peptide of the intraluminal domain of rat Syt-II in an α-helical conformation (amino acids in *orange ball* and *stick* presentation) to the saddle-like site in the distal tip of the H_{CC}-domain of BoNT/B (residues in *grey stick* presentation) (2NM1.pdb). K1192 forms an important salt bridge to E57 of rat Syt-II (dashed line). F54 corresponding to L51 in human Syt-II is highlighted in *green ball* and *stick* presentation

supposed to link synaptic vesicle fusion to Ca^{2+} influx (Sudhof 2002; Chapman 2002) (Fig. 4a; Ahnert-Hilger et al. 2012). Subsequent binding studies using recombinantly expressed Syt–I and Syt-II isoforms showed that BoNT/B exhibits a ten-times higher affinity to Syt-II (Nishiki et al. 1996a). Stable transfection of CHO cells with Syt-II identified the amino-terminal, intraluminal domain of Syt-II as binding region for BoNT/B (Fig. 4a) (Nishiki et al. 1996b). Deletion mutants of Syt-II lacking the two cytosolic Ca^{2+} binding C2 domains could still act as protein receptor of BoNT/B (Kozaki et al. 1998). The physiological role of this neurotoxin-Syt interaction was later demonstrated by means of loss-of-function and gain-of-function approaches employing PC12 cells as well as BoNT neutralisation assays in mice employing a Syt-II fragment containing the luminal and transmembrane domain and a ganglioside mix (Dong et al. 2003).

Shortly thereafter, Rummel et al. demonstrated that BoNT/G, whose cell binding domain H_C is 42 % identical to the one of BoNT/B, also interacts with both Syt-I and Syt-II in vitro, whereas neither of the remaining BoNT serotypes bound to Syt (Table 1). It was concluded that either Syt isoform act as protein receptor, as preincubation of BoNT/G with the luminal domain of Syt-I or Syt-II drastically decreased its activity at MPN preparations (Rummel et al. 2004b). In addition, the relatedness of their binding domains is reflected in the fact that binding of BoNT/G also occurs at the membrane juxtaposed 20 amino acid segment of Syt-I and Syt-II (Rummel et al. 2004b). However, their binding properties differ in one striking aspect. In contrast to BoNT/B—Syt-II, BoNT/G exhibits in vitro 5-fold and 10-fold lower affinities to Syt-II and Syt-I, respectively (Rummel et al. 2007). Final confirmation for activity-dependent uptake and Syt-mediated neuronal cell entry of BoNT/B and G was provided by mouse hippocampal Syt-I knock-out neurons and restoration of toxin sensitivity by Syt-I/Syt-II expression in those neurons (Dong et al. 2007).

Unexpectedly, BoNT/DC whose H_C-fragment is only only 22 % and 24 % identical to H_C of BoNT/B and G, respectively (Fig. 2a), precipitated Syt-I and Syt-II from rat brain detergent extracts, although to a lesser amount than BoNT/B (Peng et al. 2012). Furthermore, uptake of BoNT/B was inhibited by H_CDC and vice versa, but activity of its closest relative BoNT/C was not impaired. Knockdown of Syt-I in hippocampal neurons clearly affected activity of BoNT/DC, whereas overexpression of either Syt rescued the sensitivity towards BoNT/DC.

To identify protein receptors of the remaining BoNT serotypes synaptic vesicle proteins were preselected on the basis that their intravesicular segments exceeded 20 amino acids in size and comprised both intravesicular segments of the tetraspanin proteins synaptophysin, synaptoporin, synaptogyrin-I and -III as well as the large intravesicular domain of the SV2 isoforms A, B and C. GST-pull-down experiments employing their luminal domains discovered the interaction between the large luminal domain 4 (LD4) of SV2 and BoNT/A (Fig. 4b) (Dong et al. 2006; Mahrhold et al. 2006). The function of this integral membrane glycoprotein with 12 putative transmembrane domains (Bajjalieh et al. 1993, 1992; Janz and

Sudhof 1999; Feany et al. 1992) is likely linked to synaptic vesicles priming or rendering primed vesicles fully Ca^{2+} responsive (Chang and Sudhof 2009; Custer et al. 2006; Xu and Bajjalieh 2001) (Ahnert-Hilger et al. 2012). Recent studies showed that loss of SV2 results in elevated Ca^{2+} levels in the presynaptic terminals and also reduces the rate of compensatory membrane retrieval after synaptic vesicle release (Wan et al. 2010). Interestingly, it has been demonstrated that SV2 associates with Syt and may regulate the endocytosis of Syt (Yao et al. 2010; Nowack et al. 2010). The recombinant in *E. coli* expressed, unglycosylated, isolated LD4 of SV2C exhibits the highest affinity to BoNT/A, inhibited binding and entry of BoNT/A into hippocampal neurons and motor nerve terminals (Dong et al. 2006) and efficiently decreased BoNT/A neurotoxicity (Mahrhold et al. 2006). Loss-of-function and gain-of-function studies in hippocampal neurons (Dong et al. 2006), RNAi experiments using wild-type PC12 and Neuro-2a cells as well as transiently SV2 expressing PC12 and Neuro-2a knockdown cell lines verified that all three SV2 isoforms can act as physiological receptors for BoNT/A (Dong et al. 2006; Mahrhold 2008). Since all three SV2 isoforms are expressed in α-motoneurons (Dong et al. 2006), it is still not clear which SV2 isoform is most relevant for the physiological uptake of BoNT/A.

Analogous studies, employing co-immunoprecipitation from rat brain extracts, BoNT/E treatment of hippocampal neurons derived from SV2A/B knock-out mice and rescue of toxin sensitivity by expression of SV2 isoforms in SV2A/B knock-out hippocampal neurons revealed that the in vivo toxicity of BoNT/E predominantly relies on the binding to SV2A and to lesser extent to SV2B, but not on SV2C (Dong et al. 2008).

Barbieri et al. demonstrated that BoNT/D H_C precipitated ~ 1 % SV2A, ~10 % SV2B and 15 % Syt-II compared to BoNT/A H_C from synaptic vesicle protein lysates extracted with CHAPS and none of them from lysates extracted with Triton X-100 (Kroken et al. 2011). In contrast, Dong et al. were able to co-precipitate large amounts of SV2 and traces of Syt-I from rat brain lysates extracted with Triton X-100 (Peng et al. 2011). The SV2 precipitate could exclusively be allocated to isoform SV2B. In concordance, co-immunoprecipitation of BoNT/D H_C only yielded SV2B. Interestingly, overexpression of each SV2 isoform in SV2A/B double knock-out hippocampal neurons rescued sensitivity towards BoNT/D. In contrast to BoNT/A and E, neither the LD4 of SV2A nor other SV2 segments were identified as BoNT/D interacting segments (Peng et al. 2011). Furthermore, TeNT is suggested to employ SV2A and SV2B at central glutamatergic and GABAergic neurons thereby not excluding SV2C (Yeh et al. 2010) (Table 1). However, TeNT, BoNT/A, D and E utilise seemingly diverse SV2-binding mechanisms. Whereas BoNT/A interacts with the unglycosylated LD4 of all SV2 isoforms and N-glycosylation of N573 in SV2A only slightly enhances its uptake rate, the other three toxins do not bind to any *E. coli*-derived LD4-SV2 peptide. Biological activity of BoNT/E proved to be strictly dependent upon N–glycosylation of N573 in SV2A, the carboxyl-terminal of three conserved putative N–glycosylation sites in the LD4 of SV2 (Dong et al. 2008). In contrast, any mutation of the three putative N–glycosylation sites in SV2A like N573Q has no significant effect on the entry of BoNT/D and TeNT (Peng et al. 2011; Yeh et al. 2010). Second, the SV2-LD4 domain expressed in low density lipoprotein receptor-based or synaptogyrin-based chimeric proteins can function as the receptor for BoNT/A and E, but failed to mediate the entry of BoNT/D. Furthermore, BoNT/D does not compete for uptake in hippocampal neurons with BoNT/A and E (Peng et al. 2011). On the other hand, the cell binding domain H_C of TeNT inhibits the uptake of BoNT/E and BoNT/D at MPN preparations (Rummel et al. 2009) as well as SNAP-25 cleavage by BoNT/E in hippocampal neurons (Yeh et al. 2010). However, it has to be noted that a direct protein–protein interaction of SV2 and BoNT/D or TeNT remains to be shown. In conclusion, the employment of SV2A-C by BoNT/A as well as of SV2A and B by BoNT/E is well characterised, whereas the reported data suggest that the BoNT/D-SV2 and TeNT-SV2-binding mechanism is yet to be understood.

In addition, there is dispute about the involvement of SV2 as protein receptor for BoNT/F whose cell binding domain H_C displays 55 % and 43 % amino acid sequence identity with BoNT/E and A, respectively. As BoNT/A and E H_C are also 38 % identical, it is imaginable that BoNT/F harnesses SV2 for cell entry, too. Although BoNT/F uptake at cultured hippocampal neurons has been reported to occur independently of synaptic activity (Verderio et al. 1999), increased stimulation of motoneurons accelerates uptake of BoNT/F (Rummel et al. 2009) indicating the involvement of synaptic vesicular structures. Furthermore, BoNT/A H_C inhibits neurotoxicity of BoNT/F and BoNT/F H_C inhibits neurotoxicity of BoNT/E at MPN preparations (Rummel et al. 2009) which can be explained by competition for SV2. Moreover, two independent studies co-purified all three SV2 isoforms by BoNT/F H_C from Triton X-100 solubilised synaptic vesicle lysates (Rummel et al. 2009; Fu et al. 2009). On the other hand, cleavage of VAMP/synaptobrevin-2 by various concentrations of BoNT/F is not impaired in SV2B and SV2B/SV2A knock-out hippocampal neurons (Peng et al. 2011; Yeh et al. 2010). Residual SV2C present in hippocampal neurons and a different SV2 expression pattern in motoneurons might explain these discrepancies (Peng et al. 2011; Dong et al. 2006). However, an ultimate experiment employing pan SV2 knock-out motoneurons has not been conducted.

For BoNT/C, Kozaki and colleagues excluded proteins as receptor component, because protease treatment or boiling of solubilised rat brain synaptosomes had no effect on binding (Tsukamoto et al. 2005). These data are consistent with the observation that BoNT/C H_C failed to precipitate the synaptic vesicle proteins Syt-I/II and SV2A/B/C from Triton X-100 solubilised synaptic vesicle lysates (Rummel et al. 2009; Baldwin and Barbieri 2007). In addition, cleavage of SNAP-25 and syntaxin 1A by BoNT/C is neither impaired in SV2B and SV2B/SV2A knock-out hippocampal neurons (Peng et al. 2011) nor increased in wild-type hippocampal neurons upon high K^+ stimulation (Peng et al. 2012). However, as mentioned above the stimulation-dependent uptake of BoNT/C at motor nerve terminals points towards the involvement of synaptic vesicular structures (Rummel et al. 2009; Simpson 1982). Clearly, thorough studies are required to clarify this issue. If entry of BoNT/C turns out to depend upon proteinaceous receptors, it will be interesting to find out whether proteins different from the synaptic vesicle proteins Syt or SV2 act as receptors.

6 Mode of the BoNT-Protein Receptor Interaction

Since BoNT/A, B, E, F and G do not possess a second carbohydrate binding site, the question arises, whether protein receptors such as Syt-II for BoNT/B bind in a pocket that is homologous to the sialic acid binding site within the H_{CC}-domain of TeNT or BoNT/D. The different affinities of H_C–fragment hybrids generated of H_{CN}– and H_{CC}–domains of BoNT/B strains Okra and 111 (subtypes B1 and B2; Hill and Smith 2012), respectively, to GT1b/Syt-II endowed liposomes point to that direction (Ihara et al. 2003).

The mode of interaction of BoNT/B with its protein receptor Syt-II has been addressed by two parallel co-crystallisation approaches. In one study full-length BoNT/B was co-crystallised with a 20-mer peptide that corresponded to the proposed interacting segment of Syt-II (Chai et al. 2006). In the second study, a recombinant fusion protein consisting of the H_C-fragment of BoNT/B linked via a strep affinity tag to the complete luminal domain of Syt-II was crystallised (Jin et al. 2006). Analysis of both crystals revealed that Syt-II involving amino acids 44–60 bound to a saddle-like crevice at the distal tip of the H_{CC}-domain in the direct neighbourhood of the ganglioside binding pocket (Figs. 2b, 4c). Interestingly, the in solution unstructured luminal domain of Syt-II formed an α-helix upon binding to BoNT/B. Its interaction with the toxin relies mainly on hydrophobic interactions via two adjacent pockets on the surface of BoNT/B, but also involves important salt bridges mediated by E57 of Syt-II (Fig. 4c). Yet another parallel study starting from computer-assisted binding pocket predictions followed by mutational analyses also identified the Syt-I and Syt-II binding sites in the BoNT/B H_{CC}-domain (Rummel et al. 2007). This study also evidenced that Syt-I and Syt-II bind to the equivalent place in BoNT/G. However, only the general shape of the pocket and a few amino acids forming the surface of that binding area are conserved (Stenmark et al. 2010). As determined by isothermal titration calorimetry, Syt-I has at least two orders of magnitude lower affinity for BoNT/B compared to Syt-II (Kd = 39 nM for rat Syt-II; Jin et al. 2006), though it displays only two conservative replacements of residues at positions that were shown to be involved in the Syt-II/BoNT/B interaction, M47 and L50 instead of F55 and I58 in Syt-II. Individual conversion of M47 and L50 into phenylalanine and isoleucine, respectively, increased the binding affinity of Syt-I, and the corresponding double mutation converted Syt-I into a Syt-II-like high affinity receptor (Jin et al. 2006). This result substantiates a conserved binding mode for both Syt isoforms in BoNT/B. It is presently not solved whether Syt-I/II are bound by BoNT/G in the same configuration. Whereas the individual mutation of F47, L50, F54 or F55 of Syt-II to alanine caused loss in binding of BoNT/B (Jin et al. 2006), the Syt-II mutants F47A, L50A and F55A displayed wild-type binding to BoNT/G H_C (Stenmark et al. 2010). Other mutations in Syt-II showed comparable effects on binding to BoNT/G while mutant Syt-II-E57K, which did not bind to BoNT/B, exhibited wild-type like binding affinity in the case of BoNT/G (Rummel and Binz, manuscript in preparation). Replacement of BoNT/G-Q1200, the counterpart of BoNT/B-K1192, which forms the major salt bridge with Syt–II-E57, by the equivalent lysine residue of BoNT/B did not improve binding to Syt-II, but clearly diminished binding and neurotoxicity (Rummel et al. 2007). These findings evidenced significantly different binding modes of Syt-I and Syt-II in the similarly located protein receptor pocket of BoNT/B and BoNT/G. Also, the interaction of BoNT/DC with Syt-I and Syt-II differs from that of BoNT/B although the Syt-II mutants F47A, F54A and F55A completely lost binding to both BoNT/DC and BoNT/B in GST pull-down assays (Peng et al. 2012). But the binding constant of H_CDC with the intraluminal 20mer peptide of Syt-II is 100-fold lower (3.0 μM) than of H_CB (Berntsson et al. 2012). Superimposition of BoNT/DC H_C and H_CB-Syt-II crystal structures revealed that the surface in H_CDC homologous to the Syt-II site in H_CB greatly differs. An alternative, only partially overlapping hydrophobic area, was analysed. Here mutations Y1180K, I1264Q and P1182S/S1183Y reduced binding of both Syt–I and Syt–II, whereas L1196R and L1226K selectively diminished binding of Syt–II without affecting Syt–I binding significantly (Peng et al. 2012). Co-crystallisation of the Syt-II peptide with H_CDC revealed a novel binding site different to the one in BoNT/B and G exhibiting M1179, N1185, V1191, L1235 and I1264 as key residues (Table 1; Berntsson et al. 2012).

As mentioned above, F54 is the key hydrophobic residue in Syt-II for mediating binding to BoNT/B, DC and G. Analysis of Syt-II derived from different species revealed that the corresponding residue 51 in human and chimpanzee Syt-II is a leucine instead of phenylalanine. Introducing the corresponding mutation F54L into Syt-II of mouse or rat origin which displays identical binding affinities to BoNT/B resulted in complete loss of binding of BoNT/B, DC and G (Peng et al. 2012; Strotmeier et al. 2012). This loss cannot be rescued by the presence of gangliosides and only partially compensated by the presence of Syt-I whose corresponding residue F46 is strictly conserved in all sequenced species. Hereby the disparity in potency of BoNT/B in humans and mice as well as the 40-fold higher dosage of rimabotulinumtoxinB versus onabotulinumtoxinA is explained at the molecular level. Moreover, this study also illustrates the risk of extrapolating results of animal experiments or non-human cell line-based assays for pharmaceutical applications in humans. Furthermore, this fact might be the major reason for the rareness of botulism cases caused by BoNT/G (Strotmeier et al. 2012). A recent report classifies BoNT/DC as the most potent BoNT in mice (Nakamura et al. 2012). However, according to the lack of binding of BoNT/DC to human Syt-II this classification will not apply to humans.

So far no SV2 binding site has been identified. However, according to the data known, it is conceivable that BoNT/A like BoNT/B and G appears to interact with non-glycosylated segments of its protein ligand SV2C at a site within the H_{CC}-

domain similar to the Syt-II binding site in BoNT/B. The requirement of N–glycosylation of the SV2A-LD4 for binding BoNT/E raises the question whether the interaction occurs exclusively via the oligosaccharide chain or is a mixture of peptidic and carbohydrate interactions with BoNT/E.

As outlined above there is multiple evidence from crystallisation and mutagenesis studies that TeNT exhibits a second binding site for sialic acid. This sialic acid binding site was later on also shown to accommodate the tripeptide YEW (Jayaraman et al. 2005). It is located in the area that corresponds to the Syt binding site of BoNT/B and G, but smaller and chemically different containing several hydrophilic residues and a central arginine (R1226). A parallel situation was described for BoNT/D (Strotmeier et al. 2010). As both BoNT/D and TeNT were reported to employ glycosylated SV2 (Peng et al. 2011; Yeh et al. 2010), it is imaginable that the sialic acid binding site interacts with terminal sialic acids of the N-glycosyl branches as well as with amino acid residues of the SV2 backbone which would mediate the neuronal specificity. On the other hand, BoNT/E which clearly requires *N*-glycosylation of SV2A and SV2B does not display such a sialic acid binding pocket.

However, neither the exact binding affinities of the three SV2 isoforms to BoNT/A, D, E and TeNT nor the interacting segments and residues, respectively, of the corresponding LD4 have been determined. Moreover, the SV2 binding pockets in the respective BoNT H_C-fragments are awaiting their identification, too.

7 The BoNT Membrane Approach

Although ganglioside and protein receptor binding sites are in close proximity as demonstrated for BoNT/B and G (Fig. 2b) they function independently and do not require preformation of a ganglioside/protein receptor complex (Rummel et al. 2007). Moreover, mutants of BoNT/B with both the ganglioside and Syt binding sites (individually or simultanously) deactivated, do not exhibit appreciable toxicity excluding any significant contributions of other cell surface molecules to binding and entry of BoNT/B and G (Rummel et al. 2007).

The pH-dependent membrane insertion of the neurotoxins is the least understood step of the intoxication process primarily because it takes place inside vesicles within the cytosol. A recent study investigated this step by making it to occur at the surface of neurons due to blockade of the vesicular H^+-ATPase by bafilomycin. The neurotoxin, bound to the plasma membrane in the cold, was exposed to a warm low pH extracellular medium and the entry of the LC was monitored by measuring its specific metalloprotease activity. In conclusion, the anchorage via two receptors is a strict prerequisite for a productive low pH-induced conformational change followed by membrane translocation, and TeNT, BoNT/B, C and D change structure and interact with the membrane in the same range of pH values as of the SV lumen pH (Pirazzini et al. 2011). In contrast, Chapman et al. discovered that binding to GT1b enables BoNT/B to sense low pH, undergo a significant change in secondary structure, and transform into a hydrophobic oligomeric membrane protein. Imaging of the toxin on lipid bilayers using atomic force microscopy revealed donut-shaped channel-like structures that resemble other protein translocation assemblies (Sun et al. 2011). However, BoNT/E translocated much more rapidly than BoNT/B and required only GT1b, but no low pH, to oligomerise. In further contrast to BoNT/B, low pH alone altered the secondary structure of BoNT/E to some degree and resulted in its premature inactivation (Sun et al. 2012).

The exact position of the toxin on the membrane may be important for the subsequent endocytosis and translocation steps. Geometric restrictions are imposed by the simultaneous adherence to the ganglioside and the protein receptor (Jin et al. 2006). Different binding modes are conceivable. It has been suggested that the predominant negatively charged molecular surfaces of BoNT/B H_N favour a perpendicular orientation of the translocation domain and thus bend Syt into a membrane tangential direction. In this scenario, four solvent-exposed lysine residues conserved among Syt-I and Syt-II might interact with the phospholipid head groups of the membrane (Jin et al. 2006). This view has recently been supported by modelling GT1b into the Syt-II/BoNT/B complex based on its binding to the conserved ganglioside binding pocket of BoNT/A (Stenmark et al. 2008). This model further predicts an extended loop of BoNT/B that contains a very hydrophobic tip (G1246–F1249) and extends out between the ganglioside and Syt binding pockets to be additionally involved in the interaction with the membrane. The H_CD crystal structure revealed a corresponding loop whose hydrophobic segment F1242–Y1246 is likely to be involved in membrane association because its mutation leads to a clear loss in toxicity of BoNT/D. Also, the WY-loop of BoNT/C might play such a role. Furthermore, the monoclonal antibody 4E17 binding a defined epitope at the tip of the α-helices in the H_N domain potently neutralises BoNT/B and E indicating that this epitope plays an important role in the intoxication mechanism (Garcia-Rodriguez et al. 2011). Recently, it was demonstrated that the H_C-fragment of BoNT/A can reorientate by 180° in complex with NTNHA and/or at pH < 6.5 (Gu et al. 2012). If the H_C-fragment is now anchored to the membrane surface via double receptor interactions the remaining LH_N has to turn upside down upon pH drop prior to membrane insertion.

The results of future experiments have to link the known membrane anchorage of the BoNT with the mode of membrane contact and insertion of the translocation domain H_N to form LC translocating channels.

8 Outlook

The physiological dependence of TeNT and BoNT on gangliosides, their prefered types as well as the mode and site of binding have been exhibited in recent years thoroughly for all serotypes. Moreover, the majority of synaptic protein receptors have been elucidated. A non-gangliosidic, proteinaceous receptor molecule has not yet been ascribed to BoNT/C and TeNT which display the highest affinity to neuronal membranes and are taken up less dependently upon neuronal stimulation. Whereas all Syt-I/II binding sites are identified and partially characterised in great detail, nothing is still known for the interaction of any SV2 with BoNT/A, E and D. The value and importance of such knowledge was recently illustrated by the exhibition that human Syt-II is not a high affinity receptor for BoNT/B, DC and G which can evoke serious consequences in the use of BoNT/B as therapeutic. Furthermore, the BoNT family currently consists of 35 sero- and subtypes (variants, "native mutants"). Besides the seven well characterised serotypes the binding properties of only four subtypes (BoNT/A2, BoNT/B2, BoNT/CD and BoNT/DC) have been partially investigated and are already bearing unexpected results. Hence, all BoNT variants should be characterised thoroughly in the future to avoid dangerous surprises as well as exploit the putatively new characteristics therapeutically.

Acknowledgments The author is thankful to Dr. Thomas Binz for critically reviewing the manuscript and to his colleagues and collaborators for contributing to this research. Special thanks are due to Drs. Jasmin Strotmeier, Johannes Bigalke and Stefan Mahrhold, and to Nadja Krez and Anna Magdalena Kruel.

References

Ahnert-Hilger G, Münster-Wandowski A, Höltje M (2012) Synaptic vesicle proteins: targets and routes for botulinum neurotoxins. doi:10.1007/978-3-642-33570-9_8

Angstrom J, Teneberg S, Karlsson KA (1994) Delineation and comparison of ganglioside-binding epitopes for the toxins of Vibrio cholerae, Escherichia coli, and Clostridium tetani: evidence for overlapping epitopes. Proc Natl Acad Sci USA 91(25):11859–11863

Bajjalieh SM, Peterson K, Shinghal R, Scheller RH (1992) SV2, a brain synaptic vesicle protein homologous to bacterial transporters. Science (New York) 257(5074):1271–1273

Bajjalieh SM, Peterson K, Linial M, Scheller RH (1993) Brain contains two forms of synaptic vesicle protein 2. Proc Natl Acad Sci USA 90(6):2150–2154

Baldwin MR, Barbieri JT (2007) Association of botulinum neurotoxin serotypes a and B with synaptic vesicle protein complexes. Biochemistry 46(11):3200–3210

Benson MA, Fu Z, Kim JJ, Baldwin MR (2011) Unique ganglioside recognition strategies for clostridial neurotoxins. J Biol Chem 286(39):34015–34022

Bercsenyi K, Giribaldi F, Schiaro G (2012) The elusive compass of clostridial neurotoxins: deciding when and where to go?. doi:10.1007/978-3-642-33570-9_5

Berntsson RP-A, Peng L, Svensson LM, Dong M, Stenmark P (2012) Botulinum Neurotoxin Receptor Binding. Abstract 49th IBRCC, BS-1

Bigalke H, Muller H, Dreyer F (1986) Botulinum A neurotoxin unlike tetanus toxin acts via a neuraminidase sensitive structure. Toxicon 24(11–12):1065–1074

Binz T (2012) Clostridial neurotoxin light chains: devices for SNARE cleavage mediated blockade of neurotransmission. doi:10.1007/978-3-642-33570-9_7

Black JD, Dolly JO (1986) Interaction of 125I-labeled botulinum neurotoxins with nerve terminals. II. Autoradiographic evidence for its uptake into motor nerves by acceptor-mediated endocytosis. J Cell Biol 103(2):535–544

Blum FC, Chen C, Kroken AR, Barbieri JT (2012) Tetanus toxin and botulinum toxin a utilize unique mechanisms to enter neurons of the central nervous system. Infect Immun 80(5): 1662–1669

Brunger AT, Rummel A (2009) Receptor and substrate interactions of clostridial neurotoxins. Toxicon 54(5):550–560

Bullens RW, O'Hanlon GM, Wagner E, Molenaar PC, Furukawa K, Furukawa K, Plomp JJ, Willison HJ (2002) Complex gangliosides at the neuromuscular junction are membrane receptors for autoantibodies and botulinum neurotoxin but redundant for normal synaptic function. J Neurosci 22(16):6876–6884

Chai Q, Arndt JW, Dong M, Tepp WH, Johnson EA, Chapman ER, Stevens RC (2006) Structural basis of cell surface receptor recognition by botulinum neurotoxin B. Nature 444(7122): 1096–1100

Chang WP, Sudhof TC (2009) SV2 renders primed synaptic vesicles competent for Ca2+ -induced exocytosis. J Neurosci 29(4):883–897

Chapman ER (2002) Synaptotagmin: a Ca(2+) sensor that triggers exocytosis? Nat Rev Mol Cell Biol 3(7):498–508

Chen C, Baldwin MR, Barbieri JT (2008) Molecular basis for tetanus toxin coreceptor interactions. Biochemistry 47(27):7179–7186

Chen C, Fu Z, Kim JJ, Barbieri JT, Baldwin MR (2009) Gangliosides as high affinity receptors for tetanus neurotoxin. J Biol Chem 284(39):26569–26577

Critchley DR, Habig WH, Fishman PH (1986) Reevaluation of the role of gangliosides as receptors for tetanus toxin. J Neurochem 47(1):213–222

Custer KL, Austin NS, Sullivan JM, Bajjalieh SM (2006) Synaptic vesicle protein 2 enhances release probability at quiescent synapses. J Neurosci 26(4):1303–1313

Deinhardt K, Berninghausen O, Willison HJ, Hopkins CR, Schiavo G (2006) Tetanus toxin is internalized by a sequential clathrin-dependent mechanism initiated within lipid microdomains and independent of epsin1. J Cell Biol 174(3):459–471

Dolly JO, Black J, Williams RS, Melling J (1984) Acceptors for botulinum neurotoxin reside on motor nerve terminals and mediate its internalization. Nature 307(5950):457–60

Dong M, Richards DA, Goodnough MC, Tepp WH, Johnson EA, Chapman ER (2003) Synaptotagmins I and II mediate entry of botulinum neurotoxin B into cells. J Cell Biol 162(7):1293–1303

Dong M, Yeh F, Tepp WH, Dean C, Johnson EA, Janz R, Chapman ER (2006) SV2 is the protein receptor for botulinum neurotoxin A. Science (New York) 312(5773):592–596

Dong M, Tepp WH, Liu H, Johnson EA, Chapman ER (2007) Mechanism of botulinum neurotoxin B and G entry into hippocampal neurons. J Cell Biol 179(7):1511–1522

Dong M, Liu H, Tepp WH, Johnson EA, Janz R, Chapman ER (2008) Glycosylated SV2A and SV2B mediate the entry of botulinum neurotoxin E into neurons. Mol Biol Cell 19(12):5226–5237

Emsley P, Fotinou C, Black I, Fairweather NF, Charles IG, Watts C, Hewitt E, Isaacs NW (2000) The structures of the H_C-fragment of tetanus toxin with carbohydrate subunit complexes provide insight into ganglioside binding. J Biol Chem 275(12):8889–8894

Eswaramoorthy S, Kumaran D, Swaminathan S (2001) Crystallographic evidence for doxorubicin binding to the receptor-binding site in Clostridium botulinum neurotoxin B. Acta Crystallogr D Biol Crystallogr 57(Pt 11):1743–1746

Evinger C, Erichsen JT (1986) Transsynaptic retrograde transport of fragment C of tetanus toxin demonstrated by immunohistochemical localization. Brain Res 380(2):383–388

Feany MB, Lee S, Edwards RH, Buckley KM (1992) The synaptic vesicle protein SV2 is a novel type of transmembrane transporter. Cell 70(5):861–867

Figueiredo D, Turcotte C, Frankel G, Li Y, Dolly O, Wilkin G, Marriott D, Fairweather N, Dougan G (1995) Characterization of recombinant tetanus toxin derivatives suitable for vaccine development. Infect Immun 63(8):3218–3221

Fischer A (2012) Synchronized chaperone function of botulinum neurotoxin domains mediates light chain translocation into neurons. doi:10.1007/978-3-642-33570-9_6

Fischer A, Mushrush DJ, Lacy DB, Montal M (2008) Botulinum neurotoxin devoid of receptor binding domain translocates active protease. PLoS Pathog 4(12):e1000245

Fishman PS, Carrigan DR (1987) Retrograde transneuronal transfer of the C-fragment of tetanus toxin. Brain Res 406(1–2):275–279

Fotinou C, Emsley P, Black I, Ando H, Ishida H, Kiso M, Sinha KA, Fairweather NF, Isaacs NW (2001) The crystal structure of tetanus toxin H_C-fragment complexed with a synthetic GT1b analogue suggests cross-linking between ganglioside receptors and the toxin. J Biol Chem 276(34):32274–32281

Fu Z, Chen C, Barbieri JT, Kim JJ, Baldwin MR (2009) Glycosylated SV2 and gangliosides as dual receptors for botulinum neurotoxin serotype F. Biochemistry 48(24):5631–5641

Garcia-Rodriguez C, Geren IN, Lou J, Conrad F, Forsyth C, Wen W, Chakraborti S, Zao H, Manzanarez G, Smith TJ, Brown J, Tepp WH, Liu N, Wijesuriya S, Tomic MT, Johnson EA, Smith LA, Marks JD (2011) Neutralizing human monoclonal antibodies binding multiple serotypes of botulinum neurotoxin. Protein Eng Des Sel 24(3):321–331

Geppert M, Archer BT 3rd, Sudhof TC (1991) Synaptotagmin II. A novel differentially distributed form of synaptotagmin. J Biol Chem 266(21):13548–13552

Ginalski K, Venclovas C, Lesyng B, Fidelis K (2000) Structure-based sequence alignment for the beta-trefoil subdomain of the clostridial neurotoxin family provides residue level information about the putative ganglioside binding site. FEBS Lett 482(1–2):119–124

Gu S, Rumpel S, Zhou J, Strotmeier J, Bigalke H, Perry K, Shoemaker CB, Rummel A, Jin R (2012) Botulinum neurotoxin is shielded by NTNHA in an interlocked complex. Science 335(6071):977–981

Herreros J, Lalli G, Schiavo G (2000a) C-terminal half of tetanus toxin fragment C is sufficient for neuronal binding and interaction with a putative protein receptor. Biochem J 347 (Pt 1):199–204

Herreros J, Lalli G, Montecucco C, Schiavo G (2000b) Tetanus toxin fragment C binds to a protein present in neuronal cell lines and motoneurons. J Neurochem 74(5):1941–1950

Herreros J, Ng T, Schiavo G (2001) Lipid rafts act as specialized domains for tetanus toxin binding and internalization into neurons. Mol Biol Cell 12(10):2947–2960

Hill KK and Smith TJ (2012) Genetic diversity within *clostridium botulinum* serotypes, botulinum neurotoxin gene clusters and toxin subtypes. doi:10.1007/978-3-642-33570-9_1

Holmgren J, Elwing H, Fredman P, Strannegard O, Svennerholm L (1980) Gangliosides as receptors for bacterial toxins and sendai virus. Adv Exp Med Biol 125:453–470

Hughes R, Whaler BC (1962) Influence of nerve-ending activity and of drugs on the rate of paralysis of rat diaphragm preparations by Cl. botulinum type A toxin. J Physiol 160:221–233

Ihara H, Kohda T, Morimoto F, Tsukamoto K, Karasawa T, Nakamura S, Mukamoto M, Kozaki S (2003) Sequence of the gene for Clostridium botulinum type B neurotoxin associated with infant botulism, expression of the C-terminal half of heavy chain and its binding activity. Biochim Biophys Acta 1625(1):19–26

Janz R, Sudhof TC (1999) SV2C is a synaptic vesicle protein with an unusually restricted localization: anatomy of a synaptic vesicle protein family. Neuroscience 94(4):1279–1290

Jayaraman S, Eswaramoorthy S, Kumaran D, Swaminathan S (2005) Common binding site for disialyllactose and tri-peptide in C-fragment of tetanus neurotoxin. Proteins 61(2):288–295

Jin R, Rummel A, Binz T, Brunger AT (2006) Botulinum neurotoxin B recognizes its protein receptor with high affinity and specificity. Nature 444(7122):1092–1095

Kamata Y, Kozaki S, Sakaguchi G, Iwamori M, Nagai Y (1986) Evidence for direct binding of Clostridium botulinum type E derivative toxin and its fragments to gangliosides and free fatty acids. Biochem Biophys Res Commun 140(3):1015–1019

Karalewitz AP, Kroken AR, Fu Z, Baldwin MR, Kim JJ, Barbieri JT (2010) Identification of a unique ganglioside binding loop within botulinum neurotoxins C and D-SA. Biochemistry 49(37):8117–8126

Keller JE, Cai F, Neale EA (2004) Uptake of botulinum neurotoxin into cultured neurons. Biochemistry 43(2):526–532

Kitamura M, Iwamori M, Nagai Y (1980) Interaction between Clostridium botulinum neurotoxin and gangliosides. Biochim Biophys Acta 628(3):328–335

Kitamura M, Takamiya K, Aizawa S, Furukawa K (1999) Gangliosides are the binding substances in neural cells for tetanus and botulinum toxins in mice. Biochim Biophys Acta 1441(1):1–3

Kitamura M, Igimi S, Furukawa K, Furukawa K (2005) Different response of the knockout mice lacking b-series gangliosides against botulinum and tetanus toxins. Biochim Biophys Acta 1741(1–2):1–3

Kozaki S, Ogasawara J, Shimote Y, Kamata Y, Sakaguchi G (1987) Antigenic structure of Clostridium botulinum type B neurotoxin and its interaction with gangliosides, cerebroside, and free fatty acids. Infect Immun 55(12):3051–3056

Kozaki S, Kamata Y, Watarai S, Nishiki T, Mochida S (1998) Ganglioside GT1b as a complementary receptor component for Clostridium botulinum neurotoxins. Microb Pathog 25(2):91–99

Kroken AR, Karalewitz AP, Fu Z, Kim JJ, Barbieri JT (2011) Novel ganglioside-mediated entry of botulinum neurotoxin serotype D into neurons. J Biol Chem 286(30):26828–26837

Kumaran D, Eswaramoorthy S, Furey W, Navaza J, Sax M, Swaminathan S (2009) Domain organization in Clostridium botulinum neurotoxin type E is unique: its implication in faster translocation. J Mol Biol 386(1):233–245

Lalli G, Herreros J, Osborne SL, Montecucco C, Rossetto O, Schiavo G (1999) Functional characterisation of tetanus and botulinum neurotoxins binding domains. J Cell Sci 112(Pt 16):2715–2724

Lawrence G, Wang J, Chion CK, Aoki KR, Dolly JO (2007) Two protein trafficking processes at motor nerve endings unveiled by botulinum neurotoxin E. J Pharmacol Exp Ther 320(1): 410–418

Lazarovici P, Yavin E (1986) Affinity-purified tetanus neurotoxin interaction with synaptic membranes: properties of a protease-sensitive receptor component. Biochemistry 25(22): 7047–7054

Lightstone FC, Prieto MC, Singh AK, Piqueras MC, Whittal RM, Knapp MS, Balhorn R, Roe DC (2000) Identification of novel small molecule ligands that bind to tetanus toxin. Chem Res Toxicol 13(5):356–362

Louch HA, Buczko ES, Woody MA, Venable RM, Vann WF (2002) Identification of a binding site for ganglioside on the receptor binding domain of tetanus toxin. Biochemistry 41(46):13644–13652

Mahrhold S (2008) Identifizierung neuronaler proteinrezeptoren clostridieller neurotoxine. Leibniz Universität Hannover, Hannover

Mahrhold S, Rummel A, Bigalke H, Davletov B, Binz T (2006) The synaptic vesicle protein 2C mediates the uptake of botulinum neurotoxin A into phrenic nerves. FEBS Lett 580(8):2011–2014

Marxen P, Bigalke H (1989) Tetanus toxin: inhibitory action in chromaffin cells is initiated by specified types of gangliosides and promoted in low ionic strength solution. Neurosci Lett 107(1–3):261–266

Marxen P, Fuhrmann U, Bigalke H (1989) Gangliosides mediate inhibitory effects of tetanus and botulinum A neurotoxins on exocytosis in chromaffin cells. Toxicon 27(8):849–859

Marxen P, Erdmann G, Bigalke H (1991) The translocation of botulinum A neurotoxin by chromaffin cells is promoted in low ionic strength solution and is insensitive to trypsin. Toxicon 29(2):181–189

Matteoli M, Verderio C, Rossetto O, Iezzi N, Coco S, Schiavo G, Montecucco C (1996) Synaptic vesicle endocytosis mediates the entry of tetanus neurotoxin into hippocampal neurons. Proc Natl Acad Sci USA 93(23):13310–13315

Montecucco C (1986) How do tetanus and botulinum toxins bind to neuronal membranes? Trends Biochem Sci 11(8):314–317

Moriishi K, Koura M, Abe N, Fujii N, Fujinaga Y, Inoue K, Ogumad K (1996) Mosaic structures of neurotoxins produced from Clostridium botulinum types C and D organisms. Biochim Biophys Acta 1307(2):123–126

Munro P, Kojima H, Dupont JL, Bossu JL, Poulain B, Boquet P (2001) High sensitivity of mouse neuronal cells to tetanus toxin requires a GPI-anchored protein. Biochem Biophys Res Commun 289(2):623–629

Muraro L, Tosatto S, Motterlini L, Rossetto O, Montecucco C (2009) The N-terminal half of the receptor domain of botulinum neurotoxin A binds to microdomains of the plasma membrane. Biochem Biophys Res Commun 380(1):76–80

Nakamura K, Kohda T, Shibata Y, Tsukamoto K, Arimitsu H, Hayashi M, Mukamoto M, Sasakawa N, Kozaki S (2012) Unique biological activity of botulinum D/C mosaic neurotoxin in murine species. Infect Immun 80(8):2886–2893

Niemann H, Blasi J, Jahn R (1994) Clostridial neurotoxins: new tools for dissecting exocytosis. Trends Cell Biol 4(5):179–185

Nishiki T, Kamata Y, Nemoto Y, Omori A, Ito T, Takahashi M, Kozaki S (1994) Identification of protein receptor for Clostridium botulinum type B neurotoxin in rat brain synaptosomes. J Biol Chem 269(14):10498–10503

Nishiki T, Tokuyama Y, Kamata Y, Nemoto Y, Yoshida A, Sato K, Sekiguchi M, Takahashi M, Kozaki S (1996a) The high-affinity binding of Clostridium botulinum type B neurotoxin to synaptotagmin II associated with gangliosides GT1b/GD1a. FEBS Lett 378(3):253–257

Nishiki T, Tokuyama Y, Kamata Y, Nemoto Y, Yoshida A, Sekiguchi M, Takahashi M, Kozaki S (1996b) Binding of botulinum type B neurotoxin to Chinese hamster ovary cells transfected with rat synaptotagmin II cDNA. Neurosci Lett 208(2):105–108

Nowack A, Yao J, Custer KL, Bajjalieh SM (2010) SV2 regulates neurotransmitter release via multiple mechanisms. Am J Physiol Cell Physiol 299(5):C960–C967

Nuemket N, Tanaka Y, Tsukamoto K, Tsuji T, Nakamura K, Kozaki S, Yao M, Tanaka I (2011) Structural and mutational analyses of the receptor binding domain of botulinum D/C mosaic neurotoxin: insight into the ganglioside binding mechanism. Biochem Biophys Res Commun 411(2):433–439

Ochanda JO, Syuto B, Ohishi I, Naiki M, Kubo S (1986) Binding of Clostridium botulinum neurotoxin to gangliosides. J Biochem (Tokyo) 100(1):27–33

Peng L, Tepp WH, Johnson EA, Dong M (2011) Botulinum neurotoxin D uses synaptic vesicle protein SV2 and gangliosides as receptors. PLoS Pathog 7(3):e1002008

Peng L, Berntsson RP, Tepp WH, Pitkin RM, Johnson EA, Stenmark P, Dong M (2012) Botulinum neurotoxin D-C uses synaptotagmin I/II as receptors and human synaptotagmin II is not an effective receptor for type B, D-C, and G toxins. J Cell Sci 125(Pt 13):3233–3242

Perin MS, Fried VA, Mignery GA, Jahn R, Sudhof TC (1990) Phospholipid binding by a synaptic vesicle protein homologous to the regulatory region of protein kinase C. Nature 345(6272):260–263

Pierce EJ, Davison MD, Parton RG, Habig WH, Critchley DR (1986) Characterization of tetanus toxin binding to rat brain membranes. Evidence for a high-affinity proteinase-sensitive receptor. Biochem J 236(3):845–852

Pirazzini M, Rossetto O, Bolognese P, Shone CC, Montecucco C (2011) Double anchorage to the membrane and intact inter-chain disulfide bond are required for the low pH induced entry of tetanus and botulinum neurotoxins into neurons. Cell Microbiol 13(11):1731–1743

Rummel A, Bade S, Alves J, Bigalke H, Binz T (2003) Two carbohydrate binding sites in the H_{CC}-domain of tetanus neurotoxin are required for toxicity. J Mol Biol 326(3):835–847

Rummel A, Mahrhold S, Bigalke H, Binz T (2004a) The H_{CC}-domain of botulinum neurotoxins A and B exhibits a singular ganglioside binding site displaying serotype specific carbohydrate interaction. Mol Microbiol 51(3):631–643

Rummel A, Karnath T, Henke T, Bigalke H, Binz T (2004b) Synaptotagmins I and II act as nerve cell receptors for botulinum neurotoxin G. J Biol Chem 279(29):30865–30870

Rummel A, Eichner T, Weil T, Karnath T, Gutcaits A, Mahrhold S, Sandhoff K, Proia RL, Acharya KR, Bigalke H, Binz T (2007) Identification of the protein receptor binding site of botulinum neurotoxins B and G proves the double-receptor concept. Proc Natl Acad Sci USA 104(1):359–364

Rummel A, Häfner K, Mahrhold S, Darashchonak N, Holt M, Jahn R, Beermann S, Karnath T, Bigalke H, Binz T (2009) Botulinum neurotoxins C, E and F bind gangliosides via a conserved binding site prior to stimulation-dependent uptake with botulinum neurotoxin F utilising the three isoforms of SV2 as second receptor. J Neurochem 110(6):1942–1954

Schengrund CL, DasGupta BR, Ringler NJ (1991) Binding of botulinum and tetanus neurotoxins to ganglioside GT1b and derivatives thereof. J Neurochem 57(3):1024–1032

Schmitt A, Dreyer F, John C (1981) At least three sequential steps are involved in the tetanus toxin-induced block of neuromuscular transmission. Naunyn Schmiedebergs Arch Pharmacol 317(4):326–330

Schmitt J, Karalewitz A, Benefield DA, Mushrush DJ, Pruitt RN, Spiller BW, Barbieri JT, Lacy DB (2010) Structural analysis of botulinum neurotoxin type G receptor binding. Biochemistry 49(25):5200–5205

Shapiro RE, Specht CD, Collins BE, Woods AS, Cotter RJ, Schnaar RL (1997) Identification of a ganglioside recognition domain of tetanus toxin using a novel ganglioside photoaffinity ligand. J Biol Chem 272(48):30380–30386

Shoemaker CB and Oyler GA (2012) Persistence of *botulinum* neurotoxin inactivation of nerve function. doi:10.1007/978-3-642-33570-9_9

Simpson LL (1980) Kinetic studies on the interaction between botulinum toxin type A and the cholinergic neuromuscular junction. J Pharmacol Exp Ther 212(1):16–21

Simpson LL (1982) A comparison of the pharmacological properties of Clostridium botulinum type C1 and C2 toxins. J Pharmacol Exp Ther 223(3):695–701

Simpson LL (1984a) Botulinum toxin and tetanus toxin recognize similar membrane determinants. Brain Res 305(1):177–180

Simpson LL (1984b) The binding fragment from tetanus toxin antagonizes the neuromuscular blocking actions of botulinum toxin. J Pharmacol Exp Ther 229(1):182–187

Simpson LL (1985) Pharmacological experiments on the binding and internalization of the 50,000 dalton carboxyterminus of tetanus toxin at the cholinergic neuromuscular junction. J Pharmacol Exp Ther 234(1):100–105

Simpson LL, Rapport MM (1971a) Ganglioside inactivation of botulinum toxin. J Neurochem 18(7):1341–1343

Simpson LL, Rapport MM (1971b) The binding of botulinum toxin to membrane lipids: sphingolipids, steroids and fatty acids. J Neurochem 18(9):1751–1759

Sinha K, Box M, Lalli G, Schiavo G, Schneider H, Groves M, Siligardi G, Fairweather N (2000) Analysis of mutants of tetanus toxin Hc-fragment: ganglioside binding, cell binding and retrograde axonal transport properties. Mol Microbiol 37(5):1041–1051

Stenmark P, Dupuy J, Imamura A, Kiso M, Stevens RC (2008) Crystal structure of botulinum neurotoxin type A in complex with the cell surface co-receptor GT1b-insight into the toxin-neuron interaction. PLoS Pathog 4(8):e1000129

Stenmark P, Dong M, Dupuy J, Chapman ER, Stevens RC (2010) Crystal structure of the botulinum neurotoxin type G binding domain: insight into cell surface binding. J Mol Biol 397(5):1287–1297

Strotmeier J, Lee K, Völker AK, Mahrhold S, Zong Y, Zeiser J, Zhou J, Pich A, Bigalke H, Binz T, Rummel A, Jin R (2010) Botulinum neurotoxin serotype D attacks neurons via two carbohydrate-binding sites in a ganglioside-dependent manner. Biochem J 431(2):207–216

Strotmeier J, Gu S, Jutzi S, Mahrhold S, Zhou J, Pich A, Eichner T, Bigalke H, Rummel A, Jin R, Binz T (2011) The biological activity of botulinum neurotoxin type C is dependent upon novel types of ganglioside binding sites. Mol Microbiol. 81(1):143–156

Strotmeier J, Willjes G, Binz T, Rummel A (2012) Human synaptotagmin-II is not a high affinity receptor for botulinum neurotoxin B and G: increased therapeutic dosage and immunogenicity. FEBS Lett 586(4):310–313

Sudhof TC (2002) Synaptotagmins: why so many? J Biol Chem 277(10):7629–7632

Sun S, Suresh S, Liu H, Tepp WH, Johnson EA, Edwardson JM, Chapman ER (2011) Receptor binding enables botulinum neurotoxin B to sense low pH for translocation channel assembly. Cell Host Microbe 10(3):237–247

Sun S, Tepp WH, Johnson EA, Chapman ER (2012) Botulinum neurotoxins B and E translocate at different rates and exhibit divergent responses to GT1b and low pH. Biochemistry 50(14):2711–2713
Sutton JM, Chow-Worn O, Spaven L, Silman NJ, Hallis B, Shone CC (2001) Tyrosine-1290 of tetanus neurotoxin plays a key role in its binding to gangliosides and functional binding to neurones. FEBS Lett 493(1):45–49
Svennerholm L (1994) Designation and schematic structure of gangliosides and allied glycosphingolipids. Progress in brain research. 101:XI–XIV
Swaminathan S, Eswaramoorthy S (2000) Structural analysis of the catalytic and binding sites of Clostridium botulinum neurotoxin B. Nat Struct Biol 7(8):693–699
Takamizawa K, Iwamori M, Kozaki S, Sakaguchi G, Tanaka R, Takayama H, Nagai Y (1986) TLC immunostaining characterization of Clostridium botulinum type A neurotoxin binding to gangliosides and free fatty acids. FEBS Lett 201(2):229–232
Tsukamoto K, Kohda T, Mukamoto M, Takeuchi K, Ihara H, Saito M, Kozaki S (2005) Binding of Clostridium botulinum type C and D neurotoxins to ganglioside and phospholipid. Novel insights into the receptor for clostridial neurotoxins. J Biol Chem 280(42):35164–35171
Tsukamoto K, Kozai Y, Ihara H, Kohda T, Mukamoto M, Tsuji T, Kozaki S (2008) Identification of the receptor-binding sites in the carboxyl-terminal half of the heavy chain of botulinum neurotoxin types C and D. Microb Pathog 44(6):484–493
van Heyningen WE (1958) Identity of the tetanus toxin receptor in nervous tissue. Nature 182(4652):1809
van Heyningen WE (1959) Tentative Identification of the tetanus toxin receptor in nervous tissue. J Gen Microbiol 20:810–820
van Heyningen WE, Miller PA (1961) The fixation of tetanus toxin by ganglioside. J Gen Microbiol 24:107–119
Verderio C, Coco S, Rossetto O, Montecucco C, Matteoli M (1999) Internalization and proteolytic action of botulinum toxins in CNS neurons and astrocytes. J Neurochem 73(1): 372–379
Wan QF, Zhou ZY, Thakur P, Vila A, Sherry DM, Janz R, Heidelberger R (2010) SV2 acts via presynaptic calcium to regulate neurotransmitter release. Neuron 66(6):884–895
Williamson LC, Bateman KE, Clifford JC, Neale EA (1999) Neuronal sensitivity to tetanus toxin requires gangliosides. J Biol Chem 274(35):25173–25180
Xu T, Bajjalieh SM (2001) SV2 modulates the size of the readily releasable pool of secretory vesicles. Nat Cell Biol 3(8):691–698
Yao J, Nowack A, Kensel-Hammes P, Gardner RG, Bajjalieh SM (2010) Cotrafficking of SV2 and synaptotagmin at the synapse. J Neurosci 30(16):5569–5578
Yeh FL, Dong M, Yao J, Tepp WH, Lin G, Johnson EA, Chapman ER (2010) SV2 mediates entry of tetanus neurotoxin into central neurons. PLoS Pathog 6(11):e1001207
Yokosawa N, Kurokawa Y, Tsuzuki K, Syuto B, Fujii N, Kimura K, Oguma K (1989) Binding of Clostridium botulinum type C neurotoxin to different neuroblastoma cell lines. Infect Immun 57(1):272–277
Yowler BC, Schengrund CL (2004) Botulinum neurotoxin A changes conformation upon binding to ganglioside GT1b. Biochemistry 43(30):9725–9731
Yowler BC, Kensinger RD, Schengrund CL (2002) Botulinum neurotoxin A activity is dependent upon the presence of specific gangliosides in neuroblastoma cells expressing synaptotagmin I. J Biol Chem 277(36):32815–32819
Zhang Y, Buchko GW, Qin L, Robinson H, Varnum SM (2010) Structural analysis of the receptor binding domain of botulinum neurotoxin serotype D. Biochem Biophys Res Commun 401(4):498–503
Zhang Y, Buchko GW, Qin L, Robinson H, Varnum SM (2011) Crystal structure of the receptor binding domain of the botulinum C-D mosaic neurotoxin reveals potential roles of lysines 1118 and 1136 in membrane interactions. Biochem Biophys Res Commun 404(1):407–412

The Elusive Compass of Clostridial Neurotoxins: Deciding When and Where to Go?

Kinga Bercsenyi, Francesco Giribaldi and Giampietro Schiavo

Abstract Axonal transport ensures long-range delivery of essential components and signals between proximal and distal areas of the neuron, and it is crucial for neuronal homeostasis and survival. Several pathogens and virulence factors use this route to gain access to the central nervous system, exploiting the complex and still poorly understood trafficking mechanisms that regulate the dynamics of their cellular receptors. Studying the intracellular transport of neurotropic pathogens is therefore instrumental to glean new insights into these important molecular events. Botulinum (BoNT) and tetanus (TeNT) neurotoxins bind with high affinity to a variety of neurons and are internalised by specialised endocytic pathways leading to specific intracellular fates. Whereas BoNT trafficking is largely confined to the neuromuscular junction, TeNT is internalised in signalling endosomes shared with neurotrophins and their receptors, which are recruited to the fast axonal retrograde transport pathway. Recently, important paradigms regarding the mechanisms by which BoNT and TeNT interact with their cellular targets and are transported in neurons have been challenged. In this review, we summarise new findings concerning the uptake and intracellular trafficking of these neurotoxins, and discuss their implications in terms of the physiological effects of BoNT and TeNT in the central nervous system.

Keywords Axonal transport · Botulinum neurotoxin · Endocytosis motor neurons · Tetanus toxin

K. Bercsenyi · G. Schiavo (✉)
Molecular NeuroPathobiology Laboratory, Cancer Research UK London Research Institute, 44 Lincoln's Inn Fields, London WC2A 3LY, UK
e-mail: giampietro.schiavo@cancer.org.uk

F. Giribaldi
Department of Experimental Medicine, Section of Pharmacology and Toxicology, viale Cembrano 4, 16148 Genoa, Italy

A. Rummel and T. Binz (eds.), *Botulinum Neurotoxins*, Current Topics in Microbiology and Immunology 364, DOI: 10.1007/978-3-642-33570-9_5,

Abbreviations

$A\beta$	Amyloid beta
AP2	Adaptor protein 2
BAR	Bin–amphiphysin–rvs domain
BDNF	Brain-derived neurotrophic factor
BoNT	Botulinum neurotoxins
CAR	Coxsackie- and adenovirus receptor
CAV2	Canine adenovirus 2
CCP	Clathrin coated pits
CCV	Clathrin coated vesicles
ChAT	Choline acetyltransferase
CHO	Chinese hamster ovary
CME	Clathrin-mediated endocytosis
CNS	Central nervous system
CNT	Clostridial neurotoxins
DRG	Dorsal root ganglia
ERK	Extracellular signal-regulated kinase
GPI	Glycosylphosphatidylinositol
HC	Heavy chain
H_CT	Binding fragment of tetanus toxin
LC	Light chain
Kidins220/ARMS	Kinase-D-interacting substrate of 220 kDa/ankyrin-rich membrane spanning
MT	Microtubules
NMJ	Neuromuscular junction
NT	Neurotrophins
$p75^{NTR}$	p75 neurotrophin receptor
PC12	Pheochromocytoma cells
$PLC\gamma 1$	Phospholipase $C\gamma 1$
PrP	Prion protein
$PtdIns(4,5)P_2$	Phosphatidylinositol(4,5)bisphosphate
PV	Poliovirus
SC	Superior colliculus
SNAP-25	Synaptosomal associated protein of 25 kDa
SNARE	Soluble NSF attachment protein receptor
SV2	Synaptic vesicle glycoprotein 2
SV40	Simian virus 40
Syt	Synaptotagmin
TeNT	Tetanus neurotoxin
Trk	Tropomyosin-receptor-kinase
VAMP	Vesicle associated membrane protein

Contents

1 Introduction 93
2 Clostridial Neurotoxins are Multi-Domain Proteins 94
3 Binding of Clostridial Neurotoxins to the Cell Surface 95
4 Clostridial Neurotoxin Endocytosis 98
5 Fate Decision in the Trafficking of Clostridial Neurotoxins 100
6 Axonal Transport of Tetanus Neurotoxin 101
7 Tetanus Toxin Shares Transport Compartments with Neurotrophins and Their Receptors 101
8 Shared Pathways with Pathogens 103
9 Botulinum Neurotoxins and Their Long-Range Transport Mechanisms 104
10 Future Perspectives 106
References 107

1 Introduction

Efficient mechanisms ensuring the reliable exchange of information between cells and their environment are essential for all organisms. The outer surface of the plasma membrane provides a platform for a wide range of receptors, which sense extracellular signals and inform cells about changes in their surroundings. To ensure that cells receive relevant messages and are able to assemble suitable physiological responses, both the lipid and protein compositions of the plasma membrane must be tightly regulated. A main player in this homeostatic control is the process of endocytosis, in which selective components of the plasma membrane are internalised in a highly regulated fashion and undergo intracellular trafficking leading to degradation, recycling or targeting to other membrane compartments (McMahon and Boucrot 2011). The main roles of endocytosis include nutrient uptake, receptor-mediated signalling and the turnover of membrane components and receptors (Hoeller et al. 2005). Given the key physiological role and evolutionary conservation of these mechanisms, numerous pathogens and virulence factors, including clostridial neurotoxins (CNT) exploit selective endocytic routes to gain access to host cells. Despite the plethora of different endocytic cargoes, surprisingly little information is available regarding the molecular mechanisms regulating the recruitment of ligand-receptor complexes to specific carriers, their internalisation and transport and their ultimate fate. In light of this, infectious agents represent invaluable tools shaped by millions of years of evolution to study the molecular determinants of these membrane trafficking pathways (Schiavo and van der Goot 2001).

The most extensively studied type of endocytosis is clathrin-mediated endocytosis (CME). The list of proteins involved in the regulation of CME has been continuously growing, and as a result, a detailed view of the molecular mechanisms controlling its progression is now available (McMahon and Boucrot 2011). Mechanistically, CME occurs via four main steps, the first of which is the nucleation of clathrin at specific membrane sites. This process is initiated by accessory and adaptor proteins, such as adaptor protein 2 (AP2) and epsins (Henne et al. 2010; Jackson et al. 2010; Qualmann et al. 2011; Taylor et al. 2011). Following the recruitment of membrane curvature-inducing proteins, clathrin-coated pits (CCPs) are formed. Once the budding is completed, clathrin-coated vesicles (CCV) are clipped off the membrane by dynamin, a GTPase involved in membrane remodelling (Harper et al. 2011; Ferguson and De Camilli 2012), and the clathrin basket is removed by cytoplasmic factors to form an uncoated vesicle (Bocking et al. 2011). Interestingly, CME was shown to be the main, but not the only route taken by CNT to get access to their target cells (Deinhardt et al. 2006a; Montal 2010).

2 Clostridial Neurotoxins are Multi-Domain Proteins

The CNT family comprises tetanus toxin (TeNT) and several related botulinum neurotoxins (BoNT, serotypes A to G; Swaminathan 2011). TeNT and BoNT are synthesised as single chain proteins of 150 kDa, which are cleaved by endogenous or tissue proteases, resulting in a 50 kDa light chain (LC) and a 100 kDa heavy chain (HC) linked via a disulphide bridge. The carboxy-terminal half of the HC is responsible for the binding of CNT to the neuronal surface, whilst the amino-terminal half is involved in the translocation of the LC through the endosomal membrane (Montal 2010). The LC contains the active site of the neurotoxin and displays a very specific metalloprotease activity (Schiavo et al. 2000). To date, the synaptic SNARE (soluble NSF attachment protein receptor) proteins syntaxin-1, SNAP-25 (synaptosomal-associated protein of 25 kDa) and VAMP-1–2 (vesicle-associated membrane protein) (termed also synaptobrevin-1 and -2) are the only identified substrates of TeNT and BoNT (McMahon et al. 1993; Schiavo et al. 2000), in addition to VAMP-3, syntaxin-2 and syntaxin-3. Cleavage of these synaptic SNARE proteins yields the persistent blockade of neurotransmitter release in intoxicated neurons (Schiavo et al. 1992; Blasi et al. 1993a, b; Schiavo et al. 1993a, b; Yamasaki et al. 1994).

CNT share their core structural characteristics (Fig. 1), cleave the same class of intracellular substrates and are taken up by the same neurons (Schiavo et al. 2000). However, once endocytosed, TeNT and BoNT are sorted to different membrane trafficking pathways (Fig. 1). BoNT mainly target the neuromuscular junction (NMJ) in vivo, and only a minor fraction of the toxin is transported back to the cell body of the motor neuron (Restani et al. 2012a). In contrast, TeNT efficiently reaches the inhibitory interneurons located in the spinal cord (Fig. 1). This distinct intracellular trafficking leads to a different symptomatology of the pathologies caused by BoNT and TeNT. Botulism, which is caused by BoNT, leads to flaccid paralysis, whilst tetanus is characterised by a sustained spastic paralysis. It remains to be seen whether or not this difference in sorting is due to specific receptor complexes for TeNT and BoNT, which would target them to distinct endocytic compartments, or to the recruitment of different sorting factors after internalisation.

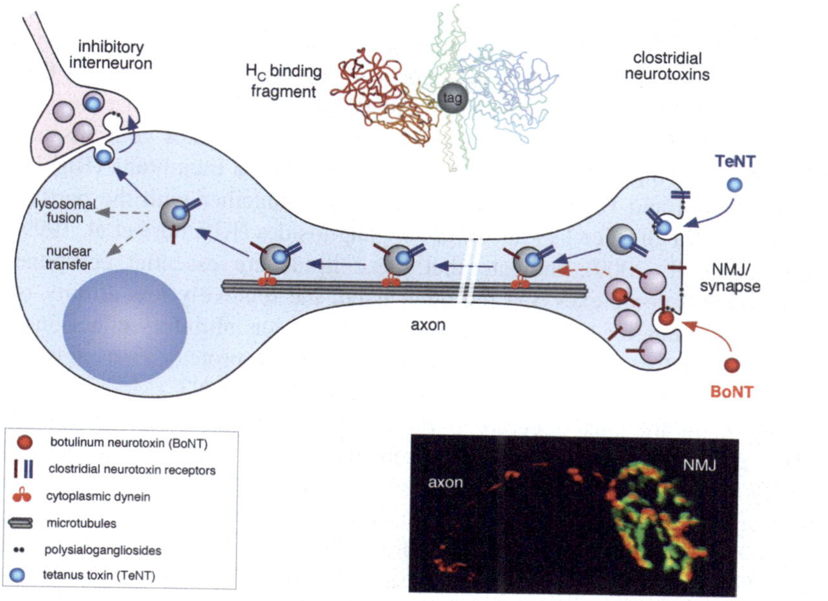

Fig. 1 Trafficking of botulinum and tetanus neurotoxins in neurons. *Top.* The three-domain structure of BoNT/A (Lacy et al. 1998), which is shared with the other CNT, is characterized by a 50 kDa LC containing the active site (shaded, on the *right*) and a 100 kDa HC responsible for membrane translocation (HN; shaded, *centre*) and binding to the neuronal surface (HC; unshaded, on the *left*). CNT fragments can be expressed as recombinant proteins bearing specific short domains that allow their labelling with fluorophores (tag) and other functional groups (Deinhardt et al. 2006b). Due to its binding and sorting characteristics and its lack of toxicity, fluorescent HCs can be used instead of the full length CNT to study binding, transport and transcytosis. *Middle.* Schematic representation of CNT trafficking in a spinal cord motor neuron. TeNT (*light blue*) and BoNT (*red*) specifically bind to neuronal membranes via receptor complex(es) formed by polysialogangliosides (*black dots*) and synaptic proteins (*blue* and *red bars*). CNT are then internalised and sorted into different intracellular compartments. Whilst BoNT remain mainly localised to the NMJ, where they enter the synaptic vesicle recycling pathway, TeNT undergoes fast axonal retrograde transport exploiting MT as tracks (*grey*) and cytoplasmic dynein as a main molecular motor (*red*). Once it has reached the soma, TeNT is trancytosed into inhibitory interneurons, where the LC reaches the cytoplasm and blocks neurotransmitter release. *Bottom.* Mouse NMJ after intramuscular administration of HCT (*red*). HCT is internalised and transported into the motor neuron shaft. The motor endplate has been counterstained with α-bungarotoxin (*green*)

3 Binding of Clostridial Neurotoxins to the Cell Surface

Complex gangliosides were shown to serve as receptors for CNT. TeNT binds to GT1b and GD1b, whereas BoNT bind to GT1b and GD1a (Habermann and Dreyer 1986; Schengrund et al. 1991; Yowler and Schengrund 2004). The affinity of CNT to immobilised polysialogangliosides is in the high nanomolar range, whilst these neurotoxins bind to synaptosomes and neurons with a much higher affinity, making their interactions with neuronal membrane almost irreversible (Habermann and Dreyer 1986). The addition of exogenous polysialogangliosides to non-neuronal cells, such as undifferentiated rat pheochromocytoma (PC12), renders them sensitive to TeNT or BoNT/A (Marxen and Bigalke 1989; Marxen et al. 1990). In agreement with this finding, neuraminidase treatment makes cells insensitive to CNT by removing sialic acid residues from the plasma membrane (Bigalke et al. 1986; Marxen and Bigalke 1989). These results, together with the partial insensitivity to CNT of mice lacking complex gangliosides (Kitamura et al. 1999, 2005; Rummel et al. 2009), indicate that these lipids are essential components of CNT receptor complexes. On the other hand, the relatively low affinity of polysialogangliosides for TeNT and BoNT and the major inhibitory effect on binding caused by pre-treatment of neurons with extracellular proteases, strongly suggests the existence of a dual protein and lipid receptor for CNT (Montecucco 1986). According to the dual receptor hypothesis (Montecucco 1986; Rummel et al. 2007), polysialogangliosides either favour the recruitment of TeNT and BoNT to specific areas of the plasma membrane enriched in protein receptors, or maintain these toxins in a preferred conformation for the receptor to bind.

The molecular identification of the protein receptors for CNT posed a challenge to the field. Both binding and internalisation of BoNT were shown to be activity-dependent (Black and Dolly 1986; Keller et al. 2004; Baldwin and Barbieri 2007) suggesting that synaptic vesicle proteins might be involved in BoNT binding and uptake. The intraluminal leaflet of the synaptic vesicle membrane becomes transiently exposed to the extracellular medium upon fusion of synaptic vesicles with the active zone during synaptic stimulation, potentially allowing the intravesicular domains of synaptic vesicle proteins to interact with the binding domain of CNT and act as their protein receptors. Pioneering work from Kozaki's group has demonstrated that BoNT/B binds simultaneously to polysialogangliosides and synaptotagmin-I (Syt-I), the main calcium sensor for synaptic vesicle fusion, and the closely related isoform synaptotagmin-II (Syt-II), with ten times higher affinity than polysialogangliosides alone (Ochanda et al. 1986; Nishiki et al. 1994; 1996a). However, direct evidence that the luminal domain of both Syt-I and Syt-II binds BoNT/B and mediates the internalisation of the toxin was provided only a decade later (Dong et al. 2003, 2007). Hippocampal neurons lacking Syt-I are insensitive to both BoNT/B and BoNT/G, but can be rendered sensitive by overexpression of Syt-I or -II (Dong et al. 2007). Similarly, the cytoplasm of PC12 and Chinese hamster ovary (CHO) cells became accessible to BoNT/B upon the surface expression of the intraluminal domain of Syt-I/II (Nishiki et al. 1996b; Pirazzini et al. 2011). However, Syt are not universal protein receptors for BoNT, as only BoNT/B, BoNT/G and the mosaic toxin BoNT/DC bind to Syt-I/II (Rummel et al. 2004, 2007; Peng et al. 2012). Based on the hypothesis that the luminal portion of synaptic vesicle proteins may serve as protein receptor of the remaining BoNT serotypes, two groups independently demonstrated that synaptic vesicle glycoprotein 2 (SV2) acts as a protein receptor for BoNT/A (Dong et al. 2006; Mahrhold et al. 2006) and several other serotypes, such as BoNT/E (Dong et al. 2008), BoNT/F (Rummel et al. 2009) and BoNT/D (Peng et al. 2011).

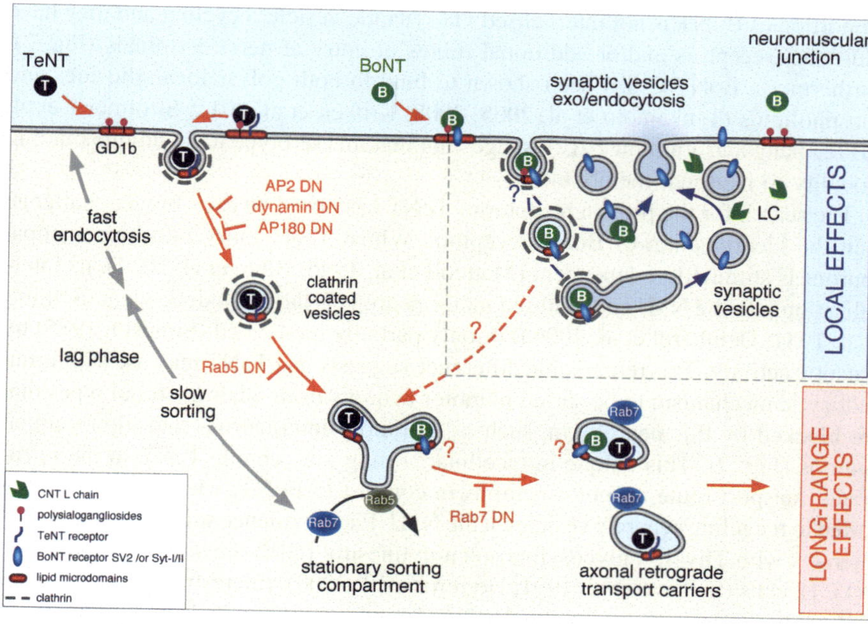

Fig. 2 Uptake of TeNT and BoNTs in cultured motor neurons. Synaptic vesicle exo/endocytosis accounts for the majority of endocytic events at the presynaptic terminal and may involve multiple clathrin-dependent step(s). Several BoNTs exploit this pathway for their internalization into neurons by binding to the luminal domain of specific vesicle proteins, such as SV2 and Syt-I/II. High affinity interaction to these proteins requires polysialogangliosides and may be regulated by lipid microdomains. Once internalised into the synaptic vesicle lumen, acidification driven by the vATPase triggers the translocation of LC into the nerve terminal cytoplasm that leads to the cleavage of SNARE proteins essential for membrane fusion. This process determines the long-lasting inhibition of neurotransmitter release at the NMJ (*local effects*). At very high doses, TeNT enters this pathway and induces a flaccid paralysis similar to that caused by BoNT. In contrast, at physiological doses TeNT exploits a pathway requiring lipid microdomains and the clathrin machinery that is largely independent of synaptic vesicle exo/endocytosis. At the NMJ, TeNT binds to a lipid-protein receptor complex containing polysialogangliosides such as GD1b, and is then laterally sorted into clathrin-coated vesicles. During this sorting event, GD1b is excluded from the toxin receptor complex (Deinhardt et al. 2006a). Internalisation of TeNT is dependent on dynamin, AP-2 and AP180, but does not require epsin1. Once internalised, TeNT is targeted to a stationary (or oscillating) early sorting compartment positive for the small GTPase Rab5, to which other endocytic routes may converge. Some of these routes may be responsible for the entry of specific BoNT serotypes (e.g. BoNT/A) to sorting endosomes, their targeting to the axonal retrograde transport pathway and their transport together with TeNT and neurotrophin receptor to the soma of motor neurons (*long-range effects*). Fast retrograde transport of these organelles requires Rab7 activity. Question marks indicate molecular events that have not been conclusively demonstrated to date and DN stands for dominant-negative mutants

Whilst it became clear that synaptic vesicle proteins play a role in the binding and uptake of BoNT, independent lines of evidence suggested that at least a fraction of these neurotoxins are internalised in an activity-independent fashion (Verderio et al. 1999; Restani et al. 2012a). These findings strongly suggest that a proportion of BoNT is not internalised via synaptic vesicle recycling and may have additional receptors and/or additional routes of entry at nerve terminals (Fig. 2). Furthermore, BoNT/C has been shown to bind to both polysialogangliosides and phospholipids (Tsukamoto et al. 2005, 2008; Kroken et al. 2011; Strotmeier et al. 2011; Zhang and Varnum 2012), suggesting that this serotype has a unique binding modality to neuronal membranes.

The search for the protein receptor of TeNT has posed an even bigger challenge than the identification of BoNT receptors. Whilst TeNT entry into hippocampal neurons is stimulation-dependent (Matteoli et al. 1996; Blum et al. 2012), its internalisation into the NMJ and cultured motor neurons is largely independent (Schmitt et al. 1981; Deinhardt et al. 2006a) or only partially modulated (Simpson 1985) by synaptic activity. This remarkable difference suggests that TeNT may use a different endocytic mechanism to be sorted in motor neurons from other neuronal types that are blocked by this neurotoxin, such as inhibitory interneurons and hippocampal neurons (Fig. 2). This unique intracellular sorting may couple TeNT to the retrograde transport route in motor neurons in contrast to BoNT, which are preferably sorted to recycling synaptic vesicles at the NMJ. Early evidence suggested that TeNT interacts with Thy-1, a glycosylphosphatidylinositol (GPI)-anchored glycoprotein, in PC12 cells (Schiavo et al. 1991; Herreros et al. 2000). Interestingly, one or more GPI-anchored proteins are involved in TeNT binding and intracellular activity, since treatment of PC12 cells or neurons with a phosphatidylinositol-specific phospholipase C prevents the TeNT-induced cleavage of VAMP-2 (Herreros et al. 2001; Munro et al. 2001). Disruption of the integrity of membrane microdomains, which is essential for GPI-anchored protein clustering, also prevented VAMP-2 cleavage by TeNT (Herreros et al. 2001; Munro et al. 2001). However, Thy-1 is unlikely to be the only protein receptor for TeNT since mice lacking Thy-1 are only slightly less sensitive to TeNT intoxication than wild-type controls (Herreros et al. 2001). SV2 was also shown to bind TeNT in central neurons (Yeh et al. 2010). However, this latter interaction has been recently disputed (Blum et al. 2012) and no evidence is presently available validating this mechanism in motor neurons or inhibitory interneurons. Last, but not least, the possibility that the interaction of TeNT with neuronal and non-neuronal cells is mediated by the binding of polysialogangliosides to two distinct sites of its binding domain (H_CT) has been proposed (Fotinou et al. 2001; Rummel et al. 2003; Chen et al. 2008, 2009). Interestingly, immobilised glycolipid complexes have been shown to display higher affinity to H_CT (Rinaldi et al. 2009), suggesting that ganglioside *cis* interactions may have important modulatory roles in the initial binding of TeNT to the neuronal membrane.

4 Clostridial Neurotoxin Endocytosis

Once CNT are bound to their surface receptors, a complex cascade of protein–protein and protein–lipid interactions trigger the recruitment of clathrin and adaptor proteins to the inner leaflet of the plasma membrane, which marks the onset of the endocytic process (Fig. 2). Although BoNT and TeNT are mainly internalised by distinct pathways (synaptic vesicle recycling and synaptic vesicle-independent clathrin-dependent endocytosis, respectively) (Deinhardt et al. 2006a; Montal 2010; Blum et al. 2012), several aspects of these mechanisms are shared at the molecular level. One of the early events of these processes is the recruitment of specific clathrin adaptors, which leads to the accumulation of effector proteins altering membrane curvature at endocytic sites. A major determinant of this process is the enrichment of specific lipids, such as phosphatidylinositol(4,5)bisphosphate (PtdIns(4,5)P_2), at these sites. This event is likely to precede membrane curvature initiation as most of the accessory proteins, such as AP2, bind to this lipid (Haucke 2005). Several factors mediating membrane curvature possess a bin–amphiphysin–rvs (BAR) domain (Qualmann et al. 2011), which enables them to initiate and maintain membrane curvature, whilst other proteins lacking the BAR domain, such as epsins, shape membranes by inserting an amphipathic helix into the inner leaflet of the lipid bilayer, making it more accessible to clathrin (Ford et al. 2002; Hinrichsen et al. 2006). Deinhardt et al. have shown that the internalisation of TeNT relies on the canonical clathrin adaptors AP2 and AP180, but is independent of epsin1, since the overexpression of an epsin1 mutant unable to bind PtdIns(4,5)P_2 did not affect the internalisation of the toxin, whilst completely abolished transferrin uptake (Deinhardt et al. 2006a). Epsin1 was shown to target ubiquitinated receptors to the late endosomal/lysosomal pathway (Le Roy and Wrana 2005), suggesting that TeNT follows an intracellular trafficking route bypassing this sorting step. Accordingly, TeNT is known to access an atypical endosomal compartment in motor neurons, which is not acidified (Lalli et al. 2003a; Bohnert and Schiavo 2005). Thus, this epsin1-independent endocytic route may prevent the membrane insertion of HC and the subsequent translocation of the active LC into the cytoplasm, a process that is triggered by low pH, allowing the delivery of TeNT to axonal retrograde carriers in an intact form (Fig. 2). At the same time, this specific sorting avoids the targeting of TeNT to lysosomes and its degradation. As to which other membrane curvature protein(s) might act as a replacement of epsin1 remains unknown.

Following the recruitment of clathrin adaptors, a clathrin basket is formed around the pit and a new CCV is ready to be born. The last step in the biogenesis of CCV is the recruitment of a member of the dynamin family, a large GTPase, which upon GTP hydrolysis, drives the fission of the CCV from the plasma membrane. Crucially, the uptake of both TeNT and BoNT/A is disrupted in the presence of dynamin inhibitors (Deinhardt et al. 2006a; Harper et al. 2011) or by overexpression of dynamin mutants (Deinhardt et al. 2006a) (Fig. 2). Electron microscopy studies have confirmed that CNT are taken up into CCV (Black and Dolly 1986; Deinhardt et al. 2006a), which undergo uncoating in the synaptic cytoplasm.

5 Fate Decision in the Trafficking of Clostridial Neurotoxins

Although several binding and endocytic determinants are shared by BoNT and TeNT, the intracellular fate of these neurotoxins is largely distinct. In motor neurons, TeNT is internalised together with neurotrophin receptors and their ligands into transport endosomes (Fig. 2), which are characterised by near-neutral pH and low degradative potential (Bohnert and Schiavo 2005), and are delivered to the cell body. Upon arrival in the soma, TeNT is sorted to a transcytotic route and gains access to inhibitory interneurons (Schiavo et al. 2000). In contrast, BoNT are mostly taken up in synaptic vesicles and remain confined at distal synapses, such as the NMJ. During reloading with neurotransmitters, the pH drop in the lumen of synaptic vesicles determines the insertion of BoNT into the lipid bilayer (Montecucco et al. 1986, 1989) which triggers the translocation of the LC into the cytoplasm, where the disulphide bond is reduced and it can specifically cleave the synaptic targets (Koriazova and Montal 2003; Fischer and Montal 2007; Fischer et al. 2008; Montal 2010). However, recent evidence suggests that BoNT/A is also retrogradely transported in several neuronal types, including hippocampal, tectal and motor neurons (Antonucci et al. 2008; Restani et al. 2012a) and undergoes transcytosis in the visual system (Restani et al. 2011, 2012b), mimicking at least in part the behaviour of TeNT.

For its long-range transport, TeNT exploits a highly specialised trafficking pathway shaped by strong evolutionary pressure. Eukaryotic cells are characterised by a highly compartmentalised organisation and efficient communication between different cellular areas is required to ensure cellular homeostasis. Transport over long distances reaches its higher specialisation in neurons, where dendritic and axonal compartments are in dynamic equilibrium via a network of highly regulated transport routes powered by molecular motors (Hirokawa et al. 2010; Soo et al. 2011; Winckler and Yap 2011). Eukaryotic cells, including neurons, rely on three superfamilies of motor proteins: kinesins and dyneins transport their cargoes on microtubules (MT), whereas myosins are F-actin-dependent (Stiess and Bradke 2011). MT and actin microfilaments extend longitudinally within neurons, with MT mainly present in axons and dendrites and actin microfilaments enriched at synaptic regions (Hirokawa et al. 2010). Both cytoskeletal elements are characterised by a highly polarised architecture. The plus, fast-growing end of MT is directed towards the periphery in axons and distal dendrites, whereas the barbed, growing end of actin microfilaments points to the plasma membrane in pre- and post-synaptic regions. Fast axonal and dendritic transport is mainly MT-dependent and relies on kinesins and cytoplasmic dynein for the distribution within different neuronal regions of a variety of cargoes, which include organelles, proteins and RNAs (Hirokawa et al. 2010). Moreover, modulation of axonal transport by varying the concentration and/or activity of individual motors constitutes a reliable size-sensing mechanism in vitro and in vivo (Rishal et al. 2012). In axons, fast axonal transport occurs both in the anterograde (from cell body towards the periphery) and retrograde (from axonal tips to cell body) directions by means of kinesins and cytoplasmic dynein, respectively (Vale et al.

1985; Hirokawa et al. 2010). Myosins also participate in axonal transport in the proximity of the synaptic regions or in areas where the distribution of MT is less uniform (Langford 2002; Lalli et al. 2003b).

6 Axonal Transport of Tetanus Neurotoxin

Kinetic analysis of axonal transport assessed in primary motor neuron cultures using either H_CT or the directly labelled full-length neurotoxin revealed a complex speed profile, which can be deconvolved in a trimodal Gaussian distribution with peaks at 1, 1.5 and 2.1 μm/s, well within the range of fast axonal transport (Lalli and Schiavo 2002; Hafezparast et al. 2003). Two different types of pleiomorphic organelles responsible for the axonal retrograde transport of TeNT were identified: vesiculo-tubular structures, that show a continuous retrograde movement and account for the faster transport component, and round vesicles characterised by a more discontinuous and slower retrograde transport (Lalli et al. 2003a). In vivo characterisation of H_CT transport in the sciatic nerve or in sensory neurons confirmed this trimodal speed distribution (Bilsland et al. 2010) (Fig. 4). Strikingly, the average velocities observed in vivo were higher than those observed in vitro (Bilsland et al. 2010). This is likely to be due to the extensive axon myelination occurring in motor neurons in the sciatic nerve, which is known to stabilise MT, and/or differences between embryonic (in vitro) and adult (in vivo) motor neurons (Bilsland et al. 2010). In addition to a major role of kinesins, cytoplasmic dynein and MT, the retrograde transport of TeNT also relies on myosin Va, an actin-associated motor protein. Whilst the fastest component of the transport is accomplished almost totally by cytoplasmic dynein and requires kinesins at equilibrium, the intermediate component is dependent on myosin Va, as demonstrated by the transport impairments observed in myosin Va-null motor neuron cultures (Lalli et al. 2003b). The requirement of both actin microfilaments and MT for fast axonal transport is in line with the tight association occurring between these two cytoskeletal elements seen by ultrastructural analysis (Bearer and Reese 1999). Furthermore, it has been proposed that the recruitment of myosins, in addition to cytoplasmic dynein, allows neurons to assure a continuous retrograde movement of organelles also during transitions between different MT (Langford 2002; Lalli et al. 2003b).

7 Tetanus Toxin Shares Transport Compartments with Neurotrophins and Their Receptors

Many different external stimuli rely on axonal retrograde transport for reliable and fast signalling from distal synapses to the cell body. Activated receptor complexes are sorted to transport organelles, called signalling endosomes (Fig. 3), that

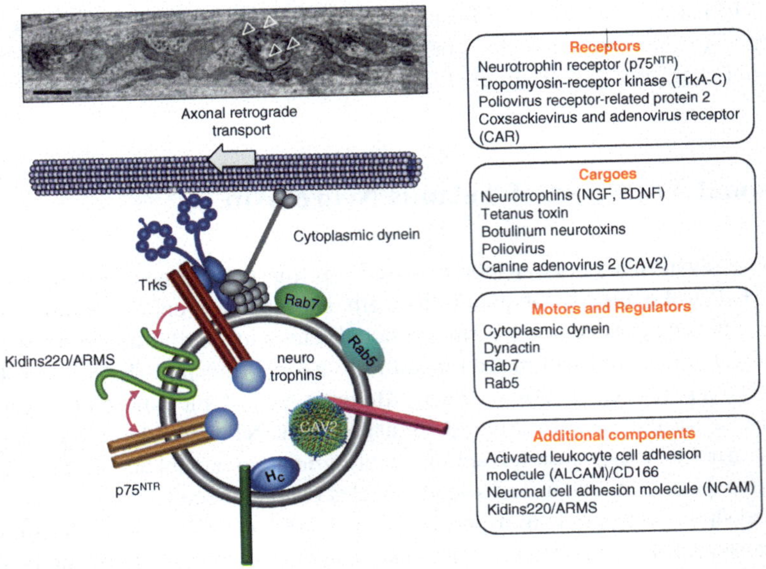

Fig. 3 Schematic representation of a TeNT-positive signalling endosome. TeNT and its binding fragment H_CT exploit the retrograde transport machinery used by physiological cargoes, such as neurotrophins (*light blue*), for their entry into the CNS. Once internalised together with its still unknown receptor(s) (*green*), H_CT enters signalling endosomes containing Trks, $p75^{NTR}$ and their interacting protein Kidins220/ARMS (kinase-D-interacting substrate of 220 kDa/ankyrin-rich membrane spanning; *green*). The progression and fast axonal transport of these signalling endosomes are dependent on the sequential recruitment of the two small GTPases Rab5 (*blue*) and Rab7 (*green*). Fast axonal retrograde transport of these signalling organelles relies on the MT-dependent motor cytoplasmic dynein (*blue*) and its associated complex dynactin (*grey*). Signalling endosomes contain several plasma membrane proteins, such as coxsackie- and adenovirus receptor (CAR, dark red), which are known to bind viruses. As such, these organelles are exploited by several viruses, such as CAV2 and poliovirus, to access the CNS. Additional components and cargoes of these axonal carriers are listed on the right. An electron microscopy image of a signalling endosome containing H_CT (empty arrows) is shown on the top. Scale bar, 200 nm

undergo fast axonal transport together with their adaptors and downstream signalling molecules, thus overcoming the limitations of signal transduction mechanisms based uniquely on diffusion (Howe 2005). Neurotrophins (NT) and their receptors, tropomyosin receptor kinase (Trk) and p75 neurotrophin receptor ($p75^{NTR}$), are among the best known examples of such retrogradely transported signalling complexes (Schecterson and Bothwell 2010). NT interact with Trks and $p75^{NTR}$ at distal sites and these activated receptor complexes are transported towards the nucleus, where they promote gene expression events controlling differentiation and survival in most types of neurons (Butowt and von Bartheld 2003; Winckler and Yap 2011). It has recently been shown that TeNT exploits the organelles used by NT receptor complexes for its journey from nerve terminals to the soma. Experiments performed in primary spinal cord motor neurons using fluorescently labelled H_CT showed that TeNT shares retrograde carriers with NGF and $p75^{NTR}$ (Lalli and Schiavo 2002). A common route for TeNT and neurotrophin receptors was further demonstrated by showing the presence in the same transported endosome of H_CT, brain-derived neurotrophic factor (BDNF), $p75^{NTR}$ and TrkB (Deinhardt et al. 2006b). This is a general mechanism, since it has been shown both in primary motor and dorsal root ganglia (DRG) neurons. Progression towards the cell body is dependent on Rab5 and Rab7, two small GTPases with multiple roles in the endocytic pathway (Deinhardt et al. 2006b) (Fig. 3). Rab5 and Rab7 act in a sequential manner: Rab5 is involved in the initial steps of the internalisation process, whereas Rab7 is required for fast progression along axons (Deinhardt et al. 2006a; Salinas et al. 2009). This unanticipated relationship between the trafficking of TeNT and NT is further supported by independent lines of evidence. BDNF was shown to increase the efficiency and kinetic of internalisation of H_CT at the murine NMJ in a dose-dependent manner (Roux et al. 2006). Other neurotrophins, such as NT4, have similar effects on the localisation and internalisation of TeNT but are less potent than BDNF (Roux et al. 2006). The functional interaction between TeNT and NT is bidirectional, since both TeNT and H_CT activate TrkA and its downstream effectors extracellular signal-regulated kinase 1/2 (ERK1/2) and phospholipase C γ1 (PLCγ1) in a dose-dependent manner (Gil et al. 2000; Gil et al. 2001, 2003).

8 Shared Pathways with Pathogens

Neurotoxins, such as TeNT and to a lesser extent BoNT, are not the only exogenous molecules gaining access to the CNS by exploiting the axonal retrograde transport pathway. Many viruses, which are endocytosed by clathrin-dependent and -independent mechanisms, rely on controlled acidification steps for their uncoating and cytoplasmic entry (Salinas et al. 2010). Once released in the cytoplasm, these virions are transported to the nucleus by binding directly motor proteins, such as cytoplasmic dynein (Greber and Way 2006; Salinas et al. 2010). However, other neurotrophic viruses, such as canine adenovirus 2 (CAV2), are retrogradely transported together with their receptors, in TeNT-positive carriers containing $p75^{NTR}$ (Salinas et al. 2009) (Fig. 3). Poliovirus (PV) is another example of a neurotrophic virus that undergoes retrograde transport in a TeNT-positive compartment in primary motor neurons (Ohka et al. 2009). Interestingly, the kinetics of these organelles seems to be controlled by the PV receptor, probably via its ability to bind directly cytoplasmic dynein (Ohka et al. 2009). Other viruses resemble TeNT dynamics in terms of their intracellular sorting and axonal trafficking. Accordingly, the first phase of influenza virus transport is largely MT-dependent with speed ranging between 1 and 4 μm/s and a pH of its carriers approaching neutrality (Lakadamyali et al. 2003). The sequential involvement of Rab5 and Rab7 is a key feature of the sorting and retrograde transport of this virus as well as of HIV (Vidricaire and Tremblay 2005), influenza virus H3N2 (Sieczkarski and Whittaker 2003) and simian virus 40 (SV40) (Vonderheit and Helenius 2005).

Similar to TeNT, SV40 and polyoma virus bind to neuronal receptors associated with lipid rafts and are internalised by a cholesterol- and glycosphingolipid-dependent mechanism (Smith et al. 2003). Although not essential for its infectivity, $p75^{NTR}$ contributes to the binding, internalisation and axonal transport of rabies virus and its subsequent transcytosis to second-order neurons (Ugolini 1995; Tuffereau et al. 2007). In addition to viruses, amyloid beta ($A\beta$), prion protein (PrP) and toxic protein aggregates have been found to be associated with axonal carriers and affect axonal transport. In particular, $A\beta$, which has been implicated in Alzheimer's disease, was found to directly interact with $p75^{NTR}$ (Yaar et al. 1997) and to impair axonal retrograde transport of BDNF in primary neurons (Poon et al. 2011). Similarly, a toxic PrP peptide has been shown to induce cell death via direct activation of $p75^{NTR}$ signalling (Della-Bianca et al. 2001), whereas the full-length protein undergoes fast axonal transport both in peripheral and central neurons (Borchelt et al. 1994; Encalada et al. 2011). Altogether, these findings suggest that the axonal retrograde transport route is a main gateway for the entry and spread into the CNS of pathological agents and virulence factors.

9 Botulinum Neurotoxins and Their Long-Range Transport Mechanisms

BoNT activity is mainly restricted to distal synapses, a feature that is exploited in the many therapeutic applications of BoNT (Schiavo et al. 2000; Hackett and Kam 2007). In spite of the overwhelming evidence supporting this peripheral mechanism of action, several reports suggest that BoNT might also have central effects in humans and animal models. More than 50 years ago, the case of a patient affected by botulism showing clear CNS effects was reported (Polley et al. 1965). Initially, these central effects of BoNT were explained by synaptic plasticity changes occurring between the central and the peripheral neuron of the network, after the latter had been silenced by BoNT intoxication. However, the suggestion of a possible, direct, central action of BoNT arose from experiments performed by Wiegand and colleagues who detected a rise in radioactivity in the ventral horn of the spinal cord after injection of radiolabelled BoNT/A in the cat gastrocnemius muscle, both ipsilaterally and contralaterally to the site of administration (Wiegand et al. 1976). Further evidence of axonal retrograde transport of BoNT/A and BoNT/B was found in mouse muscle diaphragm after intraperitoneal administration or incubation in vitro with radiolabelled neurotoxins. Although internalisation occurred with different efficiency for BoNT/A and B, both toxins were found in the axoplasm of myelinated axons within vacuolar-like structures, thus suggesting the entry of BoNT in a specific endocytic pathway (Black and Dolly 1986). Direct evidence supporting the hypothesis that BoNT/A is retrogradely transported in an active form and undergoes transcytosis in second-order neurons was provided by the detection of BoNT/A-cleaved SNAP-25 in cholinergic synapses in rat retina after injection of the neurotoxin in the superior colliculus (SC) (Antonucci et al. 2008). No BoNT/A-cleaved SNAP-25 was detected in synapses of the retina when this neurotoxin was injected in SC after MT depolymerisation, thus ruling out diffusion-based spreading mechanisms. Supporting evidence for a long-range action was also obtained by detecting changes in the activity of CA1 pyramidal neurons after injection of BoNT/A into the contralateral hippocampus (Antonucci et al. 2008).

Interestingly, transport (and transcytosis) of catalytically active BoNT/A occurs not only in the retrograde direction, but also anterogradely. This unexpected property of BoNT/A has been demonstrated by detecting the presence of neurotoxin-cleaved SNAP-25 in the SC after BoNT/A injection in the contralateral retina (Restani et al. 2011).

Interestingly, different BoNT serotypes display distinct long-range effects in vivo. Conversely to BoNT/A, BoNT/E seems unable to alter the activity of CA1 neurons after injection in the contralateral hippocampus (Antonucci et al. 2008), in spite of overwhelming evidence demonstrating its silencing activity upon ipsilateral injection (Costantin et al. 2005). This result suggests distinct sorting pathways and/or distal kinetics for different BoNT serotypes (Antonucci et al. 2008; Caleo and Schiavo 2009). This conclusion seems to be supported by overt differences in the rate of SNAP-25 cleavage induced by BoNT/A and BoNT/E in sympathetic neurons cultured in Campenot chambers (Lawrence et al. 2012). After distal application of these neurotoxins, BoNT/A-cleaved SNAP-25 was readily detectable in cell bodies, whilst the BoNT/E truncated form of this synaptic protein was only slightly detectable in the somas, despite a much higher quantity of BoNT/E being used in this experiment (Lawrence et al. 2012). The documented short-lived enzymatic activity of BoNT/E was suggested to be responsible for this lack of detection (Lawrence et al. 2012).

However, direct evidence supporting a differential rate of transport of BoNT/A and BoNT/E is still lacking. Experiments recently performed in our laboratory filled this gap by analysing the kinetic properties of the axonal transport of these serotypes once internalised in primary spinal cord motor neurons (Restani et al. 2012a). Full-length BoNT/A and BoNT/E, or their atoxic binding fragments undergo axonal retrograde transport in non-acidic organelles with speed profiles matching fast MT-dependent transport and largely overlapping with TeNT-positive carriers in living motor neurons (Restani et al. 2012a). Our analysis suggests that a lower transport efficiency of BoNT/E compared to BoNT/A may contribute to the differential in vivo effects of these two serotypes (Caleo and Schiavo 2009). Whilst BoNT/A undergoes fast and continuous retrograde transport, BoNT/E-positive organelles show a discontinuous movement towards the cell body, characterised by a higher frequency of pauses and short periods of anterograde transport (Restani et al. 2012a). Altogether, these results demonstrate that BoNT undergo axonal retrograde transport in neurons both in vitro and in vivo.

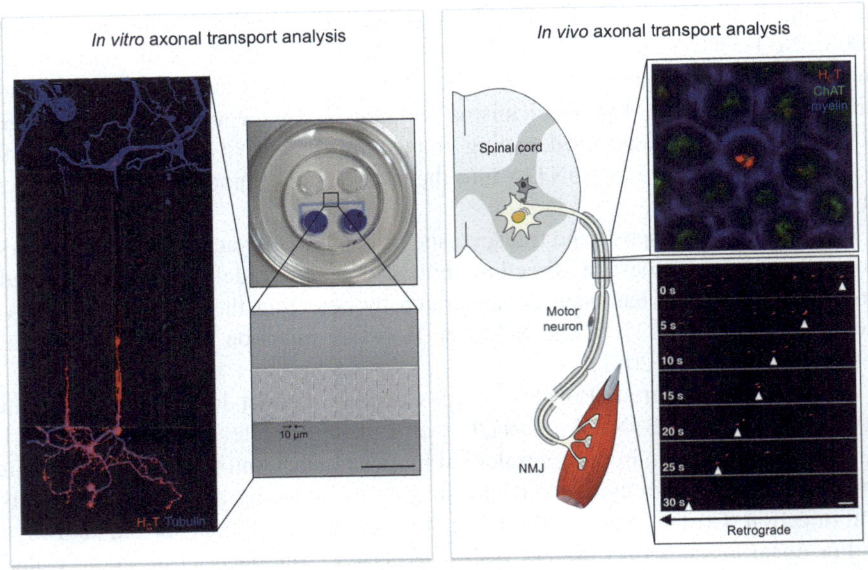

Fig. 4 Innovative strategies to monitor axonal transport in vitro and in vivo. *Left panel.* Microfluidic chambers allow the physical separation of distal axons and synapses (*bottom*) from cell bodies (*top*) in a variety of neuronal cultures, including motor neurons. HcT (*red*), added only to the axonal side, is transported towards the cell bodies located in the top compartment. The two compartments are separated by small microgrooves (10 µm width), which allow the passage of axons, but not of cell bodies. These compartments are also microfluidically separated, a feature that impedes the passive diffusion of ligands and drugs between the two sides of the device. Motor neurons are stained for β-tubulin (*blue*). Scale bar, 500 µm. *Right panel.* Axonal transport of HcT can be monitored in the intact sciatic nerve. After intramuscular injection in the mouse hindlimb, fluorescent HcT (*red*) was found within axons of the sciatic nerve (*top*), which have been stained for choline acetyl transferase (ChAT, *green*), a motor neuron marker, and myelin (*blue*). This technique allows a quantitative kinetic analysis of axonal transport of HcT in vivo (Bilsland et al. 2010). Still images from a confocal movie show an H_CT-positive endosome undergoing fast axonal retrograde transport (*arrowheads*). The cell body is out of view on the left. Scale bar, 5 µm

10 Future Perspectives

In spite of the wealth of data presently available, there are important aspects of the interaction of BoNT with the neuronal membrane that deserve further investigations. Among these the role of different isoforms of SV2 and Syt in the binding of BoNT to distinct neuronal subtypes is presently unclear. Similarly, very little is known about the role of specific posttranslational modifications on the affinity of SV2 and Syt to BoNT and on their intracellular trafficking. It is also unclear whether phospholipids and polysialogangliosides are sufficient to mediate high affinity binding and internalisation of BoNT/C and other serotypes in neurons.

Strikingly, the protein receptor(s) of TeNT at the NMJ is still unknown, although SV2 seems to play a role in TeNT binding to interneurons (Yeh et al. 2010). Previous data suggest that GPI-anchored proteins play a role in TeNT intoxication (Herreros et al. 2001; Munro et al. 2001), but it is unclear whether their contribution is direct or mediated through changes in the clustering of polysialogangliosides on the synaptic plasma membrane. Future experiments are necessary to elucidate the nature of this high affinity TeNT receptor complex and its putative role in axonal retrograde transport and intracellular sorting.

The quantitative analysis of the retrograde transport of different BoNT is still in its infancy. To date, little is known about the molecular mechanisms controlling this process, its modulators and additional cargoes sharing the same transport organelles. As significant differences between the axonal transport of BoNT/A and BoNT/E have recently been detected (Lawrence et al. 2012; Restani et al. 2012a), the CNT field is now facing the exciting prospect of better understanding the molecular determinants of this differential neuronal trafficking, and the relationship between the 3D structure of these serotypes and their intracellular fate. Furthermore, additional studies are necessary to assess whether other serotypes might be transported as well. Novel in vitro approaches, such as compartmentalized motor neuron cultures in microfluidic devices (Fig. 4) will be instrumental for the analysis of the axonal transport of these neurotoxins and the definition of the cellular machinery controlling their trafficking.

Last but not least, the molecular mechanism of transcytosis of TeNT from motor neurons into interneurons is still poorly characterised, as well as the pathway followed by BoNT/A for its transneuronal migration in the visual system (Restani et al. 2011, 2012b).

These studies will not only enable a better characterisation of the CNT at molecular level, but they will also shed light on fundamental processes essential for neuronal homeostasis and survival, providing new potential strategies for the delivery of therapeutics to the CNS.

Acknowledgments We thank Victoria Hill and the members of our laboratories for constructive comments and critical reading of the manuscript. This work was supported by Cancer Research UK (KB and GS). The authors have no conflicting financial interests.

References

Antonucci F, Rossi C, Gianfranceschi L, Rossetto O, Caleo M (2008) Long-distance retrograde effects of botulinum neurotoxin A. J Neurosci 28:3689–3696

Baldwin MR, Barbieri JT (2007) Association of botulinum neurotoxin serotypes A and B with synaptic vesicle protein complexes. Biochemistry 46:3200–3210

Bearer EL, Reese TS (1999) Association of actin filaments with axonal microtubule tracts. J Neurocytol 28:85–98

Bigalke H, Muller H, Dreyer F (1986) Botulinum A neurotoxin unlike tetanus toxin acts via a neuraminidase sensitive structure. Toxicon 24:1065–1074

Bilsland LG, Sahai E, Kelly G, Golding M, Greensmith L, Schiavo G (2010) Deficits in axonal transport precede ALS symptoms in vivo. Proc Natl Acad Sci U S A 107:20523–20528
Black JD, Dolly JO (1986) Interaction of ^{125}I-labeled botulinum neurotoxins with nerve terminals. II. Autoradiographic evidence for its uptake into motor nerves by acceptor-mediated endocytosis. J Cell Biol 103:535–544
Blasi J, Chapman ER, Link E, Binz T, Yamasaki S, De Camilli P, Südhof TC, Niemann H, Jahn R (1993a) Botulinum neurotoxin A selectively cleaves the synaptic protein SNAP-25. Nature 365:160–163
Blasi J, Chapman ER, Yamasaki S, Binz T, Niemann H, Jahn R (1993b) Botulinum neurotoxin C1 blocks neurotransmitter release by means of cleaving HPC-1/syntaxin. EMBO J 12:4821–4828
Blum FC, Chen C, Kroken AR, Barbieri JT (2012) Tetanus toxin and botulinum toxin a utilize unique mechanisms to enter neurons of the central nervous system. Infect Immun 80: 1662–1669
Bocking T, Aguet F, Harrison SC, Kirchhausen T (2011) Single-molecule analysis of a molecular disassemblase reveals the mechanism of Hsc70-driven clathrin uncoating. Nat Struct Mol Biol 18:295–301
Bohnert S, Schiavo G (2005) Tetanus toxin is transported in a novel neuronal compartment characterized by a specialized pH regulation. J Biol Chem 280:42336–42344
Borchelt DR, Koliatsos VE, Guarnieri M, Pardo CA, Sisodia SS, Price DL (1994) Rapid anterograde axonal transport of the cellular prion glycoprotein in the peripheral and central nervous systems. J Biol Chem 269:14711–14714
Butowt R, von Bartheld CS (2003) Connecting the dots: trafficking of neurotrophins, lectins and diverse pathogens by binding to the neurotrophin receptor p75NTR. Eur J Neurosci 17:673–680
Caleo M, Schiavo G (2009) Central effects of tetanus and botulinum neurotoxins. Toxicon 54:593–599
Chen C, Baldwin MR, Barbieri JT (2008) Molecular basis for tetanus toxin coreceptor interactions. Biochemistry 47:7179–7186
Chen C, Fu Z, Kim JJ, Barbieri JT, Baldwin MR (2009) Gangliosides as high affinity receptors for tetanus neurotoxin. J Biol Chem 284:26569–26577
Costantin L, Bozzi Y, Richichi C, Viegi A, Antonucci F, Funicello M, Gobbi M, Mennini T, Rossetto O, Montecucco C, Maffei L, Vezzani A, Caleo M (2005) Antiepileptic effects of botulinum neurotoxin E. J Neurosci 25:1943–1951
Deinhardt K, Berninghausen O, Willison HJ, Hopkins CR, Schiavo G (2006a) Tetanus toxin is internalized by a sequential clathrin-dependent mechanism initiated within lipid microdomains and independent of epsin1. J Cell Biol 174:459–471
Deinhardt K, Salinas S, Verastegui C, Watson R, Worth D, Hanrahan S, Bucci C, Schiavo G (2006b) Rab5 and Rab7 control endocytic sorting along the axonal retrograde transport pathway. Neuron 52:293–305
Della-Bianca V, Rossi F, Armato U, Dal-Pra I, Costantini C, Perini G, Politi V, Della Valle G (2001) Neurotrophin p75 receptor is involved in neuronal damage by prion peptide-(106-126). J Biol Chem 276:38929–38933
Dong M, Liu H, Tepp WH, Johnson EA, Janz R, Chapman ER (2008) Glycosylated SV2A and SV2B mediate the entry of botulinum neurotoxin E into neurons. Mol Biol Cell 5226–5237
Dong M, Richards DA, Goodnough MC, Tepp WH, Johnson EA, Chapman ER (2003) Synaptotagmins I and II mediate entry of botulinum neurotoxin B into cells. J Cell Biol 162:1293–1303
Dong M, Tepp WH, Liu H, Johnson EA, Chapman ER (2007) Mechanism of botulinum neurotoxin B and G entry into hippocampal neurons. J Cell Biol 179:1511–1522
Dong M, Yeh F, Tepp WH, Dean C, Johnson EA, Janz R, Chapman ER (2006) SV2 is the protein receptor for botulinum neurotoxin A. Science 312:592–596
Encalada SE, Szpankowski L, Xia CH, Goldstein LS (2011) Stable kinesin and dynein assemblies drive the axonal transport of mammalian prion protein vesicles. Cell 144:551–565
Ferguson SM, De Camilli P (2012) Dynamin, a membrane-remodelling GTPase. Nat Rev Mol Cell Biol 13:75–88

Fischer A, Montal M (2007) Crucial role of the disulfide bridge between botulinum neurotoxin light and heavy chains in protease translocation across membranes. J Biol Chem 282: 29604–29611

Fischer A, Mushrush DJ, Lacy DB, Montal M (2008) Botulinum neurotoxin devoid of receptor binding domain translocates active protease. PLoS Pathog 4:e1000245

Ford MG, Mills IG, Peter BJ, Vallis Y, Praefcke GJ, Evans PR, McMahon HT (2002) Curvature of clathrin-coated pits driven by epsin. Nature 419:361–366

Fotinou C, Emsley P, Black I, Ando H, Ishida H, Kiso M, Sinha KA, Fairweather NF, Isaacs NW (2001) The crystal structure of tetanus toxin Hc fragment complexed with a synthetic GT1b analogue suggests cross-linking between ganglioside receptors and the toxin. J Biol Chem 276:32274–32281

Gil C, Chaib-Oukadour I, Aguilera J (2003) C-terminal fragment of tetanus toxin heavy chain activates Akt and MEK/ERK signalling pathways in a Trk receptor-dependent manner in cultured cortical neurons. Biochem J 373:613–620

Gil C, Chaib-Oukadour I, Blasi J, Aguilera J (2001) H_C fragment (C-terminal portion of the heavy chain) of tetanus toxin activates protein kinase C isoforms and phosphoproteins involved in signal transduction. Biochem J 356:97–103

Gil C, Chaib-Oukadour I, Pelliccioni P, Aguilera J (2000) Activation of signal transduction pathways involving trkA, PLCgamma-1, PKC isoforms and ERK-1/2 by tetanus toxin. FEBS Lett 481:177–182

Greber UF, Way M (2006) A superhighway to virus infection. Cell 124:741–754

Habermann E, Dreyer F (1986) Clostridial neurotoxins: handling and action at the cellular and molecular level. Curr Top Microbiol Immunol 129:93–179

Hackett R, Kam PC (2007) Botulinum toxin: pharmacology and clinical developments: a literature review. Med Chem 3:333–345

Hafezparast M, Klocke R, Ruhrberg C, Marquardt A, Ahmad-Annuar A, Bowen S, Lalli G, Witherden AS, Hummerich H, Nicholson S, Morgan PJ, Oozageer R, Priestley JV, Averill S, King VR, Ball S, Peters J, Toda T, Yamamoto A, Hiraoka Y, Augustin M, Korthaus D, Wattler S, Wabnitz P, Dickneite C, Lampel S, Boehme F, Peraus G, Popp A, Rudelius M, Schlegel J, Fuchs H, Hrabe de Angelis M, Schiavo G, Shima DT, Russ AP, Stumm G, Martin JE, Fisher EM (2003) Mutations in dynein link motor neuron degeneration to defects in retrograde transport. Science 300:808–812

Harper CB, Martin S, Nguyen TH, Daniels SJ, Lavidis NA, Popoff MR, Hadzic G, Mariana A, Chau N, McCluskey A, Robinson PJ, Meunier FA (2011) Dynamin inhibition blocks botulinum neurotoxin type A endocytosis in neurons and delays botulism. J Biol Chem 286:35966–35976

Haucke V (2005) Phosphoinositide regulation of clathrin-mediated endocytosis. Biochem Soc Trans 33:1285–1289

Henne WM, Boucrot E, Meinecke M, Evergren E, Vallis Y, Mittal R, McMahon HT (2010) FCHo proteins are nucleators of clathrin-mediated endocytosis. Science 328:1281–1284

Herreros J, Lalli G, Montecucco C, Schiavo G (2000) Tetanus toxin fragment C binds to a protein present in neuronal cell lines and motoneurons. J Neurochem 74:1941–1950

Herreros J, Ng T, and Schiavo G (2001) Lipid rafts act as specialised domains for tetanus toxin binding and internalisation into neurons. Submitted

Hinrichsen L, Meyerholz A, Groos S, Ungewickell EJ (2006) Bending a membrane: how clathrin affects budding. Proc Natl Acad Sci U S A 103:8715–8720

Hirokawa N, Niwa S, Tanaka Y (2010) Molecular motors in neurons: transport mechanisms and roles in brain function, development, and disease. Neuron 68:610–638

Hoeller D, Volarevic S, Dikic I (2005) Compartmentalization of growth factor receptor signalling. Curr Opin Cell Biol 17:107–111

Howe CL (2005) Modeling the signaling endosome hypothesis: why a drive to the nucleus is better than a (random) walk. Theor Biol Med Model 2:43

Jackson LP, Kelly BT, McCoy AJ, Gaffry T, James LC, Collins BM, Honing S, Evans PR, Owen DJ (2010) A large-scale conformational change couples membrane recruitment to cargo binding in the AP2 clathrin adaptor complex. Cell 141:1220–1229

Keller JE, Cai F, Neale EA (2004) Uptake of botulinum neurotoxin into cultured neurons. Biochemistry 43:526–532

Kitamura M, Igimi S, Furukawa K (2005) Different response of the knockout mice lacking b-series gangliosides against botulinum and tetanus toxins. Biochim Biophys Acta 1741:1–3

Kitamura M, Takamiya K, Aizawa S, Furukawa K (1999) Gangliosides are the binding substances in neural cells for tetanus and botulinum toxins in mice. Biochim Biophys Acta 1441:1–3

Koriazova LK, Montal M (2003) Translocation of botulinum neurotoxin light chain protease through the heavy chain channel. Nat Struct Biol 10:13–18

Kroken AR, Karalewitz AP, Fu Z, Baldwin MR, Kim JJ, Barbieri JT (2011) Unique ganglioside binding by botulinum neurotoxins C and D-SA. FEBS J 278:4486–4496

Lacy DB, Tepp W, Cohen AC, DasGupta BR, Stevens RC (1998) Crystal structure of botulinum neurotoxin type A and implications for toxicity. Nat Struct Biol 5:898–902

Lakadamyali M, Rust MJ, Babcock HP, Zhuang X (2003) Visualizing infection of individual influenza viruses. Proc Natl Acad Sci U S A 100:9280–9285

Lalli G, Bohnert S, Deinhardt K, Verastegui C, Schiavo G (2003a) The journey of tetanus and botulinum neurotoxins in neurons. Trends Microbiol 11:431–437

Lalli G, Gschmeissner S, Schiavo G (2003b) Myosin Va and microtubule-based motors are required for fast axonal retrograde transport of tetanus toxin in motor neurons. J Cell Sci 116:4639–4650

Lalli G, Schiavo G (2002) Analysis of retrograde transport in motor neurons reveals common endocytic carriers for tetanus toxin and neurotrophin receptor p75NTR. J Cell Biol 156:233–239

Langford GM (2002) Myosin-V, a versatile motor for short-range vesicle transport. Traffic 3:859–865

Lawrence GW, Ovsepian SV, Wang J, Aoki KR, Dolly JO (2012) Extravesicular intraneuronal migration of internalized botulinum neurotoxins without detectable inhibition of distal neurotransmission. Biochem J 441:443–452

Le Roy C, Wrana JL (2005) Clathrin- and non-clathrin-mediated endocytic regulation of cell signalling. Nat Rev Mol Cell Biol 6:112–126

Mahrhold S, Rummel A, Bigalke H, Davletov B, Binz T (2006) The synaptic vesicle protein 2C mediates the uptake of botulinum neurotoxin A into phrenic nerves. FEBS Lett 580:2011–2014

Marxen P, Ahnert-Hilger G, Wellhöner HH, Bigalke H (1990) Tetanus antitoxin binds to intracellular tetanus toxin in permeabilized chromaffin cells without restoring Ca2(+)-induced exocytosis. Toxicon 28:1077–1082

Marxen P, Bigalke H (1989) Tetanus toxin: inhibitory action in chromaffin cells is initiated by specified types of gangliosides and promoted in low ionic strength solution. Neurosci Lett 107:261–266

Matteoli M, Verderio C, Rossetto O, Iezzi N, Coco S, Schiavo G, Montecucco C (1996) Synaptic vesicle endocytosis mediates the entry of tetanus neurotoxin into hippocampal neurons. Proc Natl Acad Sci U S A 93:13310–13315

McMahon HT, Boucrot E (2011) Molecular mechanism and physiological functions of clathrin-mediated endocytosis. Nat Rev Mol Cell Biol 12:517–533

McMahon HT, Ushkaryov YA, Edelmann L, Link E, Binz T, Niemann H, Jahn R, Sudhof TC (1993) Cellubrevin is a ubiquitous tetanus-toxin substrate homologous to a putative synaptic vesicle fusion protein. Nature 364:346–349

Montal M (2010) Botulinum neurotoxin: a marvel of protein design. Annu Rev Biochem 79:591–617

Montecucco C (1986) How do tetanus and botulinum toxins bind to neuronal membranes? Trends Biochem Sci 11:315–317

Montecucco C, Schiavo G, Brunner J, Duflot E, Boquet P, Roa M (1986) Tetanus toxin is labeled with photoactivatable phospholipids at low pH. Biochemistry 25:919–924

Montecucco C, Schiavo G, Dasgupta BR (1989) Effect of pH on the interaction of botulinum neurotoxins A, B and E with liposomes. Biochem J 259:47–53

Munro P, Kojima H, Dupont JL, Bossu JL, Poulain B, Boquet P (2001) High sensitivity of mouse neuronal cells to tetanus toxin requires a GPI-anchored protein. Biochem Biophys Res Commun 289:623–629

Nishiki T, Kamata Y, Nemoto Y, Omori A, Ito T, Takahashi M, Kozaki S (1994) Identification of protein receptor for *Clostridium botulinum* type B neurotoxin in rat brain synaptosomes. J Biol Chem 269:10498–10503

Nishiki T, Tokuyama Y, Kamata Y, Nemoto Y, Yoshida A, Sato K, Sekiguchi M, Takahashi M, Kozaki S (1996a) The high-affinity binding of *Clostridium botulinum* type B neurotoxin to synaptotagmin II associated with gangliosides GT1b/GD1a. FEBS Lett 378:253–257

Nishiki T, Tokuyama Y, Kamata Y, Nemoto Y, Yoshida A, Sekiguchi M, Takahashi M, Kozaki S (1996b) Binding of botulinum type B neurotoxin to Chinese hamster ovary cells transfected with rat synaptotagmin II cDNA. Neurosci Lett 208:105–108

Ochanda JO, Syuto B, Ohishi I, Naiki M, Kubo S (1986) Binding of Clostridium botulinum neurotoxin to gangliosides. J Biochem 100:27–33

Ohka S, Sakai M, Bohnert S, Igarashi H, Deinhardt K, Schiavo G, Nomoto A (2009) Receptor-dependent and -independent axonal retrograde transport of poliovirus in motor neurons. J Virol 83:4995–5004

Peng L, Berntsson RP, Tepp WH, Pitkin RM, Johnson EA, Stenmark P, Dong M (2012) Botulinum neurotoxin D-C uses synaptotagmin I/II as receptors and human synaptotagmin II is not an effective receptor for type B, D-C, and G toxins. J Cell Sci 125(Pt 13):3233–3242

Peng L, Tepp WH, Johnson EA, Dong M (2011) Botulinum neurotoxin D uses synaptic vesicle protein SV2 and gangliosides as receptors. PLoS Pathog 7:e1002008

Pirazzini M, Rossetto O, Bolognese P, Shone CC, Montecucco C (2011) Double anchorage to the membrane and intact inter-chain disulfide bond are required for the low pH induced entry of tetanus and botulinum neurotoxins into neurons. Cell Microbiol 13:1731–1743

Polley EH, Vick JA, Ciuchta HP, Fischetti DA, Macchitelli FJ, Montanarelli N (1965) Botulinum toxin, type A: effects on central nervous system. Science 147:1036–1037

Poon WW, Blurton-Jones M, Tu CH, Feinberg LM, Chabrier MA, Harris JW, Jeon NL, Cotman CW (2011) Beta-amyloid impairs axonal BDNF retrograde trafficking. Neurobiol Aging 32:821–833

Qualmann B, Koch D, Kessels MM (2011) Let's go bananas: revisiting the endocytic BAR code. EMBO J 30:3501–3515

Restani L, Antonucci F, Gianfranceschi L, Rossi C, Rossetto O, Caleo M (2011) Evidence for anterograde transport and transcytosis of botulinum neurotoxin A (BoNT/A). J Neurosci 31:15650–15659

Restani L, Giribaldi F, Manich M, Bercsenyi K, Menendez G, Rossetto O, Caleo M, and Schiavo G (2012a) Botulinum neurotoxins A and E undergo retrograde axonal transport in primary motor neurons. PLoS Pathog:in press

Restani L, Novelli E, Bottari D, Leone P, Barone I, Galli-Resta L, Strettoi E, Caleo M (2012b) Botulinum neurotoxin a impairs neurotransmission following retrograde transynaptic transport. Traffic 13:1083–1089

Rinaldi S, Brennan KM, Goodyear CS, O'Leary C, Schiavo G, Crocker PR, Willison HJ (2009) Analysis of lectin binding to glycolipid complexes using combinatorial glycoarrays. Glycobiology 19:789–796

Rishal I, Kam N, Perry RB, Shinder V, Fisher EM, Schiavo G, Fainzilber M (2012) A motor driven mechanism for cell length sensing. Cell Rep 1:608–616

Roux S, Saint Cloment C, Curie T, Girard E, Mena FJ, Barbier J, Osta R, Molgo J, Brulet P (2006) Brain-derived neurotrophic factor facilitates in vivo internalization of tetanus neurotoxin C-terminal fragment fusion proteins in mature mouse motor nerve terminals. Eur J Neurosci 24:1546–1554

Rummel A, Bade S, Alves J, Bigalke H, Binz T (2003) Two carbohydrate binding sites in the H(CC)-domain of tetanus neurotoxin are required for toxicity. J Mol Biol 326:835–847

Rummel A, Eichner T, Weil T, Karnath T, Gutcaits A, Mahrhold S, Sandhoff K, Proia RL, Acharya KR, Bigalke H, Binz T (2007) Identification of the protein receptor binding site of botulinum neurotoxins B and G proves the double-receptor concept. Proc Natl Acad Sci U S A 104:359–364

Rummel A, Hafner K, Mahrhold S, Darashchonak N, Holt M, Jahn R, Beermann S, Karnath T, Bigalke H, Binz T (2009) Botulinum neurotoxins C, E and F bind gangliosides via a conserved binding site prior to stimulation-dependent uptake with botulinum neurotoxin F utilising the three isoforms of SV2 as second receptor. J Neurochem 110:1942–1954

Rummel A, Karnath T, Henke T, Bigalke H, Binz T (2004) Synaptotagmins I and II act as nerve cell receptors for botulinum neurotoxin G. J Biol Chem 279:30865–30870

Salinas S, Bilsland LG, Henaff D, Weston AE, Keriel A, Schiavo G, Kremer EJ (2009) CAR-associated vesicular transport of an adenovirus in motor neuron axons. PLoS Pathog 5:e1000442

Salinas S, Schiavo G, Kremer EJ (2010) A hitchhiker's guide to the nervous system: the complex journey of viruses and toxins. Nat Rev Microbiol 8:645–655

Schecterson LC, Bothwell M (2010) Neurotrophin receptors: old friends with new partners. Dev Neurobiol 70:332–338

Schengrund CL, DasGupta BR, Ringler NJ (1991) Binding of botulinum and tetanus neurotoxins to ganglioside GT1b and derivatives thereof. J Neurochem 57:1024–1032

Schiavo G, Benfenati F, Poulain B, Rossetto O, Polverino de Laureto P, DasGupta BR, Montecucco C (1992) Tetanus and botulinum-B neurotoxins block neurotransmitter release by proteolytic cleavage of synaptobrevin. Nature 359:832–835

Schiavo G, Ferrari G, Rossetto O, Montecucco C (1991) Tetanus toxin receptor. Specific cross-linking of tetanus toxin to a protein of NGF-differentiated PC 12 cells. FEBS Lett 290:227–230

Schiavo G, Matteoli M, Montecucco C (2000) Neurotoxins affecting neuroexocytosis. Physiol Rev 80:717–766

Schiavo G, Santucci A, Dasgupta BR, Mehta PP, Jontes J, Benfenati F, Wilson MC, Montecucco C (1993a) Botulinum neurotoxins serotypes A and E cleave SNAP-25 at distinct COOH-terminal peptide bonds. FEBS Lett 335:99–103

Schiavo G, Shone CC, Rossetto O, Alexander FC, Montecucco C (1993b) Botulinum neurotoxin serotype F is a zinc endopeptidase specific for VAMP/synaptobrevin. J Biol Chem 268:11516–11519

Schiavo G, van der Goot FG (2001) The bacterial toxin toolkit. Nat Rev Mol Cell Biol 2:530–537

Schmitt A, Dreyer F, John C (1981) At least three sequential steps are involved in the tetanus toxin-induced block of neuromuscular transmission. Naunyn Schmiedebergs Arch Pharmacol 317:326–330

Sieczkarski SB, Whittaker GR (2003) Differential requirements of Rab5 and Rab7 for endocytosis of influenza and other enveloped viruses. Traffic 4:333–343

Simpson LL (1985) Pharmacological experiments on the binding and internalization of the 50,000 dalton carboxyterminus of tetanus toxin at the cholinergic neuromuscular junction. J Pharmacol Exp Ther 234:100–105

Smith AE, Lilie H, Helenius A (2003) Ganglioside-dependent cell attachment and endocytosis of murine polyomavirus-like particles. FEBS Lett 555:199–203

Soo KY, Farg M, Atkin JD (2011) Molecular motor proteins and amyotrophic lateral sclerosis. Int J Mol Sci 12:9057–9082

Stiess M, Bradke F (2011) Neuronal transport: myosins pull the ER. Nat Cell Biol 13:10–11

Strotmeier J, Gu S, Jutzi S, Mahrhold S, Zhou J, Pich A, Eichner T, Bigalke H, Rummel A, Jin R, Binz T (2011) The biological activity of botulinum neurotoxin type C is dependent upon novel types of ganglioside binding sites. Mol Microbiol 81:143–156

Swaminathan S (2011) Molecular structures and functional relationships in clostridial neurotoxins. FEBS J 278:4467–4485

Taylor MJ, Perrais D, Merrifield CJ (2011) A high precision survey of the molecular dynamics of mammalian clathrin-mediated endocytosis. PLoS Biol 9:e1000604

Tsukamoto K, Kohda T, Mukamoto M, Takeuchi K, Ihara H, Saito M, Kozaki S (2005) Binding of *Clostridium botulinum* type C and D neurotoxins to ganglioside and phospholipid. Novel insights into the receptor for clostridial neurotoxins. J Biol Chem 280:35164–35171

Tsukamoto K, Kozai Y, Ihara H, Kohda T, Mukamoto M, Tsuji T, Kozaki S (2008) Identification of the receptor-binding sites in the carboxyl-terminal half of the heavy chain of botulinum neurotoxin types C and D. Microb Pathog 44:484–493

Tuffereau C, Schmidt K, Langevin C, Lafay F, Dechant G, Koltzenburg M (2007) The rabies virus glycoprotein receptor p75NTR is not essential for rabies virus infection. J Virol 81:13622–13630

Ugolini G (1995) Specificity of rabies virus as a transneuronal tracer of motor networks: transfer from hypoglossal motoneurons to connected second-order and higher order central nervous system cell groups. J Comp Neurol 356:457–480

Vale RD, Schnapp BJ, Reese TS, Sheetz MP (1985) Movement of organelles along filaments dissociated from the axoplasm of the squid giant axon. Cell 40:449–454

Verderio C, Coco S, Rossetto O, Montecucco C, Matteoli M (1999) Internalization and proteolytic action of botulinum toxins in CNS neurons and astrocytes. J Neurochem 73:372–379

Vidricaire G, Tremblay MJ (2005) Rab5 and Rab7, but not ARF6, govern the early events of HIV-1 infection in polarized human placental cells. J Immunol 175:6517–6530

Vonderheit A, Helenius A (2005) Rab7 associates with early endosomes to mediate sorting and transport of Semliki forest virus to late endosomes. PLoS Biol 3:e233

Wiegand H, Erdmann G, Wellhöner HH (1976) ^{125}I-labelled botulinum A neurotoxin: pharmacokinetics in cats after intramuscular injection. Naunyn-Schemiedeberg's Archives Pharmacoloy 292:161–165

Winckler B, Yap CC (2011) Endocytosis and endosomes at the crossroads of regulating trafficking of axon outgrowth-modifying receptors. Traffic 12:1099–1108

Yaar M, Zhai S, Pilch PF, Doyle SM, Eisenhauer PB, Fine RE, Gilchrest BA (1997) Binding of beta-amyloid to the p75 neurotrophin receptor induces apoptosis. A possible mechanism for Alzheimer's disease. J Clin Invest 100:2333–2340

Yamasaki S, Baumeister A, Binz T, Blasi J, Link E, Cornille F, Roques B, Fykse EM, Sudhof TC, Jahn R et al (1994) Cleavage of members of the synaptobrevin/VAMP family by types D and F botulinal neurotoxins and tetanus toxin. J Biol Chem 269:12764–12772

Yeh FL, Dong M, Yao J, Tepp WH, Lin G, Johnson EA, Chapman ER (2010) SV2 mediates entry of tetanus neurotoxin into central neurons. PLoS Pathog 6:e1001207

Yowler BC, Schengrund CL (2004) Glycosphingolipids-sweets for botulinum neurotoxin. Glycoconj J 21:287–293

Zhang Y, Varnum SM (2012) The receptor binding domain of botulinum neurotoxin serotype C binds phosphoinositides. Biochimie 94:920–923

Synchronized Chaperone Function of Botulinum Neurotoxin Domains Mediates Light Chain Translocation into Neurons

Audrey Fischer

Abstract *Clostridium botulinum* neurotoxin (BoNT) is a multidomain protein in which the individual modules work in synchronized cooperative action in order to enter into neurons and inhibit synaptic transmission. The di-chain protein is made up of the ~50 kD light chain and the ~100 kD heavy chain. The HC can be further subdivided into the N-terminal translocation domain (H_N) and the C-terminal Receptor Binding Domain (H_C). BoNT entry into neurons requires the toxin to utilize the host cell's endocytosis pathway where it exploits the acidic environment of the endosome. Within the endosome the H_C triggers the H_N to change conformation from a soluble protein to a membrane inserted protein-conducting channel in precise timing with LC refolding. The LC must partially unfold to a translocation competent conformation in order to be translocated by the H_N channel in an N to C terminal direction. Upon completion of translocation, the LC is released from the HC and allowed to interact with its substrate SNARE protein. This article discusses the individual functions of each module as well as the mechanisms by which each domain serves as a chaperone for the others, working in concert to achieve productive intoxication.

Keywords Botulinum neurotoxin · Channels · Chaperones · Protein translocation · Membranes

Abbreviations

γ	Single channel conductance
BoNT	Botulinum neurotoxin
CD	Circular dichroism
EPR	Electron paramagnetic resonance
HC	Heavy chain

A. Fischer (✉)
Asymmetric Operations Department, Johns Hopkins University Applied Physics Laboratory, Laurel, MD 20723, USA
e-mail: Audrey.Fischer@jhuapl.edu

A. Rummel and T. Binz (eds.), *Botulinum Neurotoxins*, Current Topics in Microbiology and Immunology 364, DOI: 10.1007/978-3-642-33570-9_6,

H_C	Receptor binding domain
H_{CC}	25 kDa C-terminal subdomain of the H_C
H_{CN}	25 kDa N-terminal subdomain of the H_C
H_N	Translocation domain
$H_{N\text{-belt}}$	Beltless translocation domain
LC	Light chain
LH_N	LC + H_N
PA	Protective antigen
P_o	Probability of the channel to remain in the open state
SNAP-25	Synaptosomal-associated protein of 25 kDa
SNARE	Soluble NSF attachment protein receptor
$T_{1/2}$	Half-time for a single growing conductance event
TeNT	Tetanus neurotoxin
TSDN	Toosendanin

Contents

1 Introduction 117
2 Endosomal Entry 118
3 HC Channel Formation 119
3.1 HC Channel Characteristics 119
3.2 HC Channel Structure 121
4 Beltless H_N Represents the Minimal Channel Forming Unit 123
5 Role of the H_C in LC Translocation 124
5.1 Role of the H_C in Channel Formation 124
5.2 H_C Functions as Synchronicity Switch 124
6 HC Translocation of the LC Across Endosomal Membrane 125
6.1 Holotoxin Channel Activity Distinct From HC 125
6.2 Interpretation of Growing Conductance Event as LC Translocation 125
6.3 LC Gates HC Channel Within Endosome 127
6.4 Alternative Models for Protein Translocation 127
7 Recovery of Enzymatically Active LC Upon Completion of Translocation 128
7.1 LC Translocation Across Plasma Membrane 128
7.2 In Vitro Enzyme Activity Following Translocation Across Lipid Bilayers 128
8 Requirements for Productive LC Translocation 129
8.1 LC Unfolding is Required for Productive Translocation 129
8.2 Precisely Timed Separation of LC From HC is Required to Complete Productive Translocation 130
8.3 LC Must be Proteolytically Cleaved From HC 131
9 Chemical Inhibition of HC Channel as Potential Therapy for Botulism 131
10 Productive Intoxication Requires Progression of Inter-domain Chaperone Activity 132
References 133

1 Introduction

Botulinum neurotoxin (BoNT) inhibits synaptic exocytosis in peripheral cholinergic synapses causing the neuroparalytic illness botulism. *Clostridium botulinum* is a spore forming obligate anaerobic bacterium whose cells secrete seven BoNTs isoforms known as BoNT/A–G (Schiavo et al. 2000). All isoforms together with the related tetanus neurotoxin (TeNT) secreted by *Clostridium tetani*, are Zn^{2+}-endoproteases that block synaptic exocytosis by cleaving soluble NSF attachment protein receptor (SNARE) proteins (Schiavo et al. 2000; Blasi et al. 1993; Jahn et al. 2003; Schiavo et al. 1992; Weber et al. 1998). Assembly of the SNARE core complex is a key step preceding fusion of synaptic vesicles with the neuronal plasma membrane (Schiavo et al. 2000; Jahn et al. 2003; Jackson and Chapman 2006).

All BoNT isoforms are synthesized as a single polypeptide chain with a molecular mass of ~150 kDa before being cleaved either by clostridial or host cell proteases into two polypeptide chains. The mature di-chain holotoxin is maintained in complex by a disulfide bridge and comprises a 50 kDa light chain (LC) protease and a 100 kDa heavy chain (HC) (Fig. 1). The HC consists of the translocation domain (H_N) (the N-terminal half) and the receptor-binding domain (H_C) (the C-terminal half) (Schiavo et al. 2000; Swaminathan and Eswaramoorthy 2000; Lacy and Stevens 1999; Lacy et al. 1998).

BoNT enter neurons by receptor-mediated endocytosis, initiated by the interactions between the BoNT H_C and a ganglioside and protein receptor duo (Yowler et al. 2002; Tsukamoto et al. 2005; Nishiki et al. 1996; Ginalski et al. 2000; Rummel et al. 2004; Dong et al. 2006; Mahrhold et al. 2006; Peng et al. 2011). Exposure of the BoNT-receptor complex to the acidic environment of endosomes induces a conformational change whereby the HC inserts into the endosomal bilayer membrane (Schiavo et al. 2000; Gambale and Montal 1988; Finkelstein 1990). Under these conditions, the HC forms a protein-conducting channel that translocates the LC protease into the cytosol (Fischer and Montal 2007; Koriazova and Montal 2003). Upon release into the cytosol the LC traffics to its substrate SNARE protein.

The LC and the H_C crystal structures have been solved and their functions clearly defined (Breidenbach and Brunger 2004; Arndt et al. 2006; Arndt et al. 2005; Agarwal et al. 2004; Agarwal et al. 2005; Chai et al. 2006; Segelke et al. 2004; Hanson and Stevens 2000; Swaminathan and Eswaramoorthy 2000; Stenmark et al. 2008; Jin et al. 2007); however, the central unit of the holotoxin is less well understood. The recent crystal structures of the BoNT H_C interacting with its ganglioside and protein receptors via its C-terminal half H_{CC} raises queries regarding the purpose of the N-terminal half of the H_C (Chai et al. 2006; Swaminathan and Eswaramoorthy 2000; Stenmark et al. 2008; Jin et al. 2007). Potentially the N-terminal motif of the H_C, called H_{CN} could serve a function in priming the H_N to an insertion competent orientation with respect to the membrane (Gu et al. 2012). Although several studies have attempted to offer insight into the conformation of the H_N within the membrane, the structure of the membrane inserted, protein translocating channel has yet to be resolved. The elongated, unstructured belt

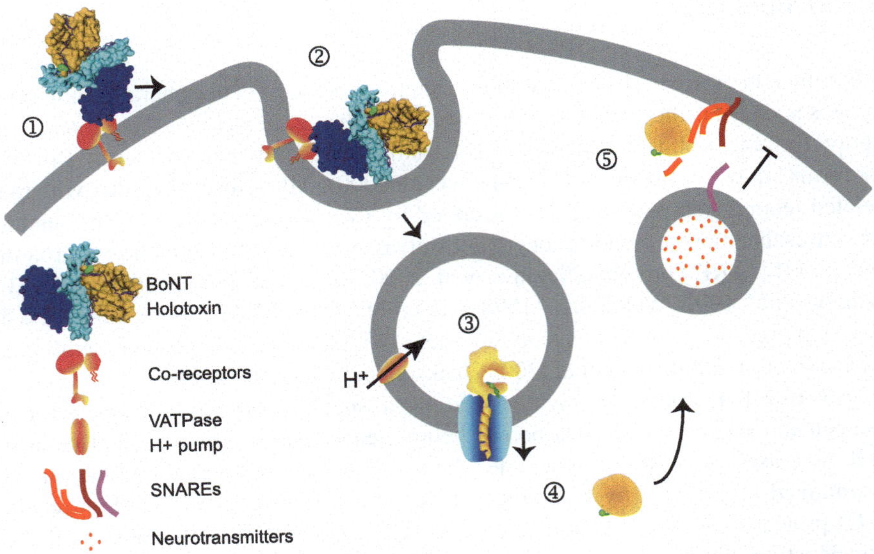

Fig. 1 Schematic of BoNT cellular intoxication at pre-synaptic membrane of neuronal cell. *1* BoNT holotoxin (N-terminal catalytic domain (LC, gold) connected to the C-terminal HC, composed of the H_N (*cyan*) and H_C (*blue*), by a belt (*purple ribbon*), and an inter-chain disulfide bridge (*green*)) binds to dual receptors on plasma membrane stimulating receptor-mediated uptake. *2* BoNT shuttles to early endosome where vATPase acidification of the vesicle lumen stimulates membrane insertion and HC channel formation as well as LC structural rearrangement to a translocation competent conformation. *3* Partially unfolded LC translocates through the HC channel before release and refolding within the cytosol. For BoNT/A in particular, the LC traffics back to the plasma membrane *4* where it cleaves its substrate SNARE protein, SNAP-25. *5* Cleavage of any component of the SNARE complex aborts synaptic vesicle fusion with the plasma membrane, precluding release of neurotransmitters and cellular signaling. The BoNT/A holotoxin image was rendered on YASARA (http://www.yasara.org) using the Protein Data Bank I.D. 3BTA (Lacy et al. 1998)

region of the H_N is another domain with an elusive function; potentially acting as a pseudo-substrate/chaperone for the LC during the majority of the intoxication process (Brunger et al. 2007). Understanding the structure as well as the individual and cooperative functions of each BoNT module may lead to design of effective inhibitors of botulism as well as novel protein-based, therapeutic delivery systems.

2 Endosomal Entry

Productive BoNT intoxication requires the toxin to exploit the host cellular machinery in order to mediate entry into sensitive neurons (Fig. 1); (Rummel 2012). BoNT and the closely related TeNT bind to protein and ganglioside receptors localized at the nerve terminal (Dolly et al. 1984). BoNT has been demonstrated to bind complex polysialogangliosides as initial cellular contacts [reviewed in

(Rummel et al. 2009)]. The protein co-receptors are unique to each serotype of toxin; however they are all part of the synaptic signaling machinery. As such these receptors are primarily localized within synaptic vesicles, residing briefly on the plasma membrane. Depolarization of the synaptic membrane stimulates synaptic vesicle fusion with the plasma membrane, thereby exposing the luminal domains to the extracellular surface and promoting BoNT dual-receptor binding (Dong et al. 2003; Habermann et al. 1980; Hughes and Whaler 1962; Simpson 1980).

Receptor binding stimulates BoNT cellular uptake into clathrin-coated synaptic vesicles (Harper et al. 2011; Petro et al. 2006; Sudhof 2004). This step has been demonstrated to require energy as intoxication can be inhibited by incubating sensitive cells at low temperatures or pretreating them with sodium azide or dinitrophenol (Dolly et al. 1984). Once internalized, BoNT is shuttled to vesicles destined for the endosome/lysosome pathway (Dong et al. 2006; Harper et al. 2011; Black and Dolly 1986; Couesnon et al. 2009; Fu et al. 2009). These vesicles mature to early endosomes where acidification of the endosomal environment is mediated by the H^+-vATPase pump. Acidification of the endosomal environment promotes the translocation step of intoxication as demonstrated by bafilomycin inhibition. Bafilomycin A1, a specific inhibitor of the H^+-vATPase, has been demonstrated to efficiently inhibit BoNT intoxication in multiple cell-based assays (Lawrence et al. 2007; Ruthel et al. 2011; Simpson et al. 1993).

3 HC Channel Formation

While the individual BoNT serotypes bind unique neuronal targets, they share a similar mechanism of protease translocation from the endosome to the cytosol. The H_C domain is responsible for targeting each serotype to its corresponding neuron; however, the H_N domain performs the protease translocation function following entry into an early endosomal compartment. This is achieved by the formation of a protein conducting channel by the HC that translocates the protease domain from the acidic endosome into the neutral cytosol.

3.1 HC Channel Characteristics

Channel activity of the toxin molecules is monitored by electrophysiological measurements; the passage of ions through a protein pore within a lipid bilayer in response to an applied voltage is measured as current. The lipid bilayer used for these types of experiments can be formed from either of two standard methods, artificial bilayers or patch clamp. Artificial bilayers are formed from a predetermined mixture of purified lipids while patch clamp recordings utilize a small piece of the plasma membrane from a neuronal cell. Neuronal membranes contain all the cofactors necessary for optimal binding; however, they contain endogenous ion channels that can complicate isolation of the toxin ion channel activity.

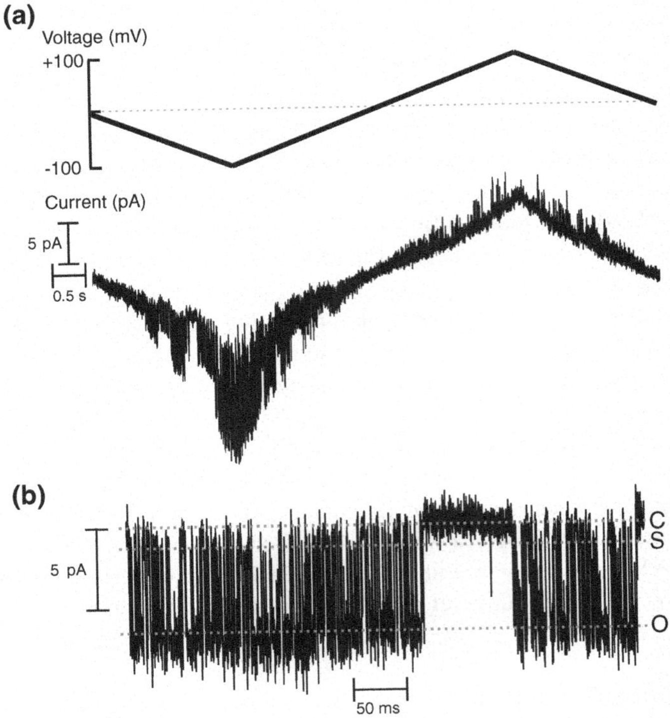

Fig. 2 BoNT/A HC channel characteristics monitored in excised patches of Neuro 2A cells (Fischer and Montal 2007; Fischer and Montal 2006). **a** applied voltage ramp cycling between −100 and +100 mV (*top panel*) stimulates BoNT/A HC channel activity measured as current resulting from Na^+ ion conductance through the channel lumen (*lower panel*). Note the occurrence of two discrete channels undergoing bursts of activity with a single channel conductance of 65 pS. Downward deflection indicates channel opening at negative voltage while upward inflection indicates channel opening at positive voltage. **b** High gain and fast time resolution of single channel recording measured at −100 mV. Transitions between the closed (*C*), sub-conductance (*S*), and open states (*O*) are clearly resolved. The channel opens to a main conductance of 65 pS and exhibits a sub-conductance of 10 pS upon entry and exit from a burst of channel activity

The HC of BoNT and TeNT form ion channels in lipid bilayers (Gambale and Montal 1988; Finkelstein 1990; Blaustein et al. 1987; Donovan and Middlebrook 1986; Hoch et al. 1985; Menestrina et al. 1989; Rauch et al. 1990). Membrane insertion and channel formation are elicited by maintaining a pH and redox gradient across the membrane, as illustrated in Fig. 2. The *cis* compartment of the apparatus mimics the endosome of a neuron with an acidic, oxidized environment while the *trans* compartment emulates the neutral pH, reduced environment of the cytosol. The HC is supplemented to the *cis* compartment, and the channel activity is monitored several minutes later.

HC channel activity exhibits characteristics similar to other bacterial toxins. The channels are cation specific, exhibiting a single channel conductance (γ) of 65 pS when both the *cis* and *trans* compartment are supplemented with 200 mM NaCl (Fischer and Montal 2007; Fischer and Montal 2006; Fischer and Montal 2007). Between variable periods of quiescence the channel enters a bursting activity state. Entry into and exit from a bursting activity state is mediated by a transition between the closed state and a subconductance state of $\gamma \sim 10$ pS (Fischer and Montal 2006). The channel then rapidly transitions between the subconductance and open states. Lipid composition regulates channel kinetics; residency within the open and closed states is significantly faster in neuronal membranes, $\leq$1 ms, than lipid bilayers, $\geq$5 ms (Koriazova and Montal 2003; Fischer and Montal 2006). Voltage dependence can also be manipulated in lipid bilayers by altering the ratio of neutral/charged lipids as demonstrated in studies with TeNT (Gambale and Montal 1988; Rauch et al. 1990). Within neurons, the channels are voltage-dependent, opening only at negative membrane potentials (Fischer and Montal 2006). In contrast, the endosomal vesicle maintains a positive membrane potential (Koivusalo et al. 2011). Evaluated together, these data suggest that the HC channel remains primarily closed within the endosome; ion conduction is minimized within this transitory compartment. LC translocation is therefore conducted without disruption of endocytic pathway, a feature that allows the LC to enter cells virtually undetected.

3.2 HC Channel Structure

The resolved crystal structures for BoNT A, B, and E visualize an intact di-chain toxin molecule soluble within aqueous solution (Lacy et al. 1998; Swaminathan and Eswaramoorthy 2000; Kumaran et al. 2009). However, the structure of the HC in its functionally active form as a channel inserted into the membrane has yet to be solved. Circular dichroism studies of the BoNT/A holotoxin, HC and LC molecules revealed ~20 % α-helical and ~40 % β-pleated sheet content for all three molecules (Singh et al. 1990; Singh and DasGupta 1989). The secondary structure of the holotoxin and HC do not undergo significant secondary structure changes upon acidification (Koriazova and Montal 2003; Fu et al. 2002). Acidic pH alone may not be sufficient to stimulate conformational changes to an active channel state. BoNT/B bound to ganglioside co-receptor GT1b underwent significant loss of helicity following acidification (Sun et al. 2011), and Electron Paramagnetic Resonance (EPR) studies of BoNT/A indicated a dynamic structure of the toxin within proteoliposomes (Mushrush et al. 2011). Conformational changes associated with channel formation therefore depend upon several factors: acidification, receptor binding, and possibly lipid association.

Evaluation of the available BoNT crystal structures reveals the core motif of the H_N to be a series of elongated α-helices ~100 Å in length. Several studies have explored the channel activity of peptides predicted to be transmembrane sequences of the channel; these studies demonstrated that peptides from amino acid region 659–681 of BoNT/A and 670–692 of TeNT form cation selective channels in lipid bilayers (Oblatt-Montal et al. 1995; Montal et al. 1992) (Fig. 3).

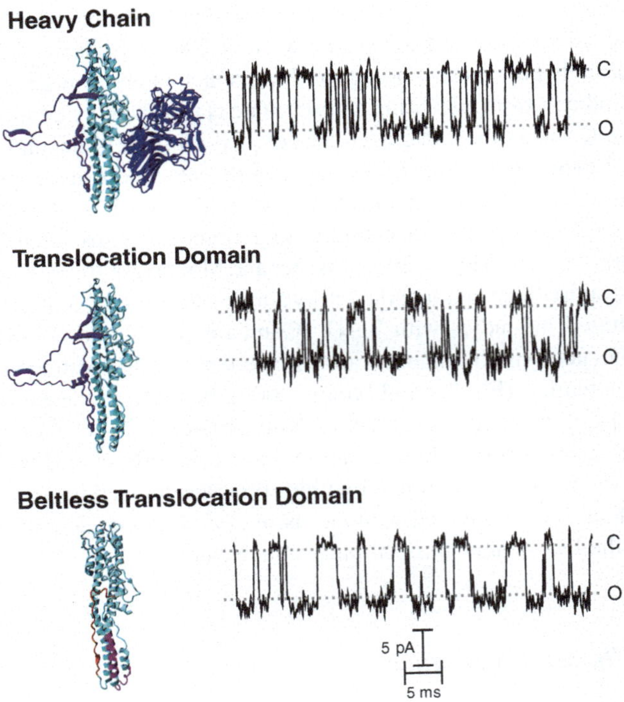

Fig. 3 BoNT HC sub-domains exhibit ion channel activity. Channel activity of the HC, H_N and $H_{N\text{-belt}}$ were measured at −100 mV applied voltage from excised patches of Neuro 2A cells (Fischer et al. 2008; Fischer et al. 2012). Channel opening is indicated by a downward deflection; *C* and *O* denote the closed and open states within a burst. $H_{N\text{-belt}}$ has been indicated with highlighted regions corresponding to predicted transmembrane regions. The peptide from region 659–681 that forms cation selective channels in lipid bilayers has been colored red (Oblatt-Montal et al. 1995; Montal et al. 1992), and the regions within residues 805–835 resistant to protease digestion following membrane insertion have been colored magenta (Mushrush et al. 2011). Other BoNT domain color conventions hold from Fig. 1

Molecular modeling of these channels revealed an amphipathic α-helical structure formed from tetramers. In contrast, a recent study by Mushrush et al. (2011) revealed two segments within the region 805–835 to be resistant to protease digestion following membrane insertion (Fig. 3). While these functional studies indicate several sections of the H_N are capable of membrane insertion and channel formation, further structural analysis of the protein translocating channel remains a critical challenge.

The oligomeric state of the HC channel is yet to be clearly defined. Chemical cross-linking studies with several serotypes indicate that the holotoxin can adopt several oligomeric states, monomer, dimer, and trimer (Ledoux et al. 1994).

EPR and Atomic Force Microscopy (AFM) studies of the toxin inserted into proteoliposomes further support the oligomeric model (Sun et al. 2011; Mushrush et al. 2011). The AFM studies visualized BoNT/B as a monomer at neutral pH that transitioned to a donut-shaped trimer upon GT1b binding and acidification. These data indicate that, like anthrax and diphtheria toxin, the protein translocating channel of clostridial neurotoxins form oligomers within the membrane.

4 Beltless H_N Represents the Minimal Channel Forming Unit

The crystal structures of BoNT holotoxins revealed the modular nature of each domain, promoting investigation into the minimal domain required to elicit ion channel activity. Evaluation of the H_N structure reveals two modules, an unstructured loop region (residues 450–544) and a set of elongated alpha helices (residues 545–870). The unstructured loop has been termed the belt region as it consists of two loops, one of which wraps tightly around the middle of the LC protease and the other runs parallel to the H_N alpha helices between the H_N and the LC. Physiochemical studies of using the H_N belt truncated at various positions revealed that the belt region perturbed membrane interaction at neutral pH; acidic pH altered the electrostatic interactions to allow efficient H_N membrane interaction and membrane insertion (Galloux et al. 2008). The belt may function as a pH trigger to promote H_N channel formation only under acidic pH.

The isolated H_N and the beltless H_N ($H_{N\text{-belt}}$) have been shown to elicit ion channels in excised patches of Neuro 2A cells (Fischer et al. 2008; Fischer et al. 2012) (Fig. 3). Under conditions which simulate the endosomal membrane of a neuron, the H_N and $H_{N\text{-belt}}$ formed ion channels with latency similar to HC. The characteristics of these ion channels were indistinguishable from that monitored for the HC, namely γ, residency within the open, closed, and subconductance states, and voltage-dependent probability to open.

The $H_{N\text{-belt}}$ serves as the minimal ion channel forming unit. The capacity of the $H_{N\text{-belt}}$ to translocate LC from the endosome to the cytosol is yet to be investigated. This is mainly resulting from the fact that within the holotoxin the belt region, and the inter-chain disulfide contained within, links the LC to the HC channel. Removal of the belt region requires novel linkage design for the LC cargo and the $H_{N\text{-belt}}$ channel. In contrast to functional analysis, structural investigation of the $H_{N\text{-belt}}$ should be more amenable than the HC or H_N. CD studies with the $H_{N\text{-belt}}$ indicate that, similar to the holotoxin and HC, minimal secondary structure rearrangements occur upon acidification (Fischer et al. 2012). Evaluation of H_N and $H_{N\text{-belt}}$ interactions with the membrane revealed minimal secondary and tertiary structure changes (Galloux et al. 2008). NMR or crystal structure analysis of the $H_{N\text{-belt}}$ inserted into lipid bicelles may be the most promising method for visualizing the toxin channel in its active state.

5 Role of the H_C in LC Translocation

5.1 Role of the H_C in Channel Formation

Removal of the H_C from the BoNT holotoxin releases the pH dependence of channel formation. HC channel activity is pH dependent; no membrane insertion or channel activity was monitored in Neuro 2A cells or lipid bilayers in which both the *cis* and *trans* compartments were maintained at neutral pH (Koriazova and Montal 2003; Fischer et al. 2008). In contrast, channel activity was monitored by patch clamp for several truncation mutants of the holotoxin over a range of acidic to neutral pH. Purified recombinant LH_N (LC + H_N), H_N, and $H_{N\text{-belt}}$ elicited channel activity in Neuro 2A cells when the *cis* compartment solution was held at pH 5, 6 or 7 (Fischer et al. 2008; Fischer et al. 2012). Ion channel activity characteristics did not vary significantly with pH; single channel conductance was determined to be pH gradient independent, remaining constant at $\gamma \sim 65$ pS. Bursting channel activity with transitions between the closed, subconductance and open states were unaffected by varying *cis* compartment pH. In contrast, latency of onset of channel activity was increased with neutral pH. Acidic pH promotes but is not required for membrane insertion and channel formation; thus removal of the H_C released the pH dependence of channel formation.

5.2 H_C Functions as Synchronicity Switch

The receptor binding domain of BoNT ensures efficient intoxication by four unique mechanisms. BoNT is secreted from the clostridium bacteria in complex with a gastro-protective non-toxic non-hemagglutinin (NTNHA) protein (Gu et al. 2012). The H_C domain is encapsulated by the three domains of the NTNHA, thereby mediating the NTNHA protection. BoNT requires the H_C to reach the peripheral nervous system. BoNT holotoxin H_C binds receptors on the mucosal surface of gut epithelial cells to mediate transcytosis to the basolateral side of the epithelial cell (Maksymowych et al. 1999; Maksymowych and Simpson 1998, 2004); (Fujinaga et al. 2012). Once released to the circulation BoNT, holotoxin reaches cholinergic nerve cells and a second round of cell entry occurs into sensitive neurons. During cell binding and internalization the H_C restricts the H_N from membrane insertion until residence within the acidic environment of the endosome. Without the H_C present, the H_N readily inserts into the plasma membrane of neuronal cells. The LC remains folded in the neutral pH environment outside the cell, incapable of translocating through the proposed ~ 15 Å pore of the H_N channel. Under these conditions, the LC remains a globular protein tethered to its carrier on the cell surface by the disulfide linkage. In this manner the H_C serves to chaperone the LC and H_N, ensuring that partial unfolding of the LC is concurrent with $H_{N\text{-belt}}$ channel formation thereby promoting productive LC translocation. Anthrax toxin utilizes a similar regulation method during its cellular intoxication. The receptor binding domain of anthrax toxin, Protective Antigen (PA), mediates the pH threshold for membrane insertion through binding of cellular receptors, TEM8 and CMG2 (Rainey et al. 2005). The receptor binding domain of PA readily binds to the receptors on the plasma membrane, and following receptor-mediated endocytosis, selectively releases binding of TEM8 at near neutral pH and CMG2 at pH 6. Receptor-binding restricts structural changes within the heptameric PA prepore that are required for membrane insertion; toxin-receptor interaction under acidic pH promotes pore formation and initiation of cargo protein translocation (Rainey et al. 2005; Pilpa et al. 2011).

6 HC Translocation of the LC Across Endosomal Membrane

6.1 *Holotoxin Channel Activity Distinct From HC*

In contrast to the steady-state ion channel activity monitored for HC channels, the activity monitored for the intact holotoxin undergoes dynamic shifts in ion conductance. For both proteins, channel activity requires the pH gradient prevalent across endosomes; however, holotoxin activity requires an additional gradient of redox potential also native to the endosome (Koriazova and Montal 2003). A family of currents recorded from BoNT/A holotoxin channels at the indicated voltages and time points is illustrated in Fig. 4 (Fischer and Montal 2007). The time course of change of γ after insertion of BoNT/A holotoxin into the membrane displays a series of discrete transient intermediate conductance states. At the onset of channel activity, small, discrete events with a $\gamma \sim 14$ pS are clearly discerned. Progressively, γ undergoes a continuous increase until reaching a stable value of 68 pS, a conductance at which they remain for the duration of the experiment. The half-time for a single growing conductance event ($T_{1/2}$) estimated from these measurements is $\sim$150 s. These results are divergent from channel activity of isolated HC where the $\gamma \sim 65$ pS is constant during the entire channel activity recording period. This invariant HC channel γ is characteristic of holotoxin channels after completion of a single growing conductance event.

6.2 *Interpretation of Growing Conductance Event as LC Translocation*

The intermediate conductance states have been interpreted as reporters of discrete transient steps during the translocation of the LC across the membrane (Fischer and Montal 2007). During protease translocation, the protein-conducting channel progressively conducts more Na^+ ions around the polypeptide chain before entering an exclusively ion-conductive state. The presence of the LC within the channel lumen obstructs Na^+ ion passage; LC localization within the channel is

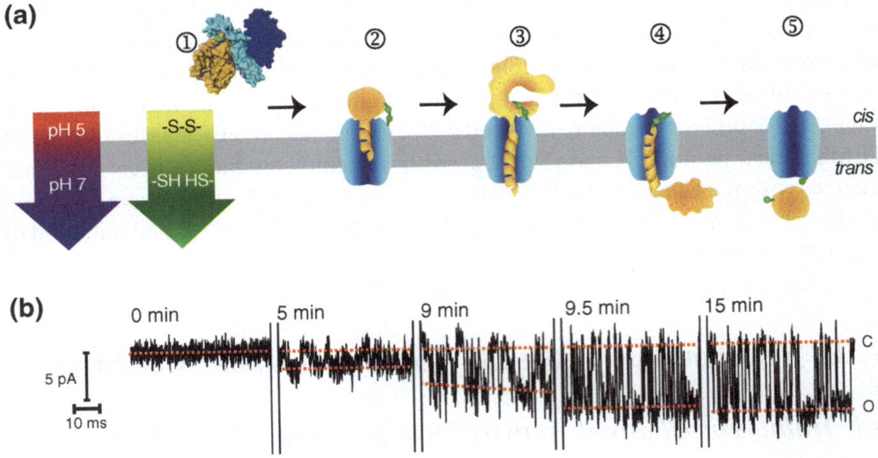

Fig. 4 BoNT LC translocation through the HC channel undergoes a series of discrete intermediates (Fischer and Montal 2007). (**a**) Schematic representation of (**b**) high-gain and fast-time resolution of BoNT/A single channel currents recorded at −100 mV in excised patches of Neuro 2A cells supplemented with the pH and redox gradient native to the endosomal membrane (*grey bar*). j initial absence of channel currents prior to exposure to BoNT/A. Subsequent recordings represent the average time course of change of channel conductance and permissive intermediate states. k represents an entry event monitored at the onset of channel activity, l indicates the series of transfer steps monitored as the partially unfolded LC translocates through the channel, and m indicates an exit event monitored immediately prior to the unoccluded channel state n. BoNT domain color conventions hold from Fig. 1

therefore detected as block of Na^+ ion conductance. The intermediate conductances may correspond to transient protein–protein interactions between the LC and the HC. This notion is consistent with the results of the analysis of the probability of channel residence in the open state (P_o) as a function of conductance for the holotoxin: the low conductance states have a lower P_o compared to the unoccluded holotoxin and HC.

Within the LC occluded state, the frequency of occurrence of the lowest conductance intermediates is higher, as determined by their individual occupancy times. The low conductance states exhibit the longest occupancy time, consistent with an entry event in which the energetic barrier associated with the unfolding of the LC must be overcome in order to initiate translocation. Conductance values between 25 pS $\leq \gamma \leq$ 50 pS have shorter lifetimes, in accordance with the occurrence of a sequence of permissible intermediate conformations, with relative fast access to the final conductance intermediate measured at $\gamma \sim$ 55 pS. This higher γ intermediate is longer lived, presumably associated with LC refolding at the exit from the channel before its final dissociation from the HC. Thus, a single LC translocation involves an entry event, a series of transfer steps, and an exit event. A single growing conductance event is monitored under these conditions rather than a series of occluded and unoccluded states. These data support several models. The channel may be formed from a holotoxin monomer that effectively translocates a single LC. Alternatively, the channel may be formed from an oligomer whose individual LC molecules interact with the entry to the channel lumen such that they are primed to follow each other through the channel. This model is supported by cargo translocation monitored for anthrax toxin in which three cargo proteins bind to a single heptameric channel (Basilio et al. 2011). Simultaneous single molecule fluorescence and ion channel measurements may reveal the functionally active state of the protein-conducting channel.

6.3 LC Gates HC Channel Within Endosome

A notable feature of the BoNT channel, both for BoNT/A and E, is that it is closed at positive voltages under conditions in which the orientation and the magnitude of the pH gradient, as well as the polarity and magnitude of the membrane potential compare fairly well with those prevailing across the endosomal membrane: pH 5.3 and positive potential on the compartment containing the BoNT and pH 7.0 and negative potential on the opposite compartment. This suggests that the BoNT HC channel would be closed in the endosome until it is gated by the LC to initiate its translocation across the membrane into the cytosol. After completion of LC translocation, the channel would also be closed, thereby precluding the passive dissipation of the ionic electrochemical gradients present across endosomal membranes.

6.4 Alternative Models for Protein Translocation

The model summarized above supports the notion that the succession of discrete intermediate states detected reflects permissive stages during LC translocation. However, a number of alternative models exist; among these, two scenarios are worth considering. First, it is plausible to envision the oligomeric channel entity undergoing self-assembly within the membrane. The intermediate states function as reporters of the oligomerization process: assembly of monomers into the developing channel progressively increases the conductance until it achieves the most stable oligomer. That the isolated HC (1) displays a unique, characteristic γ as early as detection can be resolved and (2) no intermediates are discerned indicates that this molecular entity is assembled immediately after membrane insertion. Together with the fact that conductance intermediates in BoNT/A and E holotoxin are only transient and lead to an end-point stable entity with single channel conductance and voltage-dependence features equivalent to those displayed by the HC argue against a post-membrane insertion assembly model. Furthermore, oligomerization models implicitly dictate that the LC controls the assembly: the LC would inhibit the oligomerization of the HC, and after its release the channel activity would be similar to that of isolated HC. If this were the case then one would expect to measure a growing conductance profile for isolated HC channels, albeit faster than for holotoxin; this result was never observed.

A different class of models considers that a disruption of the membrane may be generated by the interaction of the holotoxin protein with the membrane lipid bilayer. Again, the equivalency of the channel properties of isolated HC and holotoxins A and E after completion of LC translocation in terms of single channel conductance and voltage dependence are inconsistent with an irregular core breakdown of the membrane. The regularity and discreteness of γ for each of the intermediates resolved is opposite to the erratic, irregular, and spike-like current fluctuations which are typical of bilayer dielectric core disturbances produced, for example, by detergents, denatured proteins, or charged peptide-lipid interactions.

7 Recovery of Enzymatically Active LC Upon Completion of Translocation

7.1 LC Translocation Across Plasma Membrane

While electrophysiology studies directly visualize ion channel activity of BoNT, cell-based assays allow for investigation of ion channel and enzymatic activity of this class of multidomain toxins. Membrane insertion, channel formation, and recovery of enzymatic activity following cargo translocation have been demonstrated on the plasma membrane of cells. Initially this was demonstrated in vero cells with diphtheria toxin (Sandvig and Olsnes 1980; Draper and Simon 1980). The toxin was allowed to bind to cells under cold temperature which promoted binding without endocytosis. Warming to 37 °C with concurrent acidification of the extracellular solution induced membrane insertion and cargo translocation as measured by ^{35}S labeled toxin uptake, $^{22}Na^+$ influx, and cytotoxicity (Sandvig and Olsnes 1980; Draper and Simon 1980; Falnes et al. 1992). Membrane penetration by anthrax and clostridium difficile toxin B channel domains was demonstrated using $^{86}Rb^+$ release (Milne and Collier 1993; Barth et al. 2001). Recently, pH-mediated plasma membrane translocation was demonstrated for BoNT and TeNT using cerebellar granular neurons and PC12 cells (Pirazzini et al. 2011).

7.2 In Vitro Enzyme Activity Following Translocation Across Lipid Bilayers

While channel block has been the mainstay methodology for demonstrating BoNT, anthrax and diphtheria toxin cargo binding, translocation and release from the carrier protein translocating channel, recovery of an enzymatically active cargo protein following translocation across the bilayer to the *trans* compartment is a critical step that has proven difficult to quantitatively measure. BoNT is the only one of this class of AB toxins for which recovery of enzymatic activity has been monitored following translocation across lipid bilayers (Koriazova and Montal 2003). For this study, BoNT/A holotoxin was supplemented to the *cis* compartment of an artificial lipid bilayer, and following recording of macroscopic channel insertion and block, the *trans* compartment was assayed for SNAP-25 cleavage activity by ELISA. As a control, isolated LC was shown not to diffuse across the membrane. LC enzymatic activity was monitored only when the intact holotoxin was added to the *cis* compartment under conditions that simulated the pH and redox gradient prevalent across the endosome.

8 Requirements for Productive LC Translocation

8.1 LC Unfolding is Required for Productive Translocation

While the holotoxin and HC maintain a stable folded state under acidic pH, the LC undergoes significant tertiary and secondary structure rearrangements. CD studies have demonstrated that the LC reversibly shifts from a primarily α-helical structure to a largely β-sheet and β-turn structure when the pH was lowered from 7 to 5 or 4 (Koriazova and Montal 2003; Fu et al. 2002). The severity of helicity loss is also salt dependent; presence of physiological salt concentrations promoted more significant decrease in helicity. Further studies by the Singh lab indicate that the LC may adopt a molten globule-like structure under acidic pH, allowing for tertiary structure rearrangement without further loss of secondary structure (Li and Singh 2000). The results demonstrating acidic pH induced LC secondary structure refolding and tertiary structure flexibility are consistent with the model of the LC adopting a translocation competent conformation during transit through the HC channel with subsequent refolding to an enzymatically active conformation upon entry to the cytosol.

The underlying protein–protein interactions between BoNT/A LC and HC during translocation have been examined through the use of a pH resistant, stably folded, LC-specific binding protein. Fab fragments of a BoNT/A LC neutralizing antibody are ideal for these studies as they contain a single binding domain that binds an LC structural epitope with picomolar affinity (Levy et al. 2007). Removal of the constant region and separation of the two binding domains from each other generates a homogeneous mixture of 1:1 Fab fragment: holotoxin complex that can be evaluated in the Neuro 2A patch single molecule translocation assay. Pre-incubation of the Fab with BoNT/A holotoxin at pH 7 prior to the translocation assay radically transformed the pattern of channel activity, whereas no effect was monitored for that of the HC channel (Fischer and Montal 2007). Fab binding to the LC allowed channel formation and early translocation intermediates to be monitored; however, unoccluded HC channel activity was never monitored. Thus the Fab locked the channel and the LC in an irreversibly incomplete translocating conformation by maintaining the LC in folded conformation incompatible with translocation through the channel lumen.

Removal of the H_C allows for investigation of the role of pH on LC translocation. Patch clamp of Neuro 2A cells with recombinant BoNT/A LH_N over a range of pH gradient values demonstrated the necessity of LC refolding to productive LC translocation (Fischer et al. 2008). Under endosomal conditions, LH_N channel activity was similar to holotoxin; a series of multiple discrete transient intermediate conductances were visualized prior to the unoccluded state. Productive translocation by LH_N across Neuro-2A membranes was confirmed by a cellular intoxication assay which measured SNAP-25 cleavage by western blot (Fischer et al. 2008); however, drastically reduced toxicity was monitored in vivo (Chaddock et al. 2002). LH_N exhibited similar growing γ patterns at pH 6 *cis*/pH 7 *trans*, although the $T_{1/2}$ measured from the average time course of change of γ after insertion was increased from that of pH 5 *cis* conditions by ~60 s. The longer time required to complete LC translocation may be attributed to the LC conformation at pH 6; under these conditions, the LC may require more time/energy to achieve a translocation compatible conformation. Consistent with this model were the results for LH_N channel activity elicited at symmetric neutral pH; channel activity initiated and maintained at the unoccluded state of 64 pS. The conformation of the LC at pH 7 is therefore incompatible for translocation through the ~15 Å pore of the H_N channel.

8.2 Precisely Timed Separation of LC From HC is Required to Complete Productive Translocation

The role of the disulfide bridge during translocation has been examined by modulating the redox gradient during LC translocation across patch clamped neuronal membranes (Fischer and Montal 2007). In the absence of a reducing environment in the *trans* (cytosol) compartment, BoNT/A holotoxin channels were arrested in an intermediate conductance state. Without disulfide bridge reduction, the LC remained locked within the HC channel, unable to escape and refold within the cytosol. In contrast, prereduction of holotoxin before addition to the membrane resulted in HC-like channels without intermediates associated with LC translocation; the LC lost association and did not undergo translocation across the membrane. Thus the disulfide bridge is required for initiation of productive translocation. The specific step at which reduction must occur in order to complete LC translocation was investigated by reducing the disulfide at each of the permissible growing conductance states monitored during a single LC translocation event in Neuro 2A membranes. Addition of β-mercaptoethanol, a membrane permeable reductant, at either the entry (~15 pS), transfer (~20–50 pS), or exit step (~55 pS) of LC translocation precluded entrance to the unoccluded HC channel state. These results indicate that the disulfide bridge must remain intact on the *cis* (endosomal) side of the bilayer until the last stage of LC translocation; entry of the disulfide bridge into the reducing environment of the cytosol serves as the final step required for productive release. The disulfide bridge is, therefore, a crucial determinant of the BoNT toxicity, and it appears to serve a crucial role in cargo translocation.

8.3 LC Must be Proteolytically Cleaved From HC

Strains of *C. botulinum* secreting single-chain forms of BoNT/B, E, and F holotoxin belong to group II non-protoelytic *C. botulinum* (Hill and Smith 2012); these toxins have significantly reduced synaptic silencing activity as the LC must be proteolytically cleaved from the HC in order to be activated. The BoNT/E producing *C. botulinum* strains are universally non-proteolytic—trypsin is commonly used to activate BoNT/E prior to enzymatic assays of cleavage of its substrate, SNAP-25 (Sathyamoorthy and DasGupta 1985). As such, BoNT/E holotoxin represents an ideal system to investigate the role of inter-chain linkage on LC translocation. Single molecule patch clamp experiments in Neuro 2A cells under conditions which emulate the pH and redox gradients across endosomes using uncleaved BoNT/E holotoxin resulted in channels that persisted in a low conductance intermediate state (Fischer and Montal 2007). In contrast, experiments in which trypsin was supplemented to the *trans* compartment immediately following the onset of low conductance channel activity resulted in a transition through a succession of discrete transient intermediate conductance states before entering the unoccluded HC channel state of 65 pS (Fischer and Montal 2007). The intermediate conductance states monitored closely approximate most of the entry, transfer, and exit steps displayed by BoNT/A during translocation. The findings with single-chain BoNT/E therefore demonstrate that productive completion of LC translocation requires physical separation of the LC from the HC channel by both reduction of the disulfide bridge and cleavage of the scissile bond.

9 Chemical Inhibition of HC Channel as Potential Therapy for Botulism

The H_N channel represents a novel and attractive target for inhibition of BoNT neurotoxicity. Therapeutic research has primarily aimed at inhibition of the H_C and LC through small molecules, antibodies, and lectins; however, these efforts have been frustrated by the variations between serotypes and subtypes. Unique serotype specificity is imparted by the H_C and LC activity, which arises from interactions with host cellular proteins. The fact that each serotype has unique enzymatic substrates and cellular binding partners necessitates the use of a cocktail of inhibitory compounds in order to derive neutralization across the board. In contrast, the high sequence similarity of the H_N across multiple serotypes may allow for a single molecule to inhibit several serotypes. By design the H_N functions to chaperone cargo, serving as a universal carrier of partially unfolded polypeptide chains composed of a wide variety of sequences (Bade et al. 2004).

Two studies have discovered small molecules that inhibit BoNT/A channel activity with similar effects on the other serotypes (Fischer et al. 2008, 2009). Chlorpromazine and lidocaine derivative QX222, which block nicotinic acetylcholine receptors, exhibit inhibitory activity upon BoNT/A channels with μM potency (Fischer et al. 2008). These compounds significantly reduce the frequency of channel openings; stabilization of the closed channel state is a critical characteristic of channel blockers (Hille 2001). Blocking the HC channel aborts translocation of the protease and subsequently abrogates its toxicity. Additionally, this class of compounds may prove therapeutic against the synaptic silencing that is symptomatic of botulism.

Another class of antibotulismic compounds was recently discovered to derive its protection from a unique set of interactions with BoNT. Toosendanin (TSDN), a Chinese herb, has been demonstrated to protect mice, rats, and monkeys from BoNT/A, B, and E (Shi and Li 2007); however, the mechanism of protection was not known. Biophysical studies showed that TSDN directly affects BoNT channel formation through inhibition of oligomerization (Sun et al. 2011). TSDN has also been shown to exhibit bi-modal activity upon the HC channel, serving to stabilize the open channel state of the isolated HC as well as stabilizing an intermediate state of LC translocation in the occluded HC (Fischer et al. 2009). While μM concentrations of TSDN are required to modulate the unoccluded HC channel activity, LC translocation is effectively abrogated in Neuro-2A cells by ~5 nM for both BoNT/A and E. Derivatives of TSDN offer improved sensitivity with reduced cellular toxicity. One can readily envision further combinatorial development of a selective antibotulismic prophylactic compound.

10 Productive Intoxication Requires Progression of Inter-domain Chaperone Activity

Investigation of the mechanism of LC translocation through the HC channel has revealed the synchronicity of the individual BoNT domains. While each component has a unique activity that functions individually, the tight interplay between them compels each one to serve as a chaperone for the others at critical steps in the intoxication process. The H_C association with its ganglioside receptor promotes H_N channel formation through oligomerization (Sun et al. 2011). Following receptor-mediated endocytosis, the H_C ensures H_N channel formation occurs in synchrony with LC adoption of a translocation competent conformation (Sun et al. 2011; Fischer et al. 2008). The LC preserves the HC by maintaining it in a soluble conformation at neutral pH until residence within the acidic endosome where the

LC serves to chaperone the HC again. Within the endosome, the LC functions as a trigger, gating the H_N channel in order to initiate its own translocation (Fischer and Montal 2007). The H_N belt protects the LC from premature cleavage of non-specific substrates until the LC is colocalized with its specific SNARE substrate within the cytosol (Brunger et al. 2007). Although the belt is not a necessary component of the HC channel, it may facilitate LC refolding to a translocation competent conformation that would otherwise not be achieved by the isolated LC upon residence within the acidic compartment of the endosome. This model is supported by the requirement for an intact disulfide bridge between the belt and the LC until completion of LC translocation (Fischer and Montal 2007). Finally, the H_N protects the LC from misfolding and disassociating within the acidic environment of the endosome, chaperones the LC from the endosome to the cytosol, and releases an enzymatically active LC to the substrate SNARE proteins (Koriazova and Montal 2003). While it is clear that each domain can function in isolation, the elegance of cellular intoxication is only achieved by the synchronized cooperative interaction of a binding domain, pseudo-substrate, channel, and protease.

Acknowledgments This work was supported by the Johns Hopkins University Applied Physics Laboratory post-doctoral grant.

Conflict of Interest The author declares that there are no conflicts of interest.

References

Agarwal R, Eswaramoorthy S, Kumaran D, Binz T, Swaminathan S (2004) Structural analysis of botulinum neurotoxin type E catalytic domain and its mutant Glu212—Gln reveals the pivotal role of the Glu212 carboxylate in the catalytic pathway. Biochemistry 43:6637–6644

Agarwal R, Binz T, Swaminathan S (2005) Structural analysis of botulinum neurotoxin serotype F light chain: implications on substrate binding and inhibitor design. Biochemistry 44: 11758–11765

Arndt JW, Yu W, Bi F, Stevens RC (2005) Crystal structure of botulinum neurotoxin type G light chain: serotype divergence in substrate recognition. Biochemistry 44:9574–9580

Arndt JW, Chai Q, Christian T, Stevens RC (2006) Structure of botulinum neurotoxin type D light chain at 1.65 A resolution: repercussions for VAMP-2 substrate specificity. Biochemistry 45:3255–3262

Bade S, Rummel A, Reisinger C, Karnath T, Ahnert-Hilger G, Bigalke H, Binz T (2004) Botulinum neurotoxin type D enables cytosolic delivery of enzymatically active cargo proteins to neurones via unfolded translocation intermediates. J Neurochem 91:1461–1472

Barth H, Pfeifer G, Hofmann F, Maier E, Benz R, Aktories K (2001) Low pH-induced formation of ion channels by clostridium difficile toxin B in target cells. J Biol Chem 276:10670–10676

Basilio D, Kienker PK, Briggs SW, Finkelstein A (2011) A kinetic analysis of protein transport through the anthrax toxin channel. J Gen Physiol 137:521–531

Black JD, Dolly JO (1986) Interaction of 125I-labeled botulinum neurotoxins with nerve terminals. I. Ultrastructural autoradiographic localization and quantitation of distinct membrane acceptors for types A and B on motor nerves. J Cell Biol 103:521–534

Blasi J, Chapman ER, Link E, Binz T, Yamasaki S, De Camilli P, Sudhof TC, Niemann H, Jahn R (1993) Botulinum neurotoxin a selectively cleaves the synaptic protein SNAP-25. Nature 365:160–163

Blaustein RO, Germann WJ, Finkelstein A, DasGupta BR (1987) The N-terminal half of the heavy chain of botulinum type A neurotoxin forms channels in planar phospholipid bilayers. FEBS Lett 226:115–120

Breidenbach MA, Brunger AT (2004) Substrate recognition strategy for botulinum neurotoxin serotype A. Nature 432:925–929

Brunger AT, Breidenbach MA, Jin R, Fischer A, Santos JS, Montal M (2007) Botulinum neurotoxin heavy chain belt as an intramolecular chaperone for the light chain. PLoS Pathog 3:1191–1194

Chaddock JA, Herbert MH, Ling RJ, Alexander FC, Fooks SJ, Revell DF, Quinn CP, Shone CC, Foster KA (2002) Expression and purification of catalytically active, non-toxic endopeptidase derivatives of Clostridium botulinum toxin type A. Protein Expr Purif 25:219–228

Chai Q, Arndt JW, Dong M, Tepp WH, Johnson EA, Chapman ER, Stevens RC (2006) Structural basis of cell surface receptor recognition by botulinum neurotoxin B. Nature 444:1096–1100

Couesnon A, Shimizu T, Popoff MR (2009) Differential entry of botulinum neurotoxin A into neuronal and intestinal cells. Cell Microbiol 11:289–308

Dolly JO, Black J, Williams RS, Melling J (1984) Acceptors for botulinum neurotoxin reside on motor nerve terminals and mediate its internalization. Nature 307:457–460

Dong M, Richards DA, Goodnough MC, Tepp WH, Johnson EA, Chapman ER (2003) Synaptotagmins I and II mediate entry of botulinum neurotoxin B into cells. J Cell Biol 162:1293–1303

Dong M, Yeh F, Tepp WH, Dean C, Johnson EA, Janz R, Chapman ER (2006) SV2 is the protein receptor for botulinum neurotoxin A. Science 312:592–596

Donovan JJ, Middlebrook JL (1986) Ion-conducting channels produced by botulinum toxin in planar lipid membranes. Biochemistry 25:2872–2876

Draper RK, Simon MI (1980) The entry of diphtheria toxin into the mammalian cell cytoplasm: evidence for lysosomal involvement. J Cell Biol 87:849–854

Falnes PO, Madshus IH, Sandvig K, Olsnes S (1992) Replacement of negative by positive charges in the presumed membrane-inserted part of diphtheria toxin B fragment. Effect on membrane translocation and on formation of cation channels. J Biol Chem 267:12284–12290

Finkelstein A (1990) Channels formed in phospholipid bilayer membranes by diphtheria, tetanus, botulinum and anthrax toxin. J de physiol 84:188–190

Fischer A, Montal M (2006) Characterization of Clostridial botulinum neurotoxin channels in neuroblastoma cells. Neurotox Res 9:93–100

Fischer A, Montal M (2007a) Single molecule detection of intermediates during botulinum neurotoxin translocation across membranes. Proc Nat Acad Sci U S A 104:10447–10452

Fischer A, Montal M (2007b) Crucial role of the disulfide bridge between botulinum neurotoxin light and heavy chains in protease translocation across membranes. J Biol Chem 282: 29604–29611

Fischer A, Mushrush DJ, Lacy DB, Montal M (2008a) Botulinum neurotoxin devoid of receptor binding domain translocates active protease. PLoS Pathog 4:e1000245

Fischer A, Koriazova LK, Oblatt-Montal M, Montal M (2008b) Botulinum toxin: Therapeutic clinical practices and science, chapter 3, (United States of America: Elsevier Inc.)

Fischer A, Nakai Y, Eubanks LM, Clancy CM, Tepp WH, Pellett S, Dickerson TJ, Johnson EA, Janda KD, Montal M (2009) Bimodal modulation of the botulinum neurotoxin protein-conducting channel. Proc Nat Acad Sci U S A 106:1330–1335

Fischer A, Sambashivan S, Brunger AT, Montal M (2012) Beltless translocation domain of botulinum neurotoxin A embodies a minimum ion-conductive channel. J Biol Chem 287:1657–1661

Fu FN, Busath DD, Singh BR (2002) Spectroscopic analysis of low pH and lipid-induced structural changes in type A botulinum neurotoxin relevant to membrane channel formation and translocation. Biophys Chem 99:17–29

Fu Z, Chen C, Barbieri JT, Kim JJ, Baldwin MR (2009) Glycosylated SV2 and gangliosides as dual receptors for botulinum neurotoxin serotype F. Biochemistry 48:5631–5641

Fujinaga Y, Sugawara Y, Matsumura T (2012) Uptake of botulinum neurotoxin in the intestine. doi: 10.1007/978-3-642-33570-9_3

Galloux M, Vitrac H, Montagner C, Raffestin S, Popoff MR, Chenal A, Forge V, Gillet D (2008) Membrane interaction of botulinum neurotoxin A translocation (T) domain. The belt region is a regulatory loop for membrane interaction. J Biol Chem 283:27668–27676

Gambale F, Montal M (1988) Characterization of the channel properties of tetanus toxin in planar lipid bilayers. Biophys J 53:771–783

Ginalski K, Venclovas C, Lesyng B, Fidelis K (2000) Structure-based sequence alignment for the beta-trefoil subdomain of the clostridial neurotoxin family provides residue level information about the putative ganglioside binding site. FEBS Lett 482:119–124

Gu S, Rumpel S, Zhou J, Strotmeier J, Bigalke H, Perry K, Shoemaker CB, Rummel A, Jin R (2012) Botulinum neurotoxin is shielded by NTNHA in an interlocked complex. Science 335:977–981

Habermann E, Dreyer F, Bigalke H (1980) Tetanus toxin blocks the neuromuscular transmission in vitro like botulinum A toxin. Naunyn-Schmiedeberg's Arch Pharmacol 311:33–40

Hanson MA, Stevens RC (2000) Cocrystal structure of synaptobrevin-II bound to botulinum neurotoxin type B at 2.0 A resolution. Nat Struct Biol 7:687–692

Harper CB, Martin S, Nguyen TH, Daniels SJ, Lavidis NA, Popoff MR, Hadzic G, Mariana A, Chau N, McCluskey A et al (2011) Dynamin inhibition blocks botulinum neurotoxin type A endocytosis in neurons and delays botulism. J Biol Chem 286:35966–35976

Hill KK, Smith TJ (2012) Genetic diversity within *Clostridium botulinum* serotypes, botulinum neurotoxin gene clusters and toxin subtypes. doi:10.1007/978-3-642-33570-9_1

Hille B (2001) Ion channels of excitable cells. Sinauer, Sunderland, MA

Hoch DH, Romero-Mira M, Ehrlich BE, Finkelstein A, DasGupta BR, Simpson LL (1985) Channels formed by botulinum, tetanus, and diphtheria toxins in planar lipid bilayers: relevance to translocation of proteins across membranes. Proc Nat Acad Sci U S A 82: 1692–1696

Hughes R, Whaler BC (1962) Influence of nerve-ending activity and of drugs on the rate of paralysis of rat diaphragm preparations by C. botulinum type A toxin. J Physiol 160:221–233

Jackson MB, Chapman ER (2006) Fusion pores and fusion machines in Ca2 + -triggered exocytosis. Annu Rev Biophys Biomol Struct 35:135–160

Jahn R, Lang T, Sudhof TC (2003) Membrane fusion. Cell 112:519–533

Jin R, Sikorra S, Stegmann CM, Pich A, Binz T, Brunger AT (2007) Structural and biochemical studies of botulinum neurotoxin serotype C1 light chain protease: implications for dual substrate specificity. Biochemistry 46:10685–10693

Koivusalo M, Steinberg BE, Mason D, Grinstein S (2011) In situ measurement of the electrical potential across the lysosomal membrane using FRET. Traffic 12:972–982

Koriazova LK, Montal M (2003) Translocation of botulinum neurotoxin light chain protease through the heavy chain channel. Nat Struct Biol 10:13–18

Kumaran D, Eswaramoorthy S, Furey W, Navaza J, Sax M, Swaminathan S (2009) Domain organization in Clostridium botulinum neurotoxin type E is unique: its implication in faster translocation. J Mol Biol 386:233–245

Lacy DB, Stevens RC (1999) Sequence homology and structural analysis of the clostridial neurotoxins. J Mol Biol 291:1091–1104

Lacy DB, Tepp W, Cohen AC, DasGupta BR, Stevens RC (1998) Crystal structure of botulinum neurotoxin type A and implications for toxicity. Nat Struct Biol 5:898–902

Lawrence G, Wang J, Chion CK, Aoki KR, Dolly JO (2007) Two protein trafficking processes at motor nerve endings unveiled by botulinum neurotoxin E. J Pharmacol Exp Therap 320:410–418

Ledoux DN, Be XH, Singh BR (1994) Quaternary structure of botulinum and tetanus neurotoxins as probed by chemical cross-linking and native gel electrophoresis. Toxicon: Official J Int Soc Toxinol 32:1095–1104

Levy R, Forsyth CM, LaPorte SL, Geren IN, Smith LA, Marks JD (2007) Fine and domain-level epitope mapping of botulinum neurotoxin type A neutralizing antibodies by yeast surface display. J Mol Biol 365:196–210

Li L, Singh BR (2000) Spectroscopic analysis of pH-induced changes in the molecular features of type A botulinum neurotoxin light chain. Biochemistry 39:6466–6474

Mahrhold S, Rummel A, Bigalke H, Davletov B, Binz T (2006) The synaptic vesicle protein 2C mediates the uptake of botulinum neurotoxin A into phrenic nerves. FEBS Lett 580: 2011–2014

Maksymowych AB, Simpson LL (1998) Binding and transcytosis of botulinum neurotoxin by polarized human colon carcinoma cells. J Biol Chem 273:21950–21957

Maksymowych AB, Simpson LL (2004) Structural features of the botulinum neurotoxin molecule that govern binding and transcytosis across polarized human intestinal epithelial cells. J Pharmacol Exp Ther 310:633–641

Maksymowych AB, Reinhard M, Malizio CJ, Goodnough MC, Johnson EA, Simpson LL (1999) Pure botulinum neurotoxin is absorbed from the stomach and small intestine and produces peripheral neuromuscular blockade. Infect Immun 67:4708–4712

Menestrina G, Forti S, Gambale F (1989) Interaction of tetanus toxin with lipid vesicles. Effects of pH, surface charge, and transmembrane potential on the kinetics of channel formation. Biophys J 55:393–405

Milne JC, Collier RJ (1993) pH-dependent permeabilization of the plasma membrane of mammalian cells by anthrax protective antigen. Mol Microbiol 10:647–653

Montal MS, Blewitt R, Tomich JM, Montal M (1992) Identification of an ion channel-forming motif in the primary structure of tetanus and botulinum neurotoxins. FEBS Lett 313:12–18

Mushrush DJ, Koteiche HA, Sammons MA, Link AJ, McHaourab HS, Lacy DB (2011) Studies of the mechanistic details of the pH-dependent association of botulinum neurotoxin with membranes. J Biol Chem 286:27011–27018

Nishiki T, Tokuyama Y, Kamata Y, Nemoto Y, Yoshida A, Sato K, Sekiguchi M, Takahashi M, Kozaki S (1996) The high-affinity binding of Clostridium botulinum type B neurotoxin to synaptotagmin II associated with gangliosides GT1b/GD1a. FEBS Lett 378:253–257

Oblatt-Montal M, Yamazaki M, Nelson R, Montal M (1995) Formation of ion channels in lipid bilayers by a peptide with the predicted transmembrane sequence of botulinum neurotoxin A. Protein Sci: Publ Protein Soc 4:1490–1497

Peng L, Tepp WH, Johnson EA, Dong M (2011) Botulinum neurotoxin D uses synaptic vesicle protein SV2 and gangliosides as receptors. PLoS Pathog 7:e1002008

Petro KA, Dyer MA, Yowler BC, Schengrund CL (2006) Disruption of lipid rafts enhances activity of botulinum neurotoxin serotype A. Toxicon: Official J Int Soc Toxinol 48: 1035–1045

Pilpa RM, Bayrhuber M, Marlett JM, Riek R, Young JA (2011) A receptor-based switch that regulates anthrax toxin pore formation. PLoS Pathog 7:e1002354

Pirazzini M, Rossetto O, Bolognese P, Shone CC, Montecucco C (2011) Double anchorage to the membrane and intact inter-chain disulfide bond are required for the low pH induced entry of tetanus and botulinum neurotoxins into neurons. Cell Microbiol 13:1731–1743

Rainey GJ, Wigelsworth DJ, Ryan PL, Scobie HM, Collier RJ, Young JA (2005) Receptor-specific requirements for anthrax toxin delivery into cells. Proc Nat Acad Sci U S A 102: 13278–13283

Rauch G, Gambale F, Montal M (1990) Tetanus toxin channel in phosphatidylserine planar bilayers: conductance states and pH dependence. Eur Biophys J 18:79–83

Rummel A, Karnath T, Henke T, Bigalke H, Binz T (2004) Synaptotagmins I and II act as nerve cell receptors for botulinum neurotoxin G. J Biol Chem 279:30865–30870

Rummel A, Hafner K, Mahrhold S, Darashchonak N, Holt M, Jahn R, Beermann S, Karnath T, Bigalke H, Binz T (2009) Botulinum neurotoxins C, E and F bind gangliosides via a

conserved binding site prior to stimulation-dependent uptake with botulinum neurotoxin F utilising the three isoforms of SV2 as second receptor. J Neurochem 110:1942–1954
Rummel A (2012) Double receptor anchorage of botulinum neurotoxins accounts for their exquisite neurospecificity. doi:10.1007/978-3-642-33570-9_4
Ruthel G, Burnett JC, Nuss JE, Wanner LM, Tressler LE, Torres-Melendez E, Sandwick SJ, Retterer CJ, Bavari S (2011) Post-intoxication inhibition of Botulinum neurotoxin serotype A within neurons by small-molecule, non-peptidic inhibitors. Toxins 3:207–217
Sandvig K, Olsnes S (1980) Diphtheria toxin entry into cells is facilitated by low pH. J Cell Biol 87:828–832
Sathyamoorthy V, DasGupta BR (1985) Separation, purification, partial characterization and comparison of the heavy and light chains of botulinum neurotoxin types A, B, and E. J Biol Chem 260:10461–10466
Schiavo G, Benfenati F, Poulain B, Rossetto O, Polverino de Laureto P, DasGupta BR, Montecucco C (1992) Tetanus and botulinum-B neurotoxins block neurotransmitter release by proteolytic cleavage of synaptobrevin. Nature 359:832–835
Schiavo G, Matteoli M, Montecucco C (2000) Neurotoxins affecting neuroexocytosis. Physiol Rev 80:717–766
Segelke B, Knapp M, Kadkhodayan S, Balhorn R, Rupp B (2004) Crystal structure of Clostridium botulinum neurotoxin protease in a product-bound state: Evidence for noncanonical zinc protease activity. Proc Natl Acad Sci U S A 101:6888–6893
Shi YL, Li MF (2007) Biological effects of toosendanin, a triterpenoid extracted from Chinese traditional medicine. Prog Neurobiol 82:1–10
Simpson LL (1980) Kinetic studies on the interaction between botulinum toxin type A and the cholinergic neuromuscular junction. J Pharmacol Exp Ther 212:16–21
Simpson LL, Lautenslager GT, Kaiser II, Middlebrook JL (1993) Identification of the site at which phospholipase A2 neurotoxins localize to produce their neuromuscular blocking effects. Toxicon: Official J Int Soc Toxinol 31:13–26
Singh BR, DasGupta BR (1989) Structure of heavy and light chain subunits of type A botulinum neurotoxin analyzed by circular dichroism and fluorescence measurements. Mol Cell Biochem 85:67–73
Singh BR, Wasacz FM, Strand S, Jakobsen RJ, DasGupta BR (1990) Structural analysis of botulinum neurotoxin types A and E in aqueous and nonpolar solvents by fourier transform infrared, second derivative UV absorption, and circular dichroic spectroscopies. J Protein Chem 9:705–713
Stenmark P, Dupuy J, Imamura A, Kiso M, Stevens RC (2008) Crystal structure of botulinum neurotoxin type A in complex with the cell surface co-receptor GT1b-insight into the toxin-neuron interaction. PLoS Pathog 4:e1000129
Sudhof TC (2004) The synaptic vesicle cycle. Annu Rev Neurosci 27:509–547
Sun S, Suresh S, Liu H, Tepp WH, Johnson EA, Edwardson JM, Chapman ER (2011) Receptor binding enables botulinum neurotoxin B to sense low pH for translocation channel assembly. Cell Host Microbe 10:237–247
Swaminathan S, Eswaramoorthy S (2000a) Structural analysis of the catalytic and binding sites of Clostridium botulinum neurotoxin B. Nat Struct Biol 7:693–699
Swaminathan S, Eswaramoorthy S (2000b) Structural analysis of the catalytic and binding sites of Clostridium botulinum neurotoxin B. Nat Struct Biol 7:693–699
Tsukamoto K, Kohda T, Mukamoto M, Takeuchi K, Ihara H, Saito M, Kozaki S (2005) Binding of Clostridium botulinum type C and D neurotoxins to ganglioside and phospholipid. Novel insights into the receptor for clostridial neurotoxins. J Biol Chem 280:35164–35171
Weber T, Zemelman BV, McNew JA, Westermann B, Gmachl M, Parlati F, Sollner TH, Rothman JE (1998) SNAREpins: minimal machinery for membrane fusion. Cell 92:759–772
Yowler BC, Kensinger RD, Schengrund CL (2002) Botulinum neurotoxin A activity is dependent upon the presence of specific gangliosides in neuroblastoma cells expressing synaptotagmin I. J Biol Chem 277:32815–32819

Clostridial Neurotoxin Light Chains: Devices for SNARE Cleavage Mediated Blockade of Neurotransmission

Thomas Binz

Abstract Seven serologically distinct botulinum neurotoxins and tetanus neurotoxin which cause the diseases botulism and tetanus constitute the clostridial neurotoxin family. Like many other bacterial protein toxins they exhibit a modular structure. One domain mediates highly specific binding to target cells and endocytosis, while the second translocates the third, a catalytic domain across the endosomal membrane to the target cell cytosol. In case of Clostridial neurotoxins (CNT), the latter acts as extremely specific Zn^{2+}-dependent metalloproteinase. The various serotypes proteolyze each one particular peptide bond in one of the three SNARE proteins, which are the core of the membrane fusion apparatus for synaptic vesicles. SNARE cleavage causes the blockade of neurotransmitter release. This chapter details the molecular basis for the highly selective substrate recognition and cleavage mechanism of CNT.

Keywords Botulinum neurotoxin · SNAP-25 · SNARE · Syntaxin · VAMP/ Synaptobrevin · Zinc protease

Abbreviations

BoNT	botulinum neurotoxin
BoNT/X	serotype X of BoNT, X = A–G
BoNT/XY	subtype Y of serotype BoNT/X, Y = 1–8
CNT	clostridial neurotoxins
HC/X	heavy chain of BoNT serotype X/TeNT
HCX	50 kDa cell binding fragment of BoNT/TeNT
HNX	50 kDa translocation domain of BoNT/TeNT

T. Binz (✉)
Institut für Biochemie, OE 4310, Medizinische Hochschule Hannover, 30623 Hannover, Germany
e-mail: binz.thomas@mh-hannover.de

A. Rummel and T. Binz (eds.), *Botulinum Neurotoxins*, Current Topics in Microbiology and Immunology 364, DOI: 10.1007/978-3-642-33570-9_7,

LC/X	light chain, catalytic domain of BoNT serotype X/TeNT
NSF	N-ethylmaleimide-sensitive factor
SNAP	soluble NSF adapter protein
SNAP-25	synaptosomal associated protein of 25 kDa
SNAP-23	23 kDa homolog of SNAP-25
SNARE	soluble N-ethylmaleimide-sensitive factor attachment protein receptor
syntaxin-X	Isoform X of syntaxin, X = 1–19
TI-VAMP	tetanus neurotoxin insensitive VAMP
TeNT	tetanus neurotoxin
VAMP	vesicle-associated membrane protein

Contents

1 Introduction 140
2 SNAREs are the Intracellular Substrates of Clostridial Neurotoxin Light Chains 141
3 Clostridial Neurotoxin Light Chains Require Extended SNARE Regions for Efficient Binding 144
3.1 Mode of SNAP-25 Recognition by LC/A 146
3.2 Mode of VAMP Recognition by LC/F 147
4 Mechanism of Peptide Bond Cleavage by Clostridial Neurotoxin Light Chains 148
4.1 Scissile Peptide Bond Alignment at the Enzyme's Active Center 148
4.2 Catalytic Mechanism of Peptide Bond Hydrolysis 149
5 Future Perspectives 152
References 153

1 Introduction

Clostridial neurotoxins (CNT), tetanus toxin (TeNT) and the seven botulinum neurotoxins (BoNT/A-G) are produced as ~150 kDa single chain proteins by *Clostridium botulinum* and some other bacteria of the genus *Clostridium*. Subsequent cleavage by specific clostridial or host proteases converts the toxin into the fully active di-chain form. This consists of an N-terminal ~50 kDa light chain (LC) and a C-terminal ~100 kDa heavy chain (HC). Both remain linked via a single disulfide bond formed between two cysteines at the C-terminus of the LC and the N-terminus of the HC. The HC comprises two subunits, a largely α–helical domain of ~50 kDa at the N-terminus, referred as H_N domain, and a ~50 kDa fragment at the C-terminus, referred as H_C (Schiavo et al. 2000).

CNTs are considered to be the most hazardous natural substances known. The 50 % lethal doses for susceptible mammals including human are in the range of one nanogram per kg of body weight (Gill 1982). This is due to the toxins' selectivity for neuronal tissue and enzymatic activity toward the neurotransmitter release apparatus (Schiavo et al. 2000).

The HC has the task to enable the catalytic LC to reach its intracellular substrate and accomplishes that via a three-step mechanism. After CNT have gained access to the host nervous system at cholinergic motor nerve terminals, H_C mediates in the first step binding of the toxin to complex gangliosides. In order that the toxin can enter the neuron, the toxin has to associate with a second receptor. Present data suggest that those are generally synaptic vesicle resident proteins (Rummel 2012). Association with both receptors initiates the second step, endocytosis via the synaptic vesicle recycling pathway. In the course of vesicle transport the toxins become exposed to an acidic environment which initiates the third step. Acidification triggers membrane insertion of H_N to form a cation selective channel (Fischer 2012). This probably coincides with partial unfolding of the LC, allowing its entry into and transit through the H_N channel followed by its refolding in the cytosol (Fischer and Montal 2007; Koriazova and Montal 2003). Reduction of the disulfide bridge finally releases the LC from the HC. In contrast to BoNTs, TeNT is directed to the transcytosis route in the course of endocytosis. As a consequence, the mechanism leading to LC delivery to the cytosol, proceeds only at the inhibitory interneuron in the spinal cord (Schiavo et al. 2012).

Once in the cytosol the LCs exert their proteolytic activity. This is directed against those SNAREs that fuel the fusion of neurotransmitter laden vesicles with the presynaptic membrane. The proteolytic mechanism of the LCs is the subject of the following sections. Substrate cleavage by BoNT and TeNT blocks neurotransmitter release, but due to different sites of action leads to opposite clinical symptoms. BoNTs cause flaccid paralysis, which can result in respiratory failure and death, whereas TeNT leads to spastic paralysis (Schiavo et al. 2000). Although representing life-threatening toxins, BoNTs have become widely used therapeutics for the treatment of a variety of neurological disorders with still increasing numbers of indications (Davletov et al. 2005) (Bigalke 2012). Precise insights into the LCs' substrate cleavage mechanisms combined with directed modifications may contribute to expand the area of BoNT applications in the future.

2 SNAREs are the Intracellular Substrates of Clostridial Neurotoxin Light Chains

The intracellular mode of action of CNTs remained unsettled until the early 1990s. A major contribution toward its elucidation became the identification of the Zn^{2+} binding motif, His-Glu-X-X-His, in the primary sequence of TeNT (Jongeneel et al. 1989) and the fact that this motif proved to be present also in BoNT/A whose primary structure was established in 1990 (Binz et al. 1990; Thompson et al. 1990). The demonstration that various BoNTs bind zinc endorsed the idea of an intracellular mode of action as zinc-dependent proteases (Schiavo et al. 1992a).

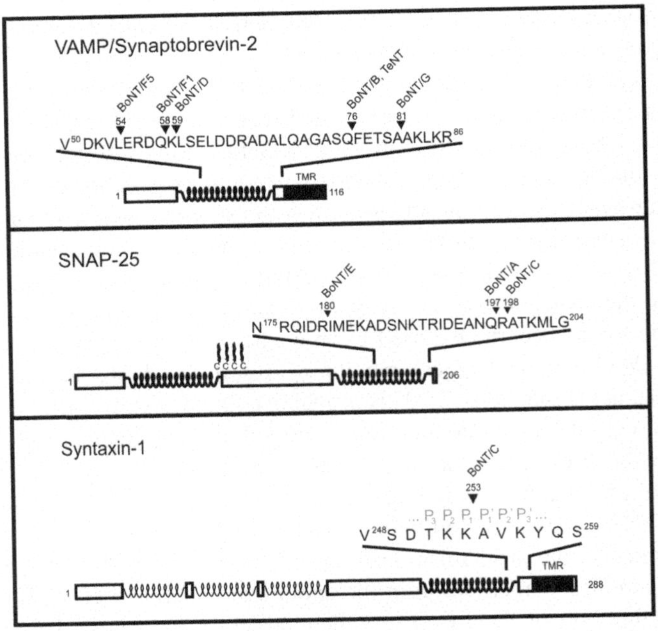

Fig. 1 Schematic drawing of the substrates of CNTs and illustration of the individual cleavage sites. VAMP/Synaptobrevin-2 (*top*) is a membrane protein of synaptic vesicles, which provides its central α-helical region for the formation of the coiled-coil four helix bundle of the SNARE complex. SNAP-25 (*middle*) is anchored in the presynaptic membrane via cysteine bound palmitoyl groups. Both its α-helical segments become constituent parts of the SNARE complex. Syntaxin-1 (*bottom*) is also a presynaptic membrane resident protein. Its C-terminal α-helical region is involved in SNARE complex formation. The individual peptide bonds hydrolyzed by CNTs are indicated by *arrow heads* above the primary structure sections and the N-terminal amino acid positions are specified. The substrate positions (P3– P3′) around the scissile peptide bond as defined by Schechter and Berger (1967) are shown above the primary structure section of syntaxin-1

Experimental evidence for a metalloproteolytic activity was finally provided by the abolishment of TeNT evoked blockade of neurotransmitter release by classical zinc endoprotease inhibitors in neurons of the sea-slug *Aplysia californica* (Schiavo et al. 1992b).

Contemporaneous to these findings on CNTs, the characterization of proteins involved in vesicular transport and synaptic function had advanced much (Ahnert-Hilger 2012). This allowed to specifically analyze synaptic proteins as potential substrates and led to the finding that BoNT/B and TeNT treatment of synaptic vesicles converted the electrophoretic mobility of the integral synaptic vesicle membrane protein vesicle associated membrane protein [VAMP (Trimble et al. 1988)], which is also termed synaptobrevin (Baumert et al. 1989), into the same two fragments (Schiavo et al. 1992c). Corresponding experiments using BoNT/D, F, and G showed that those toxins also cleave VAMP/Synaptobrevin, but generate different products (Schiavo et al. 1993a; 1993b; Yamasaki et al. 1994a). N-terminal microsequencing of the cleavage products subsequent to separation confirmed that TeNT and BoNT/B hydrolyze the same peptide bond, Gln76-Phe77, whereas BoNT/D and F cut the neighboring peptide bonds Gln58-Lys59 and Lys59-Leu60, respectively, almost in the middle of the 116-amino acids sized VAMP-2 (Fig. 1, (Schiavo et al. 1992c, 1993a, b). In contrast, the scissile bond for BoNT/G, Ala81–Ala82 is located close to the C-terminal transmembrane domain, five residues downstream from that of BoNT/B and TeNT [Fig. 1, (Yamasaki et al. 1994a)]. Many BoNT serotypes are today further distinguished into subtypes if their amino acid sequences differ by more than 2.6 %. Investigation of the various subtypes' enzymatic activity revealed that they all shared the scissile bond with their prototypical serotype, except for the very recently described subtype 5 of BoNT/F (BoNT/F5) (Kalb et al. 2012; Raphael et al. 2010). It cleaves a unique position, Leu54-Glu55, in VAMP-2, i.e., four residues upstream of the bond hydrolyzed by the prototypical BoNT/F1. This finding was less surprising, if it is considered that LC/F5 exhibits merely 46–49 % amino acid sequence identity to the presently known six other F subtypes (Kalb et al. 2012). This is the range of LC sequence conservation among the various BoNT serotypes.

The proteolytic activity of the remaining BoNTs was found to be directed against different proteins, which are localized at the presynaptic membrane. BoNT/A, C, and E proteolyze synaptosomal associated protein of 25 kDa (SNAP-25) (Schiavo et al. 1993a; Blasi et al. 1993a; Foran et al. 1996; Osen-Sand et al. 1996; Oyler et al. 1989; Williamson et al. 1996). Again different peptide bonds are attacked. BoNT/A and C hydrolyze neighboring peptide bonds, 9 or 8 amino acids, respectively, away from the C-terminus of the 206 residues comprising SNAP-25, while BoNT/E cleavage releases the 26 C-terminal residues (Binz et al. 1994; Schiavo et al. 1993c; Vaidyanathan et al. 1999). The substrate specificity of BoNT/C is unique. It is so far the only CNT capable of cleaving a second substrate, the presynaptic membrane protein, HPC-1/syntaxin-1 (Bennett et al. 1992; Blasi et al. 1993b; Inoue et al. 1992). The hydrolyzable peptide bond is located in the C-terminal region, 12 amino acids upstream of the C-terminal transmembrane domain [Fig. 1, (Schiavo et al. 1995)]. So far there are no reports available showing that BoNT/A, C, and E subtypes cleave other than the established scissile peptide bonds in SNAP-25 or syntaxin.

Each of the three CNT substrates belong to a larger protein family and they are evolutionary conserved from yeast to mammals. Due to the fact that they represent the membrane-anchored receptors for SNAPs (isoforms α, β, and γ), soluble adaptor proteins of the ATPase NSF (N-ethylmaleimide-sensitive factor), which is essential for vesicular trafficking, they are collectively termed soluble NSF attachment protein receptors (SNAREs). This together with the observation that defects in their yeast counterparts provoke defects in secretion provided independent additional evidence that proteolytic SNARE damage by CNTs indeed affects synaptic vesicle fusion and thus neurotransmitter release. The common feature of all SNAREs is the presence of one (or as an exception two) α-helical coiled-coil domain(s) of $\sim$70 amino acids, denoted SNARE domain. Four such SNARE domains, anchored in two opposing membranes, are required for each individual intracellular membrane fusion process. They interact with each other in a zipper-like fashion to form one superhelix, a parallel four-helix coiled-coil bundle. The zippering process of individual sets of such SNARE domains is assumed to fuel membrane fusion (Jahn and Scheller 2006; Sutton et al. 1998). E.g., the mammalian neuronal SNARE complex consists of syntaxin-1, SNAP-25 (comprising two SNARE domains) and VAMP-2, and mediates fusion of neurotransmitter laden synaptic vesicles with the presynaptic membrane. Consequently, their cleavage results in the blockade of neurotransmission at the neuromuscular junction.

In view of current activities to establish BoNTs with re-engineered cell specificity for clinical application (Chaddock 2012) the susceptibility of other mammalian SNAREs toward BoNTs becomes an important issue. Of particular interest are those governing excessive exocytosis in disease like release of chemokines, cytokines, hormones, mucin, etc. Based on low sequence conservation compared to the prototypical family members (less than 40 %) and major amino acid changes at the toxin cleavage sites the majority of them were ruled out as substrate of LCs. However, VAMP-1, cellubrevin/VAMP-3 as well as syntaxin-2 and syntaxin-3 exhibit more than 62 % sequence identity to the respective established substrates, VAMP-2, and syntaxin-1, and have been reported to be hydrolyzed by CNTs (Schiavo et al. 1995; McMahon et al. 1993; Yamasaki et al. 1994b). Despite the high sequence conservation among mammalian VAMP-1 the rat protein represents a poor substrate for BoNT/B and BoNT/D, unlike the human and murine (Yamasaki et al. 1994b; Yamamoto et al. 2012). Similarly, human SNAP-23 resists cleavage by BoNT/A, C, and E, whereas mouse SNAP-23 sharing 88 % identity to human SNAP-23 (both exhibit ~63 % to mouse/human SNAP-25) can be hydrolyzed by BoNT/E at higher toxin concentrations (Vaidyanathan et al. 1999; Washbourne et al. 1997).

3 Clostridial Neurotoxin Light Chains Require Extended SNARE Regions for Efficient Binding

CNT LCs are very particular proteases. There is no report that they cleave any other protein than the selected SNAREs. This raises the question in what way the proteolytic activity of CNTs is confined unlike the activity of conventional proteases that display broad specificity. First clues were reported soon after substrate identification. Cleavage assays employing various sized substrate-derived peptides revealed that extended fragments of 30 amino acids or more are required for optimal cleavage of VAMP-2 by LC/B, LC/D, LC/F, LC/G, and LC/T (Yamasaki et al. 1994a, b; Cornille et al. 1997; Foran et al. 1994; Schmidt and Stafford 2005; Shone et al. 1993) and SNAP-25 by LC/A, LC/C, and LC/E (Vaidyanathan et al. 1999; Washbourne et al. 1997). In addition, the substrate-length requirements vary

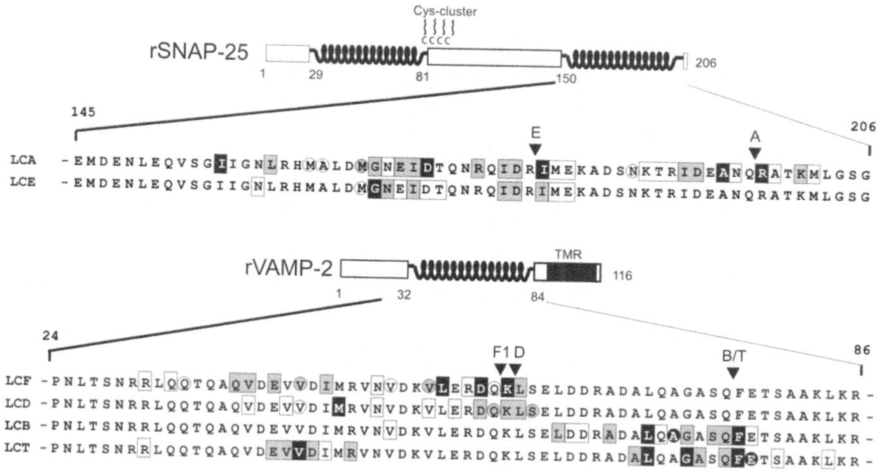

Fig. 2 Individual substrate requirements for CNT LCs. Amino acid sequence (*single letter code*) of rat SNAP-25 (*upper panel*) as well as that of rat VAMP-2 (*lower panel*) are shown below as schematic drawings. Peptide bonds cleaved by CNT LCs are indicated by *arrow heads*. Residues shown in *white* on *black* background, in a box, or on *gray* background indicate that removal of their side group results in a strong, a lesser, or an intermediate reduction, respectively, of cleavage by the respective LC (indicated on the *left*). Residues highlighted by *circles* using the same labeling depict to what extent introduction or reversal of charges affects cleavage. Data from the following publications were incorporated into this scheme: (Vaidyanathan et al. 1999; Schmidt and Stafford 2005; Jin et al. 2007; Pellizzari et al. 1997; Agarwal et al. 2009; Breidenbach and Brunger 2004; Chen and Barbieri 2006; Chen et al. 2008; Chen and Wan 2011; Fang et al. 2006; Schmidt and Bostian 1997; Shone and Roberts 1994; Sikorra et al. 2008). An averaged value is depicted, if divergent data were published

with the toxin serotype. The most striking example is provided by LC/B and LC/T. Both hydrolyze the same peptide bond, but LC/T is dependent upon a 62 residues long substrate fragment, whereas LC/B is satisfied with a substrate comprising 40 residues for optimal cleavage to occur (Foran et al. 1994; Hua and Charlton 1999). These findings were then confirmed and refined utilizing substrate point mutants. Many mutations in positions remote from the scissile peptide bond affected the cleavability (Fig. 2). Altogether, the data implied that the formation of high affinity LC-substrate complexes is dependent upon LC interactions with remote substrate residues rather than that extended substrates adopt a structure that is requisite for substrate binding (Vaidyanathan et al. 1999; Yamasaki et al. 1994b; Evans et al. 2005; Jin et al. 2007; Pellizzari et al. 1996, 1997; Sikorra et al. 2006; Wictome et al. 1996). On the contrary, if the as isolated protein largely unstructured target SNAREs assemble into the ternary SNARE complex they become virtually resistant to proteolytic attack of the LCs (Hayashi et al. 1994; Pellegrini et al. 1994).

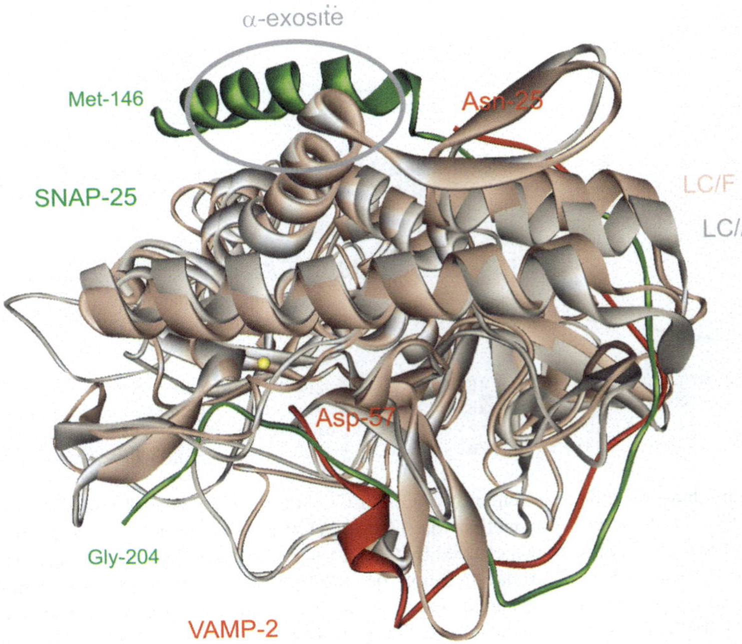

Fig. 3 Cartoon illustration of the mode of substrate binding of LC/A (pdb entry: 1XTG) and LC/F (pdb entry: 3FIE). *Ribbon* diagrams of LC/A bound to SNAP-25(146-204) and LC/F bound to VAMP-2(25-57-D-Cys58) were superimposed based on the conserved active site residues of the HExxH zinc binding motif. Zinc is shown as a *yellow sphere*

3.1 Mode of SNAP-25 Recognition by LC/A

The view of substrate/enzyme interaction via sites remote from the LCs' active site was confirmed by means of a crystal structure established for a catalytically inactive LC/A bound to the carboxyl terminal half (residues 141–204) of SNAP-25 (Breidenbach and Brunger 2004). The co-crystal revealed an extensive interface between protease and substrate. The SNAP-25 segment Gln-152 to Met-202 was found to wrap around most of the LC's circumference (Fig. 3) and 19 amino acids thereof were proposed to form side-chain/side-chain contacts with the LC. These are distributed along the whole interface, but a cumulation of interactions was noticed for the most distal 21-residue segment (residues 147–167) of SNAP-25. The hydrophobic face of this amphipathic distorted α-helical region packs against a hydrophobic patch formed at the interface of four LC α-helices, referred to as the α-exosite. The SNAP-25 α-helical region is an essential recognition area for the LC, as its absence dramatically reduces the rate of SNAP-25 cleavage (Vaidyanathan et al. 1999; Washbourne et al. 1997).

These data imply that SNAP-25 proteolysis starts via α-exosite contacts of which Ile-156, Leu-160, and Met-167 of SNAP-25 appear to be most important. After α-exosite interaction has been established, successive additional interactions via substrate amino acids located toward the C-terminal scissile peptide bond form, thereby decreasing the likelihood that LC and substrate fall apart and increasing the chance that the interaction ultimately leads to peptide bond hydrolysis. Co-crystal structure data and data of a systematic SNAP-25 mutagenesis approach suggest that major substrate residues involved are Glu-170 (interacting with LC/A N40), Ile-171 (I315; V316), Asp-172 (K41), Arg-176 (E148), Ile-178 (W118; V129), Ile-181 (V129; C134), Ile-192 (K166, methylene; F168), Asp-193 (T176). (For the sake of clarity, three-letter amino acid codes are used for substrate amino residues and single-letter codes for LC amino acids throughout this review.) Together these interactions direct the path of the substrate into the LC's active site (Breidenbach and Brunger 2004; Chen and Barbieri 2006).

Although the tertiary structure of all seven BoNT LC serotypes have been solved (Jin et al. 2007), crystallographic data on the mode of how LC/C and LC/E recognize syntaxin and/or SNAP-25, respectively, are presently not available.

3.2 *Mode of VAMP Recognition by LC/F*

Crystal structure data about the mode of VAMP-2 recognition have been achieved only for LC/F (Agarwal et al. 2009). A co-crystallization approach using wild-type LC/F1 and two N-terminal cleavage product of VAMP-2 mimicking peptide inhibitors, either 37 or 32 residues long, exhibiting a D-cysteine instead of Gln-58 at the P1 position succeeded. The direction of substrate binding was found to correspond to that determined for SNAP-25 in LC/A. In addition, the paths that the inhibitors bound along the LC/F surface are basically similar to the path of SNAP-25 on LC/A, though LC loops protrude differently and VAMP-2 exhibits one short segment of β-sheet and α-helix conformation, whereas SNAP-25 is extended in the entire alignable area (residues 168–197). A major difference in substrate binding is the longer SNAP-25-LC/A interface. A counterpart to the 21-residue distorted α-helix (residues 147–167) which joins N-terminally to the alignable segment of SNAP-25 is not required for the VAMP-2 recognition by LC/F [Fig. 3; (Agarwal et al. 2009; Breidenbach and Brunger 2004). Like the SNAP-25-LC/A interaction, amino acid side-chain contacts between VAMP-2 and LC/F were predicted to exist along the whole interface. Systematic mutagenesis studies determined the importance of individual substrate residues. Substitution of Arg-31, Gln-33, Gln-38, Val-39, Glu-41, Ile-45, Val-53, and Leu-54 showed most drastic effects (Chen and Wan 2011; Sikorra et al. 2008). In addition, several LC/F residues mediating the interaction with these amino acids were also identified (Agarwal et al. 2009; Chen and Wan 2011). These data underscored again that contacts remote from the scissile bond are critical for efficient substrate recognition. Systematic VAMP-2 mutagenesis furthermore revealed that residues downstream of the cleavage region, that roughly spans five amino acids on either side of the scissile peptide bond, did not significantly contribute to binding of LC/F (Sikorra et al. 2008). Since this is also valid for SNAP-25 (only residues down to Met-202, but not Leu-203 to Gly-206, contribute to binding (Vaidyanathan et al. 1999)], the mode of VAMP-2 binding to LC/F might occur in a similar fashion as outlined for SNAP-25 and LC/A.

Corresponding information for LC/D, LC/B, LC/T, LC/G, LC/F5 has not been worked out so far.

4 Mechanism of Peptide Bond Cleavage by Clostridial Neurotoxin Light Chains

Substrate binding as outlined in the previous chapter brings the cleavage region close to the LC's active site. In the next step substrate cleavage site residues (P5–P5′) lock into the destined LC pockets (S5–S5′) in order that the scissile peptide bond becomes exposed in proper orientation for the attack of an activated water molecule kept ready by the active site residues.

4.1 Scissile Peptide Bond Alignment at the Enzyme's Active Center

Most detailed information about substrate binding near the LCs' active site has been collected for LC/A, LC/E, and LC/F. The co-crystal structure of LC/A bound to SNAP-25 did not permit the definition of all LC S-pockets, as a double mutant of the active site had been used to render the enzyme inactive (Breidenbach and Brunger 2004). However, several co-crystal structures and docking studies with small-molecule inhibitors (Fu et al. 2006; Silvaggi et al. 2007) and SNAP-25-derived peptides (Burnett et al. 2007; Kumaran et al. 2008a; 2008b; Silvaggi et al. 2008; Zuniga et al. 2008) provided the missing information. In particular, crystals of LC/A bound to two non-cleavable short SNAP-25 peptides, one comprising Gln-197-Arg-Ala-Thr-Lys-Met-202, whereas in the second the authentic P1 residue Gln-197 was replaced with arginine, allowed the precise definition of the pockets S1–S5′ (Kumaran et al. 2008a). Parallel mutational analyses of LC residues forming these S-pockets corroborated the contribution of individual LC residues to P1–P5′ substrate residue interaction (Ahmed et al. 2008; Chen et al. 2007). Corresponding information about interactions at the active site was compiled for LC/E, too. Here, a co-crystal structure of the non-cleavable SNAP-25 peptide, Arg-180-Ile-Met-Glu-183, that represents P1– P3′, was solved to delineate the LC residues of the S1–S3′ pockets (Agarwal and Swaminathan 2008). Earlier enzymatic characterization of LC/E S-pocket mutants provided results being compatible with the co-crystal data regarding the S1′ pocket and had led to the suggestion of the residues that form the S3 and S2 pockets (Chen and Barbieri 2007).

Remarkably, these structural data could subsequently be exploited to generate an LC/E mutant whose enzymatic activity was extended to include human SNAP-23 as substrate. Earlier mutational analyses had detected that the P2 residue (Asp-179) of SNAP-25 is crucial for cleavage by LC/E and postulated it forms a salt bridge with K224 of the LC/E S2 pocket (Vaidyanathan et al. 1999; Chen and Barbieri 2007). As human SNAP-23 differs in only six amino acids within the identified LC/E interacting segment of SNAP-25 (Met-167 to Asp-186), and Asp-179 is changed to the oppositely charged lysine (position 185 in SNAP-23), it was hypothesized that the resistance of human SNAP-23 was predominantly due to repulsion of the P2 residue Lys-185. In agreement with this idea, replacement of the S2 pocket residue K224 with aspartic acid led to an LC/E that exhibited an in vitro hydrolysis rate toward human SNAP-23 in the range of that for wild-type LC/E versus SNAP-25. K_M/k_{cat} was only ninefold less (Chen and Barbieri 2009). Furthermore, LC/E-K224D decreased IL-8 and mucin secretion in TGF-α stimulated permeabilized HeLa cells (Chen and Barbieri 2009) exemplifying the potential for its future medicinal applications.

Less structural information on substrate binding at the active site is available for VAMP cleaving LCs. The co-crystals of LC/F with its substrate mimicking inhibitors provide only information on probable pockets for P5 to P2 substrate residues (Agarwal et al. 2009). However, analyses of VAMP-2 and LC/F mutants revealed that the P5, P2, P1′, and P2′ substrate residues and the appropriate S-pockets mediate fine-tuning of scissile bond (Gln-58-Lys-59) alignment to the active center of LC/F [Fig. 2; (Schmidt and Stafford 2005; Chen and Wan 2011; Sikorra et al. 2008)].

Information on LC/D, LC/B, and LC/T is exclusively limited on mutational analysis of enzyme and/or substrate. These analyses suggested that VAMP P3, P1, P1′, or P4, P2, P1′, or P4, P3, P1′ P2′ are of particular importance for cleavage by LC/D, LC/B, and LC/T, respectively (Chen et al. 2008; Sikorra et al. 2008). Complementary data on several of the appropriate S-pockets of LC/B and LC/T have been published recently (Chen et al. 2012). No data exist on LC/G and LC/F5. Characterization of the former is hampered by the requirement of the substrate transmembrane domain in in vitro cleavage assays (Sikorra and Binz, unpublished observation), whereas the scissile bond for the latter was only recently identified.

4.2 Catalytic Mechanism of Peptide Bond Hydrolysis

Clostridial neurotoxin LCs act as Zn^{2+}-endoproteases. Their Zn^{2+}-coordination mode classifies them into the gluzincin superfamily among the clan MA of metalloproteases (Hooper 1994; Rawlings and Barrett 1995). The characteristic feature of this superfamily is the coordination of Zn^{2+} by two histidine residues of the

H-E-X-X-H motif, a water molecule, which is bonded to the glutamate residue of the motif, and a second glutamic acid being located about 35 residues toward the C-terminus. Searches for proteins exhibiting a related tertiary structure revealed that the central portion is structurally similar to a portion of the zinc metalloprotease thermolysin (Lacy et al. 1998). These similarities are limited to the helix containing the H-E-X-X-H motif and a four-stranded β-sheet buttressing the helix. The loop regions connecting the conserved secondary structure elements of the catalytic core are drastically expanded compared to those of thermolysin and the CNT catalytic core is not involved in substrate binding via side-chain side-chain interactions. Substrate specificity is thus imparted by structural elements flanking the core. These contain the substrate-binding sites and appear to be structurally unique in CNTs (Breidenbach and Brunger 2005).

Thermolysin represents the well-characterized prototypical member of the gluzincin metalloprotease superfamily. Its mode of peptide bond cleavage is presumably to a large extent similar to that of CNTs. A central point is that peptide inhibitors were found to bind to thermolysin with the same directionality as SNAP-25 and VAMP-derived peptides do at LC/A, LC/E, or LC/F, respectively. In thermolysin, catalysis follows a general base-type mechanism (Hangauer et al. 1984). It was proposed that the water molecule is polarized via its interaction with Zn^{2+} and the glutamic acid carboxylate group of the H-E-X-X-H-motif and thus its oxygen becomes capable of nucleophilically attacking the carbonyl carbon of the scissile peptide bond. An oxyanion develops. Simultaneously, a proton abstracted from the attacking water is shuttled via the carboxyl group of the glutamate to the scissile peptide bond nitrogen. The glutamate probably then stabilizes the tetrahedral intermediate by forming a salt bridge with the positively charged amide nitrogen. The negative charge of the oxyanion is stabilized in the tetrahedral transition state by hydrogen bond interactions with a protonated histidine (position 231) and the hydroxyl group of a tyrosine (position 157). It is assumed that H231 is retained in proper position and protonated state through a hydrogen bonding interaction with an aspartate (position 226). The protonated amide nitrogen finally facilitates C-N-bond disruption and may receive the second proton of the water molecule. This is possibly again mediated by the glutamic acid carboxylate group (Hangauer et al. 1984; Matthews 1988).

The available data on CNT LCs support a very similar mechanism (Fig. 4). In fact, the nucleophilic attack of the scissile bond carbonyl carbon by the coordinated polarized water molecule as well as the subsequent transfer of a proton from the attacking water to the scissile peptide bond nitrogen may be ascribed to the action of the glutamate of the H-E-X-X-H motif as its mutation leads to complete deactivation of the catalytic activity of, e.g., BoNT/A (Li et al. 2000) and BoNT/E (Agarwal et al. 2004).

Primary structure alignments of all eight CNTs revealed three conserved residues within the otherwise less conserved region around the zinc coordinating residues of the LCs, i.e., a tyrosine, an arginine, and a glutamate (Fig. 4). They reside $\sim$120 amino acids downstream of the H-E-X-X-H motif and are separated from each other by 2 and 11 amino acids, respectively. The crystal structures of

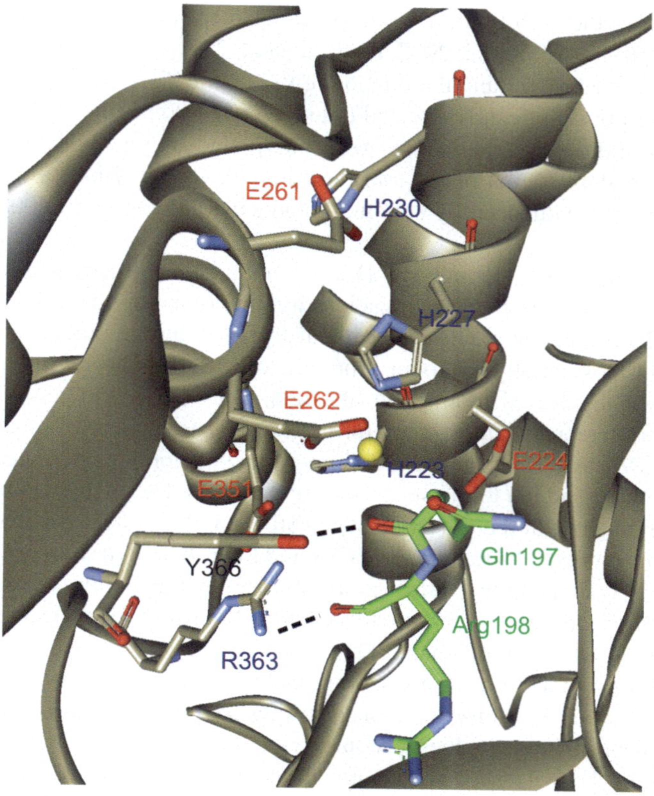

Fig. 4 Arrangement of CNT LC active site amino acids involved in peptide bond hydrolysis. Close-up view of the active center of LC/A bound to the SNAP-25-derived peptide QRATKM (pdg entry: 3DDA). LC/A is shown as a *olive ribbon* diagram. Crucial residues are depicted in stick representation. Only the P1 and P1′ residues (Gln197-Arg198) of the SNAP-25-derived hexapeptide are shown in *green stick* representation. H223, H227, and E262 coordinate the zinc ion, which is shown as a *yellow sphere*. A water molecule (not shown) held in place by E224 is the fourth zinc ligand. This water molecule nucleophilically attacks the carbonyl carbon of the scissile peptide bond. The developing oxyanion becomes stabilized through H-bond formation with Y366, and the guanidinium group of R363 forms an H-bond with the carbonyl oxygen of the P1′ amino acid. E261 and E351 interact with H227 and H230, or R363 and H223, respectively. These interactions likely stabilize the side-chain conformations of H223 and H227 allowing them to be properly oriented for zinc coordination. All residues are strictly conserved among TeNT and all BoNT serotypes and subtypes except that the counterpart of LC/A E261 is replaced in LC/C with alanine

LC/A and LC/E bound to non-cleavable short SNAP-25 peptides comprising the respective scissile bonds showed that this tyrosine (position 366 and 351 in LC/A and LC/E, respectively) lines up in close proximity to the scissile peptide bond (Kumaran et al. 2008; Agarwal and Swaminathan 2008) and might therefore be the functionally equivalent residue to Y157 of thermolysin. Its central role in substrate cleavage was demonstrated for TeNT (position 375) and BoNT/A and E. In BoNT/

A and TeNT the substitution with phenylalanine drastically decreased the hydrolytic activity (Binz et al. 2002; Rossetto et al. 2001). Kinetic analyses for LC/A-Y366F revealed a substantial reduction of k_{cat}, while Zn^{2+}-binding and substrate binding were not significantly affected (Ahmed et al. 2008; Binz et al. 2002). This finding and additional calculations were interpreted as an involvement of this amino acid in stabilization of the transition state oxyanion (Binz et al. 2002) and was later on also evidenced by structural studies (Kumaran et al. 2008; Silvaggi et al. 2008).

The counterpart of thermolysin H-231 might be the second of the above-mentioned strictly conserved amino acids, arginine (positions 363 and 348 in LC/A and LC/E, respectively). Its replacement in LC/A by various residues and by alanine in LC/E generally led to an approxmately 100-fold decrease in k_{cat}, while K_M remained unaffected (Ahmed et al. 2008; Binz et al. 2002; Agarwal et al. 2005). The difference in free energy for transition state binding of mutated versus wild-type LC/A as well as the guanidinium group positioning in the vicinity of the scissile peptide bond led to the suggestion that this residue might also be involved in oxyanion binding in the transition state (Silvaggi et al. 2008; Binz et al. 2002). However, the co-crystal structures of LC/A and LC/E bound to short non-cleavable SNAP-25 derived peptides rather propose H-bond interaction of the arginine guanidinium group and the carbonyl oxygen of the P1' position (Kumaran et al. 2008; Agarwal and Swaminathan 2008).

CNT LCs presumably also possess an analog to D226 of thermolysin. A glutamate, the third of the above-mentioned strictly conserved residues in the vicinity of the zinc coordinating residues, occupies a similar position in all CNTs (Jin et al. 2007). Removal of its negative charge dramatically diminished the hydrolytic activity of BoNT/A and E (Binz et al. 2002; Agarwal et al. 2005). In addition to keeping the guanidinium group of the conserved arginine in proper position and in protonated state, at variance with thermolysin, it appears to be important for orienting one of the Zn^{2+} coordinating histidines at the active site (Binz et al. 2002; Agarwal et al. 2005).

Knowledge of the catalytic mechanism is being exploited for the development of save vaccines, by recombinantly producing BoNTs which are rendered non-toxic by mutations of crucial amino acids of the catalytic center (Pier et al. 2008; Webb et al. 2009). Such preparations exhibit fewer side effects compared to chemically inactivated BoNT preparations.

5 Future Perspectives

BoNT/A and B are approved therapeutics for the treatment of a variety of clinical conditions relying on hyperactive neurons that innervate skeletal muscles. The range of their application has steadily increased during the past 20 years and includes today neurons that innervate smooth muscles and glands like in overactive bladder and axillary hyperhidrosis.

Further expansion of their applications is limited, as they act exclusively in neurons and preferentially cleave neuronal SNAREs. Consequently, retargeting BoNT activity to specific neurons or non-neuronal cells by replacing their cell binding subunit is one current field of scientific activity. Pioneer studies already provided proof of principle that BoNT LCs can be harnessed to interfere with therapeutically relevant secretion of non-neuronal cells (Chaddock 2012). However, exocytosis in potential target cells may be driven by non-substrate SNAREs, like syntaxin-4, SNAP-23, and TI-VAMP/VAMP-7 in human mastocytes (Chaineau et al. 2009). Hence, LCs have to be reengineered in order to be able to attack the desired non-cleavable SNARE. Based on the available detailed knowledge about LC-substrate interactions this is reasonable and has exemplarily been already achieved for LC/E. As described above (Sect. 4.1) LC/E could be endowed with the ability to cleave human SNAP-23 by a single amino acid substitution (Chen and Barbieri 2009). Similarly, based on the LC/A-SNAP-25 co-crystal structure an LC/C variant was engineered that retained its activity against syntaxin but lost it against SNAP-25 (Wang et al. 2011).

Further properties that are presumably amenable to targeted modifications comprise the catalytic efficiency and the persistency of LC activity in target cells (Shoemaker and Oyler 2012). Increased enzymatic activity and persistency could reduce toxin doses and intervals of injections, respectively, and thereby minimize the risk of triggering antibody response in patients. Achievement of these goals clearly requires further basic research efforts. Systematic biochemical characterization of the growing number of BoNT LC subtypes whose sequences are now being determined (Hill and Smith 2012) might add to this. It might quarry LCs that exhibit novel characteristics in terms of substrate specificity, enzymatic activity, intracellular longevity, etc. The so far conducted work has led to the identification of a novel cleavage site for LC/F5 in VAMP-2 [see Sect. 2; Fig. 1; (Kalb et al. 2012)], different VAMP-2 recognition requirements of BoNT/F7 (produced by *C. baratii*) (Kalb et al. 2011), and revealed different catalytic potency for LC/A3 and LC/A4 versus LC/A1, which shares 82 and 89 %, respectively, sequence identity with them (Henkel et al. 2009). Anyway, careful characterization of new subtypes will undoubtedly add to a comprehensive knowledge about the mode of CNT LC activities.

Acknowledgments I thank Andreas Rummel for critically reviewing the manuscript. This work was supported by grant BI 660/3-1from the Deutsche Forschungsgemeinschaft to T. B.

References

Agarwal R, Swaminathan S (2008) SNAP-25 substrate peptide (residues 180–183) binds to but bypasses cleavage by catalytically active Clostridium botulinum neurotoxin E. J Biol Chem 283:25944–25951

Agarwal R, Eswaramoorthy S, Kumaran D et al (2004) Structural analysis of botulinum neurotoxin type E catalytic domain and its mutant Glu212→Gln reveals the pivotal role of the Glu212 carboxylate in the catalytic pathway. Biochemistry 43:6637–6644

Agarwal R, Binz T, Swaminathan S (2005) Analysis of active site residues of botulinum neurotoxin E by mutational, functional, and structural studies: Glu335Gln is an apoenzyme. Biochemistry 44:8291–8302

Agarwal R, Schmidt JJ, Stafford RG et al (2009) Mode of VAMP substrate recognition and inhibition of Clostridium botulinum neurotoxin F. Nat Struct Mol Biol 16:789–794

Ahmed SA, Olson MA, Ludivico ML et al (2008) Identification of residues surrounding the active site of type A botulinum neurotoxin important for substrate recognition and catalytic activity. Protein J 27:151–162

Ahnert-Hilger G (2012) Synaptic vesicle proteins: targets and routes for botulinum neurotoxins. doi:10.1007/978-3-642-33570-9_8

Baumert M, Maycox PR, Navone F et al (1989) Synaptobrevin: an integral membrane protein of 18,000 daltons present in small synaptic vesicles of rat brain. EMBO J 8:379–384

Bennett MK, Calakos N, Scheller RH (1992) Syntaxin: a synaptic protein implicated in docking of synaptic vesicles at presynaptic active zones. Science 257:255–259

Bigalke H (2012) Botulinum toxin: application, safety, and limitations doi:10.1007/978-3-642-33570-9_14

Binz T, Kurazono H, Wille M et al (1990) The complete sequence of botulinum neurotoxin type A and comparison with other clostridial neurotoxins. J Biol Chem 265:9153–9158

Binz T, Blasi J, Yamasaki S et al (1994) Proteolysis of SNAP-25 by types E and A botulinal neurotoxins. J Biol Chem 269:1617–1620

Binz T, Bade S, Rummel A et al (2002) Arg(362) and Tyr(365) of the botulinum neurotoxin type A light chain are involved in transition state stabilization. Biochemistry 41:1717–1723

Blasi J, Chapman ER, Link E et al (1993a) Botulinum neurotoxin A selectively cleaves the synaptic protein SNAP-25. Nature 365:160–163

Blasi J, Chapman ER, Yamasaki S et al (1993b) Botulinum neurotoxin C1 blocks neurotransmitter release by means of cleaving HPC-1/syntaxin. EMBO J 12:4821–4828

Breidenbach MA, Brunger AT (2004) Substrate recognition strategy for botulinum neurotoxin serotype A. Nature 432:925–929

Breidenbach MA, Brunger AT (2005) 2.3 A crystal structure of tetanus neurotoxin light chain. Biochemistry 44:7450–7457

Burnett JC, Ruthel G, Stegmann CM et al (2007) Inhibition of metalloprotease botulinum serotype A from a pseudo-peptide binding mode to a small molecule that is active in primary neurons. J Biol Chem 282:5004–5014

Chaddock J (2012) Transforming the domain structure of botulinum neurotoxins into novel therapeutics. doi:10.1007/978-3-642-33570-9_13

Chaineau M, Danglot L, Galli T (2009) Multiple roles of the vesicular-SNARE TI-VAMP in post-Golgi and endosomal trafficking. FEBS Lett 583:3817–3826

Chen S, Barbieri JT (2006) Unique substrate recognition by botulinum neurotoxins serotypes A and E. J Biol Chem 281:10906–10911

Chen S, Barbieri JT (2007) Multi-pocket recognition of SNAP-25 by botulinum neurotoxin serotype E. J Biol Chem 282:25540–25547

Chen S, Barbieri JT (2009) Engineering botulinum neurotoxin to extend therapeutic intervention. Proc Natl Acad Sci U S A 106:9180–9184

Chen S, Wan HY (2011) Molecular mechanisms of substrate recognition and specificity of botulinum neurotoxin serotype F. Biochem J 433:277–284

Chen S, Kim JJ, Barbieri JT (2007) Mechanism of substrate recognition by botulinum neurotoxin serotype A. J Biol Chem 282:9621–9627

Chen S, Hall C, Barbieri JT (2008) Substrate recognition of VAMP-2 by botulinum neurotoxin B and tetanus neurotoxin. J Biol Chem 283:21153–21159

Chen S, Karalewitz AP, Barbieri JT (2012) Insights into the different catalytic activities of Clostridium neurotoxins. Biochemistry 51:3941–3947

Cornille F, Martin L, Lenoir C et al (1997) Cooperative exosite-dependent cleavage of synaptobrevin by tetanus toxin light chain. J Biol Chem 272:3459–3464

Davletov B, Bajohrs M, Binz T (2005) Beyond BOTOX: advantages and limitations of individual botulinum neurotoxins. Trends Neurosci 28:446–452

Evans ER, Sutton JM, Gravett A et al (2005) Analysis of the substrate recognition domain determinants of Botulinum type B toxin using phage display. Toxicon 46:446–453

Fang H, Luo W, Henkel J et al (2006) A yeast assay probes the interaction between botulinum neurotoxin serotype B and its SNARE substrate. Proc Natl Acad Sci U S A 103:6958–6963

Fischer A, Montal M (2007) Single molecule detection of intermediates during botulinum neurotoxin translocation across membranes. Proc Natl Acad Sci U S A 104:10447–10452

Fischer A (2012) Synchronized chaperone function of botulinum neurotoxin domains mediates light chain translocation into neurons. doi:10.1007/978-3-642-33570-9_6

Foran P, Shone CC, Dolly JO (1994) Differences in the protease activities of tetanus and botulinum B toxins revealed by the cleavage of vesicle-associated membrane protein and various sized fragments. Biochemistry 33:15365–15374

Foran P, Lawrence GW, Shone CC et al (1996) Botulinum neurotoxin C1 cleaves both syntaxin and SNAP-25 in intact and permeabilized chromaffin cells: correlation with its blockade of catecholamine release. Biochemistry 35:2630–2636

Fu Z, Chen S, Baldwin MR et al (2006) Light chain of botulinum neurotoxin serotype A: structural resolution of a catalytic intermediate. Biochemistry 45:8903–8911

Gill DM (1982) Bacterial toxins: a table of lethal amounts. Microbiol Rev 46:86–94

Hangauer DG, Monzingo AF, Matthews BW (1984) An interactive computer graphics study of thermolysin-catalyzed peptide cleavage and inhibition by N-carboxymethyl dipeptides. Biochemistry 23:5730–5741

Hayashi T, McMahon H, Yamasaki S et al (1994) Synaptic vesicle membrane fusion complex: action of clostridial neurotoxins on assembly. EMBO J 13:5051–5061

Hill KK , Smith TJ (2012) Genetic diversity within clostridium botulinum serotypes, botulinum neurotoxin gene clusters and toxin subtypes. doi:10.1007/978-3-642-33570-9_1

Henkel JS, Jacobson M, Tepp W et al (2009) Catalytic properties of botulinum neurotoxin subtypes A3 and A4. Biochemistry 48:2522–2528

Hooper NM (1994) Families of zinc metalloproteases. FEBS Lett 354:1–6

Hua SY, Charlton MP (1999) Activity-dependent changes in partial VAMP complexes during neurotransmitter release. Nat Neurosci 2:1078–1083

Inoue A, Obata K, Akagawa K (1992) Cloning and sequence analysis of cDNA for a neuronal cell membrane antigen, HPC-1. J Biol Chem 267:10613–10619

Jahn R, Scheller RH (2006) SNAREs—engines for membrane fusion. Nat Rev Mol Cell Biol 7:631–643

Jin R, Sikorra S, Stegmann CM et al (2007) Structural and biochemical studies of botulinum neurotoxin serotype C1 light chain protease: implications for dual substrate specificity. Biochemistry 46:10685–10693

Jongeneel CV, Bouvier J, Bairoch A (1989) A unique signature identifies a family of zinc-dependent metallopeptidases. FEBS Lett 242:211–214

Kalb SR, Baudys J, Egan C et al (2011) Different substrate recognition requirements for cleavage of synaptobrevin-2 by Clostridium baratii and Clostridium botulinum type F neurotoxins. Appl Environ Microbiol 77:1301–1308

Kalb SR, Baudys J, Webb RP et al (2012) Discovery of a novel enzymatic cleavage site for botulinum neurotoxin F5. FEBS Lett 586:109–115

Koriazova LK, Montal M (2003) Translocation of botulinum neurotoxin light chain protease through the heavy chain channel. Nat Struct Biol 10:13–18

Kumaran D, Rawat R, Ahmed SA et al (2008a) Substrate binding mode and its implication on drug design for botulinum neurotoxin A. PLoS Pathog 4:e1000165

Kumaran D, Rawat R, Ludivico ML et al (2008b) Structure- and substrate-based inhibitor design for Clostridium botulinum neurotoxin serotype A. J Biol Chem 283:18883–18891

Lacy DB, Tepp W, Cohen AC et al (1998) Crystal structure of botulinum neurotoxin type A and implications for toxicity. Nat Struct Biol 5:898–902

Li L, Binz T, Niemann H et al (2000) Probing the mechanistic role of glutamate residue in the zinc-binding motif of type A botulinum neurotoxin light chain. Biochemistry 39:2399–2405
Matthews BW (1988) Structural basis of the action of thermolysin and related zinc peptidases. Acc Chem Res 21:333–340
McMahon HT, Ushkaryov YA, Edelmann L et al (1993) Cellubrevin is a ubiquitous tetanus-toxin substrate homologous to a putative synaptic vesicle fusion protein. Nature 364:346–349
Osen-Sand A, Staple JK, Naldi E et al (1996) Common and distinct fusion proteins in axonal growth and transmitter release. J Comp Neurol 367:222–234
Oyler GA, Higgins GA, Hart RA et al (1989) The identification of a novel synaptosomal-associated protein, SNAP-25, differentially expressed by neuronal subpopulations. J Cell Biol 109:3039–3052
Pellegrini LL, O'Connor V, Betz H (1994) Fusion complex formation protects synaptobrevin against proteolysis by tetanus toxin light chain. FEBS Lett 353:319–323
Pellizzari R, Rossetto O, Lozzi L et al (1996) Structural determinants of the specificity for synaptic vesicle-associated membrane protein/synaptobrevin of tetanus and botulinum type B and G neurotoxins. J Biol Chem 271:20353–20358
Pellizzari R, Mason S, Shone CC et al (1997) The interaction of synaptic vesicle-associated membrane protein/synaptobrevin with botulinum neurotoxins D and F. FEBS Lett 409:339–342
Pier CL, Tepp WH, Bradshaw M et al (2008) Recombinant holotoxoid vaccine against botulism. Infect Immun 76:437–442
Raphael BH, Choudoir MJ, Luquez C et al (2010) Sequence diversity of genes encoding botulinum neurotoxin type F. Appl Environ Microbiol 76:4805–4812
Rawlings ND, Barrett AJ (1995) Evolutionary families of metallopeptidases. Methods Enzymol 248:183–228
Rossetto O, Caccin P, Rigoni M et al (2001) Active-site mutagenesis of tetanus neurotoxin implicates TYR-375 and GLU-271 in metalloproteolytic activity. Toxicon 39:1151–1159
Rummel A (2012) Double receptor anchorage of botulinum neurotoxins accounts for their exquisite neurospecificity. doi:10.1007/978-3-642-33570-9_4
Schechter I, Berger A (1967) On the size of the active site in proteases. I. Papain. Biochem Biophys Res Commun 27:157–162
Schiavo G, Rossetto O, Santucci A et al (1992a) Botulinum neurotoxins are zinc proteins. J Biol Chem 267:23479–23483
Schiavo G, Poulain B, Rossetto O et al (1992b) Tetanus toxin is a zinc protein and its inhibition of neurotransmitter release and protease activity depend on zinc. EMBO J 11:3577–3583
Schiavo G, Benfenati F, Poulain B et al (1992c) Tetanus and botulinum-B neurotoxins block neurotransmitter release by proteolytic cleavage of synaptobrevin. Nature 359:832–835
Schiavo G, Rossetto O, Catsicas S et al (1993a) Identification of the nerve terminal targets of botulinum neurotoxin serotypes A, D, and E. J Biol Chem 268:23784–23787
Schiavo G, Shone CC, Rossetto O et al (1993b) Botulinum neurotoxin serotype F is a zinc endopeptidase specific for VAMP/synaptobrevin. J Biol Chem 268:11516–11519
Schiavo G, Santucci A, Dasgupta BR et al (1993c) Botulinum neurotoxins serotypes A and E cleave SNAP-25 at distinct COOH-terminal peptide bonds. FEBS Lett 335:99–103
Schiavo G, Shone CC, Bennett MK et al (1995) Botulinum neurotoxin type C cleaves a single Lys-Ala bond within the carboxyl-terminal region of syntaxins. J Biol Chem 270:10566–10570
Schiavo G, Matteoli M, Montecucco C (2000) Neurotoxins affecting neuroexocytosis. Physiol Rev 80:717–766
Schiavo G, Bercsenyi K, Giribaldi F (2012) The Elusive Compass of Clostridial Neurotoxins: deciding when and where to go? doi:10.1007/978-3-642-33570-9_5
Schmidt JJ, Bostian KA (1997) Endoproteinase activity of type A botulinum neurotoxin: substrate requirements and activation by serum albumin. J Protein Chem 16:19–26
Schmidt JJ, Stafford RG (2005) Botulinum neurotoxin serotype F: identification of substrate recognition requirements and development of inhibitors with low nanomolar affinity. Biochemistry 44:4067–4073

Shoemaker CP, Oyler GA (2012) Persistence of botulinum neurotoxin inactivation of nerve function. doi:10.1007/978-3-642-33570-9_9

Shone CC, Roberts AK (1994) Peptide substrate specificity and properties of the zinc-endopeptidase activity of botulinum type B neurotoxin. Eur J Biochem 225:263–270

Shone CC, Quinn CP, Wait R et al (1993) Proteolytic cleavage of synthetic fragments of vesicle-associated membrane protein, isoform-2 by botulinum type B neurotoxin. Eur J Biochem 217:965–971

Sikorra S, Henke T, Swaminathan S et al (2006) Identification of the amino acid residues rendering TI-VAMP insensitive toward botulinum neurotoxin B. J Mol Biol 357:574–582

Sikorra S, Henke T, Galli T et al (2008) Substrate recognition mechanism of VAMP/synaptobrevin-cleaving clostridial neurotoxins. J Biol Chem 283:21145–21152

Silvaggi NR, Boldt GE, Hixon MS et al (2007) Structures of Clostridium botulinum neurotoxin serotype A light chain complexed with small-molecule inhibitors highlight active-site flexibility. Chem Biol 14:533–542

Silvaggi NR, Wilson D, Tzipori S et al (2008) Catalytic features of the botulinum neurotoxin A light chain revealed by high resolution structure of an inhibitory peptide complex. Biochemistry 47:5736–5745

Sutton RB, Fasshauer D, Jahn R et al (1998) Crystal structure of a SNARE complex involved in synaptic exocytosis at 2.4 A resolution. Nature 395:347–353

Thompson DE, Brehm JK, Oultram JD et al (1990) The complete amino acid sequence of the Clostridium botulinum type A neurotoxin, deduced by nucleotide sequence analysis of the encoding gene. Eur J Biochem 189:73–81

Trimble WS, Cowan DM, Scheller RH (1988) VAMP-1: a synaptic vesicle-associated integral membrane protein. Proc Natl Acad Sci U S A 85:4538–4542

Vaidyanathan VV, Yoshino K, Jahnz M et al (1999) Proteolysis of SNAP-25 isoforms by botulinum neurotoxin types A, C, and E: domains and amino acid residues controlling the formation of enzyme-substrate complexes and cleavage. J Neurochem 72:327–337

Wang D, Zhang Z, Dong M et al (2011) Syntaxin requirement for Ca2+-triggered exocytosis in neurons and endocrine cells demonstrated with an engineered neurotoxin. Biochemistry 50:2711–2713

Washbourne P, Pellizzari R, Baldini G et al (1997) Botulinum neurotoxin types A and E require the SNARE motif in SNAP-25 for proteolysis. FEBS Lett 418:1–5

Webb RP, Smith TJ, Wright P et al (2009) Production of catalytically inactive BoNT/A1 holoprotein and comparison with BoNT/A1 subunit vaccines against toxin subtypes A1, A2, and A3. Vaccine 27:4490–4497

Wictome M, Rossetto O, Montecucco C et al (1996) Substrate residues N-terminal to the cleavage site of botulinum type B neurotoxin play a role in determining the specificity of its endopeptidase activity. FEBS Lett 386:133–136

Williamson LC, Halpern JL, Montecucco C et al (1996) Clostridial neurotoxins and substrate proteolysis in intact neurons: botulinum neurotoxin C acts on synaptosomal-associated protein of 25 kDa. J Biol Chem 271:7694–7699

Yamamoto H, Ida T, Tsutsuki H et al (2012) Specificity of botulinum protease for human VAMP family proteins. Microbiol Immunol 56:245–253

Yamasaki S, Binz T, Hayashi T et al (1994a) Botulinum neurotoxin type G proteolyses the Ala81-Ala82 bond of rat synaptobrevin 2. Biochem Biophys Res Commun 200:829–835

Yamasaki S, Baumeister A, Binz T et al (1994b) Cleavage of members of the synaptobrevin/VAMP family by types D and F botulinal neurotoxins and tetanus toxin. J Biol Chem 269:12764–12772

Zuniga JE, Schmidt JJ, Fenn T et al (2008) A potent peptidomimetic inhibitor of botulinum neurotoxin serotype A has a very different conformation than SNAP-25 substrate. Structure 16:1588–1597

Synaptic Vesicle Proteins: Targets and Routes for Botulinum Neurotoxins

Gudrun Ahnert-Hilger, Agnieszka Münster-Wandowski and Markus Höltje

Abstract Synaptic vesicles (SV) are key organelles of neuronal communication. SV are responsible for the storage of neurotransmitters, which are released by Ca^{2+}-dependent exocytosis. After release and interaction with postsynaptic receptors, transmitters rapidly diffuse out of the synaptic cleft and are sequestered by plasma membrane transporters (in some cases following enzymatic conversion). SVs undergo endocytosis and are refilled by specific vesicular transmitter transporters different in the various neuronal subtypes. Besides these differences, SVs in general are equipped with a remarkable common set of proteins. Botulinum neurotoxins (BoNTs) inhibit neurotransmitter release from almost all types of neurons by cleaving proteins required for membrane fusion localized either to SVs (synaptobrevin) or to the plasma membrane (SNAP-25 and syntaxin) depending on the BoNT serotype. To enter the neuronal cytoplasm, BoNTs specifically interact with the luminal domain of SV proteins (synaptotagmin or SV2, depending on serotype) transiently exposed during exocytotic membrane fusion and occurring in almost every neuron. Thus, the highly specific interaction with luminal domains of SV proteins commonly expressed on all SV types is one reason why BoNTs exhibit such a high neuronal specificity but attack almost every neuron type.

Keywords Synaptic vesicle (SV) · SV2 · Synaptotagmin (Syt) · Synaptobrevin/ VAMP · SNAP-25 · Syntaxin

---

G. Ahnert-Hilger (✉) · A. Münster-Wandowski · M. Höltje
AG Functional Cell Biology, Institute for Integrative Neuroanatomy,
Charité-Universitätsmedizin, Philippstr 12, 10115 Berlin, Germany
e-mail: gudrun.ahnert@charite.de

A. Rummel and T. Binz (eds.), *Botulinum Neurotoxins*, Current Topics in Microbiology and Immunology 364, DOI: 10.1007/978-3-642-33570-9_8,

Abbreviations

BoNTs	Botulinum neurotoxin
ClC3	Chloride exchanger 3
Habc	N-terminal a-helical domain of syntaxin
HC	Heavy chain
LC	Light chain
SNAP	Soluble N-ethylmaleimide-sensitive factor attachment protein
SNARE	SNAP receptor) complex
SNAP-25	Synaptosomal-associated protein of 25 kDa
SV	Synaptic vesicle
SV2	Synaptic vesicle glycoprotein 2
Syt	Synaptotagmin
TI-VAMP	Tetanus toxin insensitive VMAP
VAMP	Vesicle-associated membrane protein
V-ATPase	Vacuolar proton ATPase
VGLUT	Vesicular glutamate transporter
VGAT	Vesicular GABA transporter
VMAT	Vesicular monoamine transporter

Contents

1 Introduction 160
2 Protein Equipment of Synaptic Vesicles 161
  2.1 Common and Differential Protein Makeup of SV 162
  2.2 Vesicular Transmitter Transporter and SV2 168
3 SNARE Proteins: Targets of BoNTs 170
4 Synaptotagmin and SV2: Routes for BoNTs into Nerve Terminals 171
5 Variations of Vesicular Properties: Modulation of BoNT Intoxication? 172
6 Concluding Remarks 172
References 173

1 Introduction

Communication between neurons in the central nervous system and between neurons and muscles in the periphery mainly occurs at specialized structures, the synapses. Functional variations at these sites by either modulating the postsynaptic or the presynaptic function will change input and output at synapses. In a physiological context, this describes synaptic plasticity. Damage of pre- and post-synaptic structures and functions is the initial cause of a variety of diseases affecting the central nervous system. The postsynaptic answer mainly depends on the postsynaptic receptors and ion channels and, presynaptically, on the availability and fusion competence of synaptic vesicles (SV). Furthermore, the amount of neurotransmitter stored in an individual vesicle contributes in determining the synaptic strength. Botulinum neurotoxins (BoNTs) serotypes A–G and tetanus neurotoxin produced by six *Clostridium* species specifically and exclusively attack presynaptic nerve terminals. BoNTs consist of a heavy chain (HC) responsible for internalization and a light chain (LC) that harbors protease activity. Intoxication exerts a long-lasting block of neurotransmitter release and therefore prevents communication between the pre- and the post-synaptic site. This review will summarize the abundance, structure, interaction, and function of synaptic proteins that play a role in the pathophysiology of the different BoNTs.

2 Protein Equipment of Synaptic Vesicles

SV are key organelles of neuronal communication. They are found at the end of neuronal axons in presynaptic terminals. Their main function is to concentrate, store, and release neurotransmitters into the synaptic cleft. SV undergo repeated rounds of exo- and endocytosis. From the cytoplasmic space, SV first have to dock, followed by a priming step that renders them ready for fusion. Finally, mediated by an increase of the cytoplasmic free Ca^{2+} concentration, SV fuse with the presynaptic membrane and release their transmitter content into the synaptic cleft (for review see (Südhof 2012)). Thus, vesicular exocytosis of transmitter is one of the most important steps for neuronal communication. The specific protein equipment ensures the high reliability of these processes spatiotemporally. Due to their rather uniform biophysical properties, SV can be highly purified. This has opened the possibility for proteome analysis of SV from brain by mass spectrometry. The outcome was a rather complex assembly of proteins, most of them involved in membrane trafficking (Takamori et al. 2006).

The majority of these proteins appears to be common to most SV populations irrespective of the type of neuron they are hosted. In addition to the brain, this may also apply to the peripheral nervous system at least for some of the major SV proteins. Besides these common proteins, subpopulations of SV differ in vesicular transmitter transporters, which are highly specific for the various transmitters occurring in the nervous system. The various BoNTs preferentially affect cholinergic neurons under pathophysiological conditions, but in general prevent neurotransmission from almost every type of neuron suggesting common targets present in the majority of neurons. Regarding the molecular mechanism of how BoNTs affect neurotransmission, two molecular events have to be distinguished. The first involves the traffic routes by which the high molecular weight toxins reach their intracellular target. This includes the endocytic uptake, the translocation of the LC to the cytoplasm, and a cytosolic reduction of the S–S bridge linking the heavy and the light chain. While the proteins involved in endocytic uptake have been mostly identified, the further processing of the internalized toxin is still poorly understood. Second, the intracellular target molecules that include synaptic proteins necessary for vesicular exocytosis are proteolytically cleaved directly leading to an inhibition of neurotransmission. This review will mainly focus on vesicular and synaptic proteins being either targets or relevant for BoNT uptake.

2.1 Common and Differential Protein Makeup of SV

SV are equipped with a great variety of integral and associated proteins. More than 80 different integral membrane proteins were identified by proteomics. The integral membrane proteins may be grouped into those responsible for trafficking and those required for transmitter loading and its modulation.

One of the key events in vesicular exocytosis is the formation of the core or SNARE (soluble N-ethylmaleimide-sensitive factor attachment protein (SNAP) receptor) complex consisting of the vesicular protein synaptobrevin and the preferentially plasma membrane-localized proteins syntaxin and SNAP-25 (synaptosomal-associated protein of 25 kDa) (Sutton et al. 1998).

Synaptobrevin-2 or vesicle-associated membrane protein (VAMP) -2 is one of the most abundant proteins on SV but is also present on other secretory vesicles outside the central nervous system (see also Table 1). VAMPs cover a family of proteins which are characterized by a C-terminal integral membrane domain. The N-terminus (aa 1–90 or more depending on isoform and species) faces the cytosol and comprises a SNARE motif for the interaction with SNAP-25 and syntaxin. About 70 copies of VAMP-2 are present per SV. Besides VAMP-2, other family members occur on SV including synaptobrevin-1/VAMP-1, which is the second abundant VAMP isoform (Raptis et al. 2005; Schoch et al. 2001), VAMP-4, and VAMP-7/Ti-VAMP. According to a proteomic comparison of glutamatergic (VGLUT1) and GABAergic SV isolated from whole rat brain, VAMP-1, VAMP-2, and VAMP-7 appear to be more abundant on VGLUT1 compared to VGAT expressing SV (Gronborg et al. 2010). In cholinergic terminals, both VAMP-1 and VAMP-2 have been identified (see Table 1) All VAMPs are engaged in membrane fusion processes. VAMPs interact by their N-terminal chain with syntaxin and SNAP-25 family members to form the fusion core or SNARE complex as a prerequisite for SV exocytosis. The fundamental role of VAMP for vesicle fusion has been shown when analyzing VAMP-2 knockout animals. The mutants die at birth and lack almost completely the fast Ca^{2+}-dependent fusion. Other VAMP isoforms appear not to reconstitute for the lack of VAMP-2. Interestingly, the mutants show no morphological sign of neuronal degeneration at birth and the majority of other SV proteins was not changed compared to wild-type littermates suggesting that only at the end of prenatal development will VAMP-2 take over from other VAMP isoforms (Schoch et al. 2001). Besides its interaction with syntaxin and SNAP-25, VAMP-1 and VAMP-2 have been shown to form a transient complex with synaptophysin. Both interactions of VAMP with either synaptophysin or the SNARE proteins are mutually exclusive (Becher et al. 1999; Yelamanchili et al. 2005). Other members of the VAMP family are engaged in fusion processes outside the nervous system (see Table 1).

Table 1 Vesicle-associated membrane proteins (VAMP)

	Function	Distribution	Reference
Synaptobrevin-1/VAMP-1	v-SNARE essential for SV fusion	All types of central neurons but more abundant in spinal cord compared to VAMP-2; neuromuscular junction; brain nuclei, and motor neurons	(Raptis et al. 2005; Liu et al. 2011) 2011 (Aguado et al. 1999)
Synaptobrevin-2/VAMP-2	Most abundant v-SNARE essential for SV fusion and endocytosis	All types of central neurons; chromaffin cells; endocrine and exocrine pancreatic islet cells; mast cells; platelets; parietal cells; neuromuscular junction	(Schoch et al. 2001; Deak et al. 2004)
Cellubrevin/VAMP-3	v-SNARE essential for SV fusion	Neurons and nonneuronal cells	(Yang et al. 2001)
VAMP-4	v-SNARE also found on other organelles besides SV	Subset of central neurons secretory granules, SV	(Raingo et al. 2012)
VAMP-5	v-SNARE in skeletal muscle	Skeletal muscle	(Rose et al. 2009)
TI-VAMP /VAMP-7	v-SNARE nonsensitive to TeNT	Developing central neurons; adult all brain regions	(Coco et al. 1999; Danglot et al. 2012)
VAMP-8	v-SNARE	Endosome fusion	(Antonin et al. 2000)
Synaptotagmins			
Synaptotagmin-I	Ca^{2+} sensor of SV fusion	All types of central neurons; some sensory neurons; sympathetic peripheral neurons; PC12 cells; chromaffin cells; subpopulation of neuromuscular junction; autonomic and sensory neurons	(Aguado et al. 1999; Berton et al. 1997; Li et al. 1994; Pang et al. 2006; Perin et al. 1991; Xue et al. 2010)
Synaptotagmin-II	Ca^{2+} sensor of SV fusion	Preferential in spinal cord neurons; Motor neurons, and neuromuscular junction	(Aguado et al. 1999; Berton et al. 1997; Geppert et al. 1991; Pang et al. 2006)
Synaptotagmin-IV	Ca^{2+} sensor of SV fusion preferential during development	Subsets of brain and spinal cord neurons	(Berton et al. 1997)
Synaptotagmin-VII	Ca^{2+} sensor of SV fusion	Chromaffin cells pancreatic β-cells	(Gustavsson et al. 2008)

(continued)

Table 1 (continued)

	Function	Distribution	Reference
Synaptotagmin-IX	Ca^{2+} sensor of SV fusion	Preferential in neurons of the limbic system and the striatum	(Xu et al. 2007)
Synaptotagmin-V, -XII, and -XVII	Ca^{2+} sensor of SV fusion	SV with a preference for VGLUT1	(Gronborg et al. 2010)
SV2 proteins			
SV2A	Membrane glycoprotein 12 membrane spanning domains	All brain areas, less in spinal cord neuromuscular junction according to functional analysis	(Bajjalieh et al. 1993; Dong et al. 2006)
SV2B	Membrane glycoprotein 12 membrane spanning domains	All brain areas, less in spinal cord neuromuscular junction according to functional analysis	(Dong et al. 2006; Bajjalieh et al. 1993)
SV2C	Membrane glycoprotein 12 membrane spanning domains	More distinct distribution, preferential in basal ganglia neuromuscular junction according to functional analysis	(Dardou et al. 2011; Dong et al. 2006; Rummel et al. 2009)
SNAP-25 family members			
SNAP-25A	Membrane-associated protein, member of the SNARE complex	All brain areas before birth	(Bark et al. 2004)
SNAP-25B	Splice isoform of SNAP-25A	All brain areas and peripheral nervous system	(Bark et al. 2004)
SNAP-23	Membrane-associated protein, member of the SNARE complex resistant to BoNT	Cortex preferential GABAergic neurons	(Bragina et al. 2007)
SNAP-29	Membrane-associated protein, also other intracellular membranes, inhibits dissociation of SNARE complex	Ubiquitously, including neurons	(Pan et al. 2005)
SNAP-47	Membrane-associated protein, member of the SNARE complex, also other intracellular membranes besides SV, resistant to BoNT	All brain areas	(Holt et al. 2006)
Syntaxin			

(continued)

Table 1 (continued)

	Function	Distribution	Reference
Syntaxin-1A	Member of the SNARE complex	Ubiquitously expressed in brain preferentially present in sensory neurons and autonomic fibers	(Ruiz-Montasell et al. 1996); (Aguado et al. 1999; Bennett et al. 1993)
Syntaxin-1B	Member of the SNARE complex	Ubiquitously expressed in brain preferentially present in motor neurons	(Ruiz-Montasell et al. 1996); (Aguado et al. 1999)
Syntaxin-2 (Epimorphin)	Member of SNARE complex	Present in mesenchymal cells, fibroblasts, and myofibroblast adjacent to epithelia of lung, gut liver	(Wang et al. 2006)
Syntaxin-3, Syntaxin-4	Plasma membrane	Epithelial cells	(Watson and Pessin 2001; Karvar et al. 2005)
Syntaxin-5	Cis golgi	Heart, lung, liver, spleen, kidney	(Bennett et al. 1993; Watson and Pessin 2001)

The plasma membrane SNARE proteins syntaxin and SNAP-25 also reside on SV (Holt et al. 2006). Since SNARE complexes do not dissociate in their membrane bridging (trans-) configuration, every cycling of SV leaves some syntaxin and SNAP-25 in the vesicular membrane. Syntaxin-1 consists of two closely related isoforms A and B (over 80 % of sequence homology) which are the main isoforms involved in SV exocytosis (Bennett et al. 1993). Other members of the large syntaxin family are involved in all types of membrane fusion in different cells (see Table 1). Syntaxin is composed of an N-terminal α-helical domain (Habc), a C-terminal SNARE motif H3, followed by a transmembrane region. In an unbound stage, the Habc motif folds back to the SNARE motif keeping syntaxin in a closed conformation thereby preventing an interaction with SNAP-25 and VAMP. Before entering the SNARE complex, the Habc domain frees the SNARE motif and syntaxin is in an open conformation now available for an interaction with SNAP-25. The presynaptic matrix protein munc-18—a key modulator of exocytosis—binds to syntaxin in its closed conformation which is probably necessary during sorting to the plasma membrane, thereby preventing uncontrolled interactions with other SNARE proteins on ER or Golgi membranes (Christie et al. 2012). In addition, munc-18 binds to syntaxin engaged in the SNARE complex formation and this binding is mediated by the N-terminal N-peptide of syntaxin. Probably, munc-18 either promotes the syntaxin SNARE protein interaction by a controlled release of the bound Habc motif or stabilizes the interaction with the SNARE partners (Christie et al. 2012) for review see (Han et al. 2010). Expression of syntaxin-1A and 1B overlaps in most brain areas but in some parts of the central and peripheral nervous system, the isoforms are differentially distributed. In this respect, syntaxin-1B is mainly found in fibers from motor neurons whereas syntaxin-1A appears to be preferentially expressed in perivascular fibers belonging to the autonomic nervous system (Aguado et al. 1999). Deletion of syntaxin-1A has no severe impact on mice survival indicating that syntaxin-1B probably functionally replaces syntaxin-1A (Han et al. 2010; Gerber et al. 2008). When expressing syntaxin-1B in a predominantly locked open conformation, the resulting homozygous mutants developed lethal epileptic seizures and exhibited decreased syntaxin-1B levels (Gerber et al. 2008). These data suggest that syntaxin-1 by its isoforms may gradually interfere with the various steps involved in SV exocytosis.

SNAP-25 occurs in the two splice variants SNAP-25 A and B that differ in only nine amino acids out of the 39 residues encoded by the alternatively spliced exon. The expression of the splice variants appears to be switched from "A" to "B" after birth (Bark et al. 2004). SNAP-23 represents another isoform and is also present in nonneuronal cells (see Table 1). Members of the SNAP-25 family are stably associated with the plasma membrane by palmitoylation of four cysteine residues in between the two SNARE motifs. Deletion of SNAP-25A/B in mice does not prevent prenatal development but mutants die perinatally. SNAP-25 deletion mutants do not show stimulated transmitter release indicating the importance of SNAP-25 for neuronal communication (Delgado-Martinez et al. 2007; Washbourne et al. 2002).

All three SNARE proteins (VAMP, SNAP-25, and syntaxin) are involved in vesicular exocytosis and termed v-SNARE (VAMP) or t-SNAREs (syntaxin and

SNAP-25) signifying their location on vesicular or target plasma membrane. As can be deduced from the various deletion mutants, SNARE proteins are mandatory for stimulated vesicular release from neurons. They are also targets for the different serotypes of BoNT proteases (see below). SNARE complex formation involves the N-terminal cytoplasm facing part of VAMP, the C-terminal SNARE motif of syntaxin adjacent to the transmembrane domain, and the N- and C-terminal domains of SNAP-25. These four α-helical chains tether to a supercoiled string whose formation provides the energy to overcome the rejecting forces between the vesicular and the plasma membrane, allowing them to fuse (Sutton et al. 1998).

Besides the SNARE proteins, synaptotagmins (Syt), further SV proteins are involved in membrane fusion. Syt are the Ca^{2+} sensor of vesicle exocytosis. So far, 17 isoforms have been described with Syt-I and -II mainly responsible for neuronal SV exocytosis. Syt contain a single transmembrane domain and two C2 domains (C2A and C2B) that both bind Ca^{2+} with different affinity. In addition, Syt also bind to negatively charged membrane phospholipids like the highly abundant phosphatidylserine and phosphatidyl inositol 4,5-biphosphate, thereby accelerating membrane fusion by more than four orders of magnitude (Rhee et al. 2005; Vennekate et al. 2012). Syts have been shown to be essential for exocytotic membrane fusion (Chapman 2008; de Wit et al. 2009; Xue et al. 2010). Syt-I is a presynapse-specific isoform that regulates synaptic vesicle trafficking. Syt-I is upregulated during postnatal development mainly in the brain, while Syt-II appears to be dominantly expressed in spinal cord neurons probably including motor neurons (Berton et al. 1997). The various Syt isoforms exhibit a differential distribution. Syt-I is present in all types of brain neurons as well as in terminals of the autonomic nervous system. By contrast, Syt-II that is more expressed in the spinal cord preferentially occurs in α-motor neurons and neuromuscular endplates (Li et al. 1994; Pang et al. 2006). Other isoforms also confer Ca^{2+} sensitivity to exocytotic membrane fusion in nonneuronal cells (see Table 1). Besides Syt-I and -II, Syt-IV and -V are also highly expressed in brain. Expression of Syt-IV in neurons is regulated during development and is induced by neuronal activity (Berton et al. 1997; Ibata et al. 2002). Null mutants for Syt-IV exhibit deficits in motor coordination and hippocampus-dependent memory formation (Ferguson et al. 2000). Syt-IV is the first Syt isoform that has been found to exhibit random distribution in neurons not preferentially localized to SV release sites and has only 16 amino acids facing the vesicular lumen (Ibata et al. 2002). Whether SNARE complex formation primes vesicles for fusion or executes it is still a matter of debate. Recent data with Syt-I may shed some light on this question. In an elegant study using liposome fusion approaches, it was convincingly shown that Syt overcomes the electrostatic forces and bridges the vesicular and the plasma membrane. Following an action potential, the inflowing Ca^{2+} binds to Syt thereby changing its conformation and decreasing the space between the two membranes. Now the three SNARE proteins come close enough to tether into the SNARE complex and to execute the final fusion. In this scenario Syt acts upstream of core complex formation. This effect appears to be mediated by the C2B domain and is characteristic for Syt-I (van den Bogaart et al. 2011).

SV are equipped with a great variety of membrane proteins, many of them occurring in several copy numbers. Synaptophysin is the most abundant synaptic vesicle protein with respect to protein mass. It occurs in almost every type of neuron in the central and the peripheral nervous system. With about 30 copies per vesicle, it is only outnumbered by VAMP. Synaptophysin is a member of the tetraspan family including synaptoporin and synaptogyrin that all have four transmembrane domains and an N- and C-terminus facing the cytosol. Synaptophysin is X-chromosomal linked while the other tetraspan proteins are located on other genes. Despite these facts, little is known about its physiological functions. Synaptophysin knockout animals show no severe phenotypical changes. Synaptophysin is developmentally upregulated and the synaptophysin/VAMP interaction represents a hallmark of synaptic maturation (Becher et al. 1999). In addition, the synaptophysin/VAMP interaction appears to correlate with synaptic activity. It is increased after prolonged (Hinz et al. 2001) and spent during short-term stimulation (Reisinger et al. 2004). Synaptophysin appears to be relevant for some steps during endocytosis (Kwon and Chapman 2011; Spiwoks-Becker et al. 2001) and especially for VAMP retrieval following exocytotic membrane fusion (Gordon et al. 2011), but the slowed down endocytosis does not severely impair the animals. Thus, synaptophysin appears to modulate special, higher functions and consequences of synaptophysin deletion are only obvious when analyzing complex behavioral tasks (Schmitt et al. 2009).

All SV are equipped with an oligomeric vacuolar proton ATPase (V-ATPase) that is essential for proton pumping. The resulting acidification of the lumen of SV provides the energy required for transmitter loading. Surprisingly, the copy number per vesicle has been estimated to be between one and two, making it possible that an SV may also fail to incorporate a copy of this enzyme during recycling. Such an SV would remain silent but has the chance to get a copy of the proton pump during its next excursion to the plasma membrane (Takamori et al. 2006).

2.2 Vesicular Transmitter Transporter and SV2

Despite the rather uniform protein equipment of SV regulating their traffic, exocytosis and endocytic retrieval, SV differ in the transmitter phenotype and therefore in the transporters they use to fill SV with transmitters. BoNT preferentially target cholinergic neurons but prevent neurotransmission from almost every neuron irrespective of its transmitter phenotype with varying sensitivity (Foran et al. 2003). Classical (i.e., nonpeptide) neurotransmitters are synthesized in the synaptic cytoplasm and then loaded into SV by means of vesicular transporters that are driven by a proton electrochemical gradient across the vesicle membrane built up by a V-ATPase. It appears that the vesicular transporters, in conjunction with the biosynthetic enzymes and (at least in some cases) the plasma membrane transporters, are mainly responsible for determining the neurotransmitter that is released from a particular neuron. For the main transmitters involved in neuronal communication, vesicular transporters have been identified for glutamate (VGLUT, three isoforms), GABA and glycine (VGAT or VIAAT), acetylcholine (VAChT), and catecholamines and serotonin (VMAT, two isoforms) (Hnasko et al. 2010). VAChT, VMAT, and mammalian VGLUTs are characterized by 12 membrane domains. An exception is VGAT which has nine membrane domains and the C-terminus facing the lumen of the SV (Martens et al. 2008). Generally, transport involves the exchange of one or two protons per molecule of neurotransmitter depending on the transmitter transporter type.

SV usually contain only transporters for a single neurotransmitter, however, exceptions from this rule increase. Expression of VGLUT in GABAergic or aminergic neurons probably results in stimulus-dependent corelease of glutamate but may also improve loading of the "home" transmitter by increasing the electrochemical gradient due to the negatively charged glutamate that have to be compensated by additional protons. Examples of such scenarios are VGLUT3 on cholinergic vesicles (Gras et al. 2008; Holt et al. 2006), VGLUT2 (Zander et al. 2010) or VGLUT3 (Seal et al. 2008) on subpopulations of GABAergic vesicles, or VGLUT2 on dopaminergic (Hnasko et al. 2010) and VMAT2 vesicles (Zander et al. 2010). Other proteins such as ion channels or ion exchangers may be required in addition to the transporters in order to load vesicles efficiently with high concentrations of transmitter. In this context, the vesicle-associated chloride exchanger ClC3 appears to modulate transmitter loading by increasing ΔpH (Ahnert-Hilger and Jahn 2011; Riazanski et al. 2011; Stobrawa et al. 2001).

Although overlapping expression between the various vesicular transmitter transporters can no longer be totally excluded, immunoisolation techniques using transporter-specific antibodies allowed to purify SV subpopulation and to compare their protein equipment by mass spectroscopy. Surprisingly, the differences between VGLUT1 and VGAT SV subpopulations are small suggesting only subtle differences between SV subpopulations besides the transporter equipment. One of the main differences includes MAL, a tetraspan protein distantly related to synaptophysin which is only present on VGLUT1-positive SV (Gronborg et al. 2010).

SV glycoprotein 2 is characterized by a 12 membrane spanning structure reminiscent of sugar transporter proteins. However, so far no transporter function has been described. There are three isoforms SV2A, B, and C transcribed by different genes (Bajjalieh et al. 1992; Janz and Sudhof 1999). While SV2A is present in all types of neurons, SV2B and C have a more differentiated distribution (Bajjalieh et al. 1994; Janz and Sudhof 1999). Indeed, SV2B and SV2C appear to preferentially reside on VGLUT1 or VGAT SV, respectively (Gronborg et al. 2010) All SV2 proteins directly interact with Syt-I and appear to influence its expression, trafficking, or endocytosis (Yao et al. 2010). SV2A deletion mutants have severe seizures and animals die a few weeks after birth, while SV2B knockouts are phenotypically almost normal but exhibit reduced neurotransmission at rod photoreceptor synapses (Morgans et al. 2009). SV2A knockout neurons exhibit decreased transmitter release and the epileptic phenotype in SV2A mutants is probably due to the reduced GABA release (Chang and Sudhof 2009). Another feature of SV2A is its selective interaction with the antiepileptic drug levetiracetam not shared by the other SV2 isoforms (Lynch et al. 2004). Although looking like a transporter, current models argue for SV2 proteins being mainly involved in Syt homeostasis (Yao et al. 2010).

3 SNARE Proteins: Targets of BoNTs

Neuronal SNARE proteins are cleaved by the zinc-dependent endoprotease activity of the light chains of the various clostridial neurotoxin serotypes in a SNARE protein-specific way. BoNT/A and E cleave SNAP-25, BoNT/B, D, F, G, and TeNT the SV protein VAMP, and BoNT/C hydrolyzes the integral plasma membrane protein syntaxin and at higher concentrations of the LC also SNAP-25 (Schiavo et al. 2000, Binz 2012). Generally, neurotoxin-mediated cleavage sites for VAMP and syntaxin are located between the C-terminal membrane anchors and the ionic '0' layer. Only uncomplexed or partially assembled SNARE proteins can be proteolyzed by neurotoxins while the fully assembled SNARE complex is resistant to cleavage (Sutton et al. 1998). Cleavage of VAMP 2 by BoNT/G yields the shortest and by BoNT/F the longest C-terminal peptide. Also, BoNT/C1, A, and E cleave SNAP-25 at different sites (Binz 2012). Cleavage of either of these proteins prevents neurotransmission underscoring the importance of SNARE proteins for neuronal communication. All BoNT serotypes also affect human neurons (Humeau et al. 2000).

Besides the overall effects of BoNTs on Ca^{2+}-dependent neurotransmitter release, there are distinct kinetic differences indicating that it matters which of the SNARE proteins is cleaved and at what site. Although it may be difficult to analyze these differences at conventional synapses, large terminals with several release sites exhibiting larger pools of SV may be suitable to discriminate kinetic differences. The Calyx of Held is a large glutamatergic terminal in the auditory pathway which can be voltage-clamped and manipulated with respect to its intracellular Ca^{2+} concentration. This experimental arrangement allows a direct application of the light chains of clostridial neurotoxins. By this approach, the acute effects of various toxins can be studied and kinetic differences be worked out in a limited time window. It could be shown that cleavage of syntaxin by BoNT/C yields a complete block without changing the kinetics of the remaining SV and their release site so far not affected by the toxin. The same applied to VAMP cleaved by the LC of tetanus neurotoxin and probably also for the LC of BoNT/B which uses the identical cleavage site. Manipulating the intracellular Ca^{2+} concentrations by caged Ca^{2+} (thereby mimicking Ca^{2+} influx through channels), however, revealed that cleaved VAMP modifies the coupling between Ca^{2+} channels and release-competent vesicles. This may be either due to a loss of a postulated interaction between vesicles and the special release site or by vesicles which are delayed in endocytosis and therefore occlude release sites (Sakaba et al. 2005). By contrast, BoNT/A produced a strong reduction in the Ca^{2+} sensitivity of neurotransmitter release which can be overcome by increasing the presynaptic free Ca^{2+} concentration (Sakaba et al. 2005). Given the extension of medical applications for BoNTs, deeper insight into the differences of their effects on neurotransmission at individual synapses are helpful to shape their therapeutic profiles.

Generally, BoNTs affect all types of neurons with the highest preference for cholinergic ones. Regarding central excitatory and inhibitory neurons, it appears that especially both BoNT/A and BoNT/E differ in their potency in glutamatergic versus GABAergic neurons. The reason for this phenomenon may be the reduced or almost absent expression of SNAP-25 in GABAergic neurons, which is replaced by another member of the SNAP-25 family, i.e., SNAP-23 especially in the adult nervous system. Overexpression of SNAP-25 increases the sensitivity of GABAergic neurons for BoNT/A suggesting that resistance may be primarily caused by the target and not mediated by the protein receptor required for internalization (Matteoli et al. 2009; Verderio et al. 2006; Verderio et al. 2007). Expression of BoNT/A cleavage-resistant isoforms of the SNAP-25 family in these neurons may be an explanation. These include SNAP-23 (Ravichandran et al. 1996; Galli et al. 1998), SNAP-47 (Holt et al. 2006), or SNAP-29 (Schiavo et al. 2000; Steegmaier et al. 1998). Probably, resistance to BoNT/A relies more on the reduced amount of SNAP-25 which may be overcome by increased SNAP-25 expression (even in its BoNT/A cleaved from) as well as on the special Ca^{2+} dynamics in GABAergic terminals than on the presence of a different SNAP-25 isoform (Grumelli et al. 2010). In contrast to these reports, peripheral neurons which differ only in their amount of SNAP-25 exhibit an almost identical sensitivity toward BoNT/A treatment with respect to fast synaptic transmission (Gibbins et al. 2003). In addition, it is also conceivable that subpopulations of inhibitory neurons differ in their BoNT/A resistance either on the basis of SNAP-25 family member isoforms or on the equipment with gangliosides and vesicular proteins necessary for internalization (Rummel 2012).

4 Synaptotagmin and SV2: Routes for BoNTs into Nerve Terminals

Internalization of the various BoNTs involves binding and subsequent transfer into the neuron by the toxins' HC (Rummel 2012). A double receptor concept has been developed which applies to all BoNTs. It is worth notifying that the two closely interacting proteins, SV2 and Syt-I and -II, also represent the tracking molecules for almost all BoNTs. Thus, it may be speculated that the protein receptor for BoNT/C that escaped detection so far could also be SV2, Syt, or represent complexed heterodimers of these SV proteins (Yao et al. 2010).

5 Variations of Vesicular Properties: Modulation of BoNT Intoxication?

All BoNTs are taken up via SV that upon cycling transiently expose their luminal face to the extracellular surface allowing the heavy chains to bind via their C-terminal half to their protein receptor, thereby initiating the uptake of the whole toxin. Thus, variation in the equipment with these protein receptors or their interaction with other SV proteins may have an impact on the efficiency of BoNTs. The trapped BoNT forms a pore by its N-terminal part of the heavy chain, which allows the LC to enter the cytosol (Fischer 2012). This translocation step depends on acidic pH inside the vesicle and the preceding/progression of intoxination can be stopped by inhibiting the vacuolar proton pump, i.e., via bafilomycin. Immediately after endocytosis, the SV is empty and has an increased pH compared to a ready-to-fuse SV. The low H^+ concentration is then increased by the proton pump, and once the luminal pH is low enough it then allows the translocation of the LC into the cytosol. As mentioned above, the vacuolar proton pump resides with one or two copies on average per SV opening the possibility that proton pump-free SV will be endocytosed. In these, trapped BoNT will not have the chance to be translocated to the cytosol and eventually released again by another round of exocytosis or end up in the endosomal compartment. Given the variation between different neuronal subtypes, SV depending on their equipment with transmitter transporter and additional exchangers may differ in the building up and maintenance of a low luminal pH. Besides the putative absence of a proton pump, other examples of such scenario are the chloride exchangers or VGLUTs on GABAergic SV, which slightly change luminal pH and consequently transmitter storage. Depending on the changes in the luminal pH, the effects of BoNTs may be either slowed down or accelerated. Although these effects may be small, they may add to the differences observed for the different neuronal subtypes and their sensitivity toward the various BoNTs.

6 Concluding Remarks

All BoNTs are taken up by neurons. Their internalization is best described by a double-receptor concept that involves an initial binding to neuron-specific gangliosides. A second protein receptor appears during neuronal stimulation. SV fuse with the plasma membrane and transiently expose their intravesicular luminal site to the extracellular surface. These protein receptors include the SV proteins Syt and SV2. The acidification of SV mediated by the vacuolar proton pump and sustained by vesicular transporters and ion exchangers mediate translocation of the BoNT LC to the cytosol. The light chains of all BoNTs harbor a protease specific for one of the SNARE proteins. Cleavage of SNARE proteins prevents SV fusion with the plasma membrane and arrests neuronal communication (Fig. 1). This has

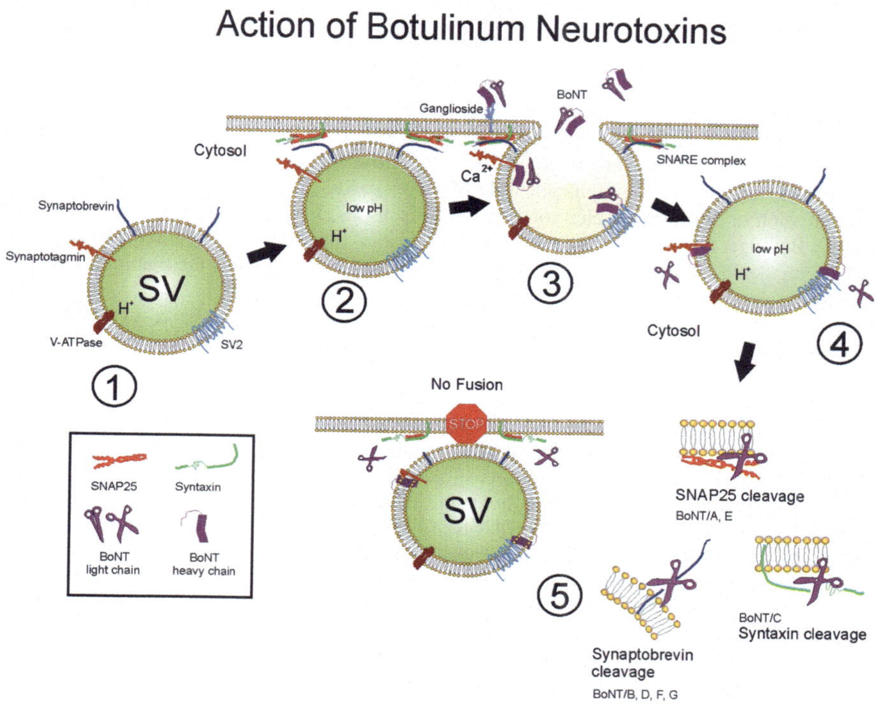

Fig. 1 Botulinum neurotoxins (BoNT) prevent exocytosis by cleaving synaptic proteins. Synaptic vesicles (SV) approach the synaptic membrane (*1*, *2*) and form the SNARE complex as a prerequisite for the fusion pore and exocytotic transmitter release (Geppert et al. 1991). During this step, BoNT bound to gangliosides at the extracellular side can be taken up into the vesicle by binding of the heavy chain to specific target proteins synaptotagmin or SV2. Following retrieval from the plasma membrane, recycling vesicles become acidified again by means of the vacuolar proton pump (V-ATPase). This results in incorporation of the N-terminus of the heavy chain into the vesicular membrane, partial unfolding, and translocation of the light chain into the cytosol (*4*) to target the specific synaptic proteins involved in the exocytotic machinery (*5*)

deleterious effects on the organism without severely damaging the toxin hosting neuron. Variable equipment of SV with protein isoforms responsible for either internalization or SNARE complex formation (see Table 1) may explain the differences between neuron subpopulations regarding variations in their sensitivity toward BoNT serotypes.

References

Aguado F, Majo G, Ruiz-Montasell B, Llorens J, Marsal J, Blasi J (1999) Syntaxin 1A and 1B display distinct distribution patterns in the rat peripheral nervous system. Neuroscience 88:437–446

Ahnert-Hilger G, Jahn R (2011) CLC-3 spices up GABAergic synaptic vesicles. Nat Neurosci 14:405–407

Antonin W, Holroyd C, Tikkanen R, Honing S, Jahn R (2000) The R-SNARE endobrevin/VAMP-8 mediates homotypic fusion of early endosomes and late endosomes. Mol Biol Cell 11:3289–3298

Bajjalieh SM, Peterson K, Shinghal R, Scheller RH (1992) SV2, a brain synaptic vesicle protein homologous to bacterial transporters. Science 257:1271–1273

Bajjalieh SM, Peterson K, Linial M, Scheller RH (1993) Brain contains two forms of synaptic vesicle protein 2. Proc Natl Acad Sci USA 90:2150–2154

Bajjalieh SM, Frantz GD, Weimann JM, McConnell SK, Scheller RH (1994) Differential expression of synaptic vesicle protein 2 (SV2) isoforms. J Neurosci 14:5223–5235

Bark C, Bellinger FP, Kaushal A, Mathews JR, Partridge LD, Wilson MC (2004) Developmentally regulated switch in alternatively spliced SNAP-25 isoforms alters facilitation of synaptic transmission. J Neurosci 24:8796–8805

Becher A, Drenckhahn A, Pahner I, Margittai M, Jahn R, Ahnert-Hilger G (1999) The synaptophysin-synaptobrevin complex: a hallmark of synaptic vesicle maturation. J Neurosci 19:1922–1931

Bennett MK, Garcia-Arraras JE, Elferink LA, Peterson K, Fleming AM, Hazuka CD, Scheller RH (1993) The syntaxin family of vesicular transport receptors. Cell 74:863–873

Berton F, Iborra C, Boudier JA, Seagar MJ, Marqueze B (1997) Developmental regulation of synaptotagmin I, II, III, and IV mRNAs in the rat CNS. J Neurosci 17:1206–1216

Binz T (2012) Clostridial neurotoxin light chains: devices for snare cleavage mediated blockade of neurotransmission. doi:10.1007/978-3-642-33570-9_7

Bragina L, Candiracci C, Barbaresi P, Giovedi S, Benfenati F, Conti F (2007) Heterogeneity of glutamatergic and GABAergic release machinery in cerebral cortex. Neuroscience 146: 1829–1840

Chang WP, Sudhof TC (2009) SV2 renders primed synaptic vesicles competent for Ca^{2+}-induced exocytosis. J Neurosci 29:883–897

Chapman ER (2008) How does synaptotagmin trigger neurotransmitter release? Annu Rev Biochem 77:615–641

Christie MP, Whitten AE, King GJ (2012) Low-resolution solution structures of Munc18: syntaxin protein complexes indicate an open binding mode driven by the Syntaxin N-peptide. Proc Natl Acad Sci USA 109:9816–9821

Coco S, Raposo G, Martinez S (1999) Subcellular localization of tetanus neurotoxin-insensitive vesicle-associated membrane protein (VAMP)/VAMP7 in neuronal cells: evidence for a novel membrane compartment. J Neurosci 19:9803–9812

Danglot L, Zylbersztejn K, Petkovic M (2012) Absence of TI-VAMP/Vamp7 leads to increased anxiety in mice. J Neurosci 32:1962–1968

Dardou D, Dassesse D, Cuvelier L, Deprez T, De Ryck M, Schiffmann SN (2011) Distribution of SV2C mRNA and protein expression in the mouse brain with a particular emphasis on the basal ganglia system. Brain Res 1367:130–145

de Wit H, Walter AM, Milosevic I, Gulyas-Kovacs A, Riedel D, Sorensen JB, Verhage M (2009) Synaptotagmin-1 docks secretory vesicles to syntaxin-1/SNAP-25 acceptor complexes. Cell 138:935–946

Deak F, Schoch S, Liu X, Sudhof TC, Kavalali ET (2004) Synaptobrevin is essential for fast synaptic-vesicle endocytosis. Nat Cell Biol 6:1102–1108

Delgado-Martinez I, Nehring RB, Sorensen JB (2007) Differential abilities of SNAP-25 homologs to support neuronal function. J Neurosci 27:9380–9391

Dong M, Yeh F, Tepp WH, Dean C, Johnson EA, Janz R, Chapman ER (2006) SV2 is the protein receptor for botulinum neurotoxin A. Science 312:592–596

Ferguson GD, Anagnostaras SG, Silva AJ, Herschman HR (2000) Deficits in memory and motor performance in synaptotagmin IV mutant mice. Proc Natl Acad Sci USA 97:5598–5603

Fischer A (2012) Synchronized chaperone function of botulinum neurotoxin domains mediates light chain translocation into neurons. doi:10.1007/978-3-642-33570-9_6

Foran PG, Mohammed N, Lisk GO, Nagwaney S, Lawrence GW, Johnson E, Smith L, Aoki KR, Dolly JO (2003) Evaluation of the therapeutic usefulness of botulinum neurotoxin B, C1, E, and F compared with the long lasting type A. Basis for distinct durations of inhibition of exocytosis in central neurons. J Biol Chem 278:1363–1371

Galli T, Zahraoui A, Vaidyanathan VV, Raposo G, Tian JM, Karin M, Niemann H, Louvard D (1998) A novel tetanus neurotoxin-insensitive vesicle-associated membrane protein in SNARE complexes of the apical plasma membrane of epithelial cells. Mol Biol Cell 9: 1437–1448

Geppert M, Archer BT 3rd, Sudhof TC (1991) Synaptotagmin II a novel differentially distributed form of synaptotagmin. J Biol Chem 266:13548–13552

Gerber SH, Rah JC, Min SW (2008) Conformational switch of syntaxin-1 controls synaptic vesicle fusion. Science 321:1507–1510

Gibbins IL, Jobling P, Teo EH, Matthew SE, Morris JL (2003) Heterogeneous expression of SNAP-25 and synaptic vesicle proteins by central and peripheral inputs to sympathetic neurons. J Comp Neurol 459:25–43

Gordon SL, Leube RE, Cousin MA (2011) Synaptophysin is required for synaptobrevin retrieval during synaptic vesicle endocytosis. J Neurosci 31:14032–14036

Gras C, Amilhon B, Lepicard EM (2008) The vesicular glutamate transporter VGLUT3 synergizes striatal acetylcholine tone. Nat Neurosci 11:292–300

Gronborg M, Pavlos NJ, Brunk I, Chua JJ, Münster-Wandowski A, Riedel D, Ahnert-Hilger G, Urlaub H, Jahn R (2010) Quantitative comparison of glutamatergic and GABAergic synaptic vesicles unveils selectivity for few proteins including MAL2, a novel synaptic vesicle protein. J Neurosci 30:2–12

Grumelli C, Corradini I, Matteoli M, Verderio C (2010) Intrinsic calcium dynamics control botulinum toxin a susceptibility in distinct neuronal populations. Cell Calcium 47:419–424

Gustavsson N, Lao Y, Maximov A (2008) Impaired insulin secretion and glucose intolerance in synaptotagmin-7 null mutant mice. Proc Natl Acad Sci USA 105:3992–3997

Han GA, Malintan NT, Collins BM, Meunier FA, Sugita S (2010) Munc18-1 as a key regulator of neurosecretion. J Neurochem 115:1–10

Hinz B, Becher A, Mitter D, Schulze K, Heinemann U, Draguhn A, Ahnert-Hilger G (2001) Activity-dependent changes of the presynaptic synaptophysin-synaptobrevin complex in adult rat brain. Eur J Cell Biol 80:615–619

Hnasko TS, Chuhma N, Zhang H, Goh GY, Sulzer D, Palmiter RD, Rayport S, Edwards RH (2010) Vesicular glutamate transport promotes dopamine storage and glutamate corelease in vivo. Neuron 65:643–656

Holt M, Varoqueaux F, Wiederhold K, Takamori S, Urlaub H, Fasshauer D, Jahn R (2006) Identification of SNAP-47, a novel Qbc-SNARE with ubiquitous expression. J Biol Chem 281:17076–17083

Humeau Y, Doussau F, Grant NJ, Poulain B (2000) How botulinum and tetanus neurotoxins block neurotransmitter release. Biochimie 82:427–446

Ibata K, Hashikawa T, Tsuboi T, Terakawa S, Liang F, Mizutani A, Fukuda M, Mikoshiba K (2002) Non-polarized distribution of synaptotagmin IV in neurons: evidence that synaptotagmin IV is not a synaptic vesicle protein. Neurosci Res 43:401–406

Janz R, Sudhof TC (1999) SV2C is a synaptic vesicle protein with an unusually restricted localization: anatomy of a synaptic vesicle protein family. Neuroscience 94:1279–1290

Karvar S, Zhu L, Crothers J Jr, Wong W, Turkoz M, Forte JG (2005) Cellular localization and stimulation-associated distribution dynamics of syntaxin-1 and syntaxin-3 in gastric parietal cells. Traffic 6:654–666

Kwon SE, Chapman ER (2011) Synaptophysin regulates the kinetics of synaptic vesicle endocytosis in central neurons. Neuron 70:847–854

Li JY, Jahn R, Dahlstrom A (1994) Synaptotagmin I is present mainly in autonomic and sensory neurons of the rat peripheral nervous system. Neuroscience 63:837–850

Liu Y, Sugiura Y, Lin W (2011) The role of synaptobrevin1/VAMP1 in Ca^{2+}-triggered neurotransmitter release at the mouse neuromuscular junction. J Physiol 589:1603–1618

Lynch BA, Lambeng N, Nocka K, Kensel-Hammes P, Bajjalieh SM, Matagne A, Fuks B (2004) The synaptic vesicle protein SV2A is the binding site for the antiepileptic drug levetiracetam. Proc Natl Acad Sci USA 101:9861–9866

Martens H, Weston MC, Boulland JL (2008) Unique luminal localization of VGAT-C terminus allows for selective labeling of active cortical GABAergic synapses. J Neurosci 28: 13125–13131

Matteoli M, Pozzi D, Grumelli C, Condliffe SB, Frassoni C, Harkany T, Verderio C (2009) The synaptic split of SNAP-25: different roles in glutamatergic and GABAergic neurons? Neuroscience 158:223–230

Morgans CW, Kensel-Hammes P, Hurley JB, Burton K, Idzerda R, McKnight GS, Bajjalieh SM (2009) Loss of the Synaptic Vesicle Protein SV2B results in reduced neurotransmission and altered synaptic vesicle protein expression in the retina. PLoS ONE 4:e5230

Pan PY, Cai Q, Lin L, Lu PH, Duan S, Sheng ZH (2005) SNAP-29-mediated modulation of synaptic transmission in cultured hippocampal neurons. J Biol Chem 280:25769–25779

Pang ZP, Melicoff E, Padgett D, Liu Y, Teich AF, Dickey BF, Lin W, Adachi R, Sudhof TC (2006) Synaptotagmin-2 is essential for survival and contributes to Ca^{2+} triggering of neurotransmitter release in central and neuromuscular synapses. J Neurosci 26:13493–13504

Perin MS, Brose N, Jahn R, Sudhof TC (1991) Domain structure of synaptotagmin (p65). J Biol Chem 266:623–629

Raingo J, Khvotchev M, Liu P (2012) VAMP4 directs synaptic vesicles to a pool that selectively maintains asynchronous neurotransmission. Nat Neurosci 15:738–745

Raptis A, Torrejon-Escribano B, de Gomez Aranda I, Blasi J (2005) Distribution of synaptobrevin/VAMP 1 and 2 in rat brain. J Chem Neuroanat 30:201–211

Ravichandran V, Chawla A, Roche PA (1996) Identification of a novel syntaxin- and synaptobrevin/VAMP-binding protein, SNAP-23, expressed in non-neuronal tissues. J Biol Chem 271:13300–13303

Reisinger C, Yelamanchili SV, Hinz B, Mitter D, Becher A, Bigalke H, Ahnert-Hilger G (2004) The synaptophysin/synaptobrevin complex dissociates independently of neuroexocytosis. J Neurochem 90:1–8

Rhee JS, Li LY, Shin OH, Rah JC, Rizo J, Sudhof TC, Rosenmund C (2005) Augmenting neurotransmitter release by enhancing the apparent Ca^{2+} affinity of synaptotagmin 1. Proc Natl Acad Sci USA 102:18664–18669

Riazanski V, Deriy LV, Shevchenko PD, Le B, Gomez EA, Nelson DJ (2011) Presynaptic CLC-3 determines quantal size of inhibitory transmission in the hippocampus. Nat Neurosci 14: 487–494

Rose AJ, Jeppesen J, Kiens B, Richter EA (2009) Effects of contraction on localization of GLUT4 and v-SNARE isoforms in rat skeletal muscle. Am J Physiol Regul Integr Comp Physiol 297:R1228–R1237

Ruiz-Montasell B, Aguado F, Majo G, Chapman ER, Canals JM, Marsal J, Blasi J (1996) Differential distribution of syntaxin isoforms 1A and 1B in the rat central nervous system. Eur J Neurosci 8:2544–2552

Rummel A (2012) Double receptor anchorage of botulinum neurotoxins accounts for their exquisite neurospecificity. doi:10.1007/978-3-642-33570-9_4

Rummel A, Hafner K, Mahrhold S (2009) Botulinum neurotoxins C, E and F bind gangliosides via a conserved binding site prior to stimulation-dependent uptake with botulinum neurotoxin F utilising the three isoforms of SV2 as second receptor. J Neurochem 110:1942–1954

Sakaba T, Stein A, Jahn R, Neher E (2005) Distinct kinetic changes in neurotransmitter release after SNARE protein cleavage. Science 309:491–494

Schiavo G, Matteoli M, Montecucco C (2000) Neurotoxins affecting neuroexocytosis. Physiol Rev 80:717–766

Schmitt U, Tanimoto N, Seeliger M, Schaeffel F, Leube RE (2009) Detection of behavioral alterations and learning deficits in mice lacking synaptophysin. Neuroscience 162:234–243

Schoch S, Deak F, Konigstorfer A, Mozhayeva M, Sara Y, Sudhof TC, Kavalali ET (2001) SNARE function analyzed in synaptobrevin/VAMP knockout mice. Science 294:1117–1122

Seal RP, Akil O, Yi E (2008) Sensorineural deafness and seizures in mice lacking vesicular glutamate transporter 3. Neuron 57:263–275

Spiwoks-Becker I, Vollrath L, Seeliger MW, Jaissle G, Eshkind LG, Leube RE (2001) Synaptic vesicle alterations in rod photoreceptors of synaptophysin-deficient mice. Neuroscience 107:127–142

Steegmaier M, Yang B, Yoo JS, Huang B, Shen M, Yu S, Luo Y, Scheller RH (1998) Three novel proteins of the syntaxin/SNAP-25 family. J Biol Chem 273:34171–34179

Stobrawa SM, Breiderhoff T, Takamori S (2001) Disruption of ClC-3, a chloride channel expressed on synaptic vesicles, leads to a loss of the hippocampus. Neuron 29:185–196

Südhof TC (2012) The presynaptic active zone. Neuron 75:11–25

Sutton RB, Fasshauer D, Jahn R, Brunger AT (1998) Crystal structure of a SNARE complex involved in synaptic exocytosis at 2.4 a resolution. Nature 395:347–353

Takamori S, Holt M, Stenius K (2006) Molecular anatomy of a trafficking organelle. Cell 127:831–846

van den Bogaart G, Thutupalli S, Risselada JH (2011) Synaptotagmin-1 may be a distance regulator acting upstream of SNARE nucleation. Nat Struct Mol Biol 18:805–812

Vennekate W, Schroder S, Lin CC, van den Bogaart G, Grunwald M, Jahn R, Walla PJ (2012) Cis- and trans-membrane interactions of synaptotagmin-1. Proc Natl Acad Sci USA 109:11037–11042

Verderio C, Rossetto O, Grumelli C, Frassoni C, Montecucco C, Matteoli M (2006) Entering neurons: botulinum toxins and synaptic vesicle recycling. EMBO Rep 7:995–999

Verderio C, Grumelli C, Raiteri L (2007) Traffic of botulinum toxins A and E in excitatory and inhibitory neurons. Traffic 8:142–153

Wang Y, Wang L, Iordanov H, Swietlicki EA, Zheng Q, Jiang S, Tang Y, Levin MS, Rubin DC (2006) Epimorphin(-/-) mice have increased intestinal growth, decreased susceptibility to dextran sodium sulfate colitis, and impaired spermatogenesis. J Clin Invest 116:1535–1546

Washbourne P, Thompson PM, Carta M (2002) Genetic ablation of the t-SNARE SNAP-25 distinguishes mechanisms of neuroexocytosis. Nat Neurosci 5:19–26

Watson RT, Pessin JE (2001) Transmembrane domain length determines intracellular membrane compartment localization of syntaxins 3, 4, and 5. Am J Physiol Cell Physiol 281:C215–C223

Xu J, Mashimo T, Sudhof TC (2007) Synaptotagmin-1, -2, and -9: Ca^{2+} sensors for fast release that specify distinct presynaptic properties in subsets of neurons. Neuron 54:567–581

Xue M, Craig TK, Shin OH, Li L, Brautigam CA, Tomchick DR, Sudhof TC, Rosenmund C, Rizo J (2010) Structural and mutational analysis of functional differentiation between synaptotagmins-1 and -7. *PLoS One* 5

Yang C, Mora S, Ryder JW, Coker KJ, Hansen P, Allen LA, Pessin JE (2001) VAMP3 null mice display normal constitutive, insulin- and exercise-regulated vesicle trafficking. Mol Cell Biol 21:1573–1580

Yao J, Nowack A, Kensel-Hammes P, Gardner RG, Bajjalieh SM (2010) Cotrafficking of SV2 and synaptotagmin at the synapse. J Neurosci 30:5569–5578

Yelamanchili SV, Reisinger C, Becher A, Sikorra S, Bigalke H, Binz T, Ahnert-Hilger G (2005) The C-terminal transmembrane region of synaptobrevin binds synaptophysin from adult synaptic vesicles. Eur J Cell Biol 84:467–475

Zander JF, Munster-Wandowski A, Brunk I, Pahner I, Gomez-Lira G, Heinemann U, Gutierrez R, Laube G, Ahnert-Hilger G (2010) Synaptic and vesicular coexistence of VGLUT and VGAT in selected excitatory and inhibitory synapses. J Neurosci 30:7634–7645

Persistence of *Botulinum* Neurotoxin Inactivation of Nerve Function

Charles B. Shoemaker and George A. Oyler

Abstract The extraordinary persistence of intoxication occurring after exposure to some Botulinum neurotoxin (BoNT) serotypes is both a therapeutic marvel and a biodefense nightmare. Understanding the mechanisms underlying BoNT persistence will offer new strategies for improving the efficacy and extending the applications of BoNT therapeutic agents as well as for treating the symptoms of botulism. Research indicates that the persistence of BoNT intoxication can be influenced both by the ability of the toxin protease or its cleaved soluble N-ethylmaleimide-sensitive factor attachment protein receptor (SNARE) protein substrate to resist turnover. Protease turnover seems to be mediated in part by the ubiquitin-proteasome system (UPS) and efforts to manipulate the UPS may prove to be an effective strategy for improving therapeutic utility of BoNT products and in the development of botulism antidotes.

Keywords Persistence · Polyubiquitination · TRAF2 · E3-ligase · Botulinum neurotoxin light chain

Abbreviations

BoNT	Botulinum neurotoxin
UPS	Ubiquitin-proteasome system
LAS	Lysosome-autophagy system
GFP	Green fluorescent protein
LC/X	Light chain of BoNT/A-G
PMA	Phorbol 12-myristate 13-acetate

C. B. Shoemaker (✉) · G. A. Oyler
Division of Infectious Diseases, Department of Biomedical Sciences,
Tufts Cummings School of Veterinary Medicine, Westboro Road 200,
North Grafton, MA 01536, USA
e-mail: charles.shoemaker@tufts.edu
URL: http://www.tufts.edu/vet/facpages/shoemaker_c.html

A. Rummel and T. Binz (eds.), *Botulinum Neurotoxins*, Current Topics in Microbiology and Immunology 364, DOI: 10.1007/978-3-642-33570-9_9,

SNAP-25	synaptosomal-associated protein of 25 kDa
SNARE	Soluble N-ethylmaleimide-sensitive factor attachment protein receptor
TFB	Targeted F-box
TNF	Tumor necrosis factor
VAMP	Vesicle-associated membrane protein
VHH	Camelid heavy-chain-only V_H domains

Contents

1 Introduction 180
2 Mechanisms of BoNT Persistence 181
2.1 Role of SNARE Proteins in the Persistence of BoNT Intoxication 182
2.2 Role of BoNT Proteases in the Persistence of BoNT Intoxication 184
3 Exploiting BoNT Persistence for Therapeutic Applications 190
3.1 Increasing BoNT Persistence for Improved Therapeutic Benefit 190
3.2 Reversing Paralysis by Accelerated Turnover of BoNT Protease 191
4 Conclusions 193
References 194

1 Introduction

Cellular proteins turn over with half-lives that vary from seconds to several days in a highly regulated manner, influencing critical processes such as the cell cycle, apoptosis, stress responses, and immunity (Varshavsky 1996). Neurons, like all cells, possess elaborate systems dedicated to the maintenance of a healthy proteome in which improperly folded proteins are either refolded or shunted for degradation, most commonly via the ubiquitin-proteasome system (UPS) or lysosome-autophagy system (LAS) (Wong and Cuervo 2010). Botulinum neurotoxins (BoNTs) entering a neuron will surely encounter these same systems and be degraded at some rate. Rapid turnover of intraneuronal BoNTs would reduce the timeframe of toxin action and mitigate the morbidity and mortality associated with intoxication. Because it is assumed that death of the intoxicated animal is an outcome that benefits *Clostridium botulinum* in nature, evolutionary pressures result in BoNTs having greater persistence of action. This pressure has apparently led to BoNT proteases that resist degradation within the cytosol of their targeted neuronal cells. This chapter describes the remarkable persistence of some BoNT proteases and considers the mechanisms whereby the proteases promote their own stability. Strategies for perturbing BoNT protease persistence, both positively and negatively, to generate more effective therapeutic agents or to develop strategies to accelerate patient recovery from pathogenic intoxication are described.

C. botulinum strains produce a number of different neurotoxin serotypes and subtypes (Hill and Smith 2012) which show variable potency and persistence in different animals, presumably reflecting evolutionary selection for strains that are lethal to a broader range of target species. The mechanisms responsible for BoNT toxicity and persistence may vary widely between the seven known BoNT serotypes and within the huge range of susceptible animal species, yet have been studied in detail for only a small subset. Therefore, the current understanding of BoNT persistence mechanisms, which is primarily based on human and rodent studies employing BoNT/A and BoNT/E serotypes, may not prove broadly applicable in other species and BoNT serotypes. Nevertheless, substantial progress has been made in our understanding of persistence in humans and this may translate into improved therapeutic BoNT agents and in the development of antidotes for reversal of unwanted intoxication.

The persistence of muscle paralysis in humans following treatment with different BoNT serotypes varies dramatically from 4 to 6 months with BoNT/A and BoNT/C1 to a 1–4 weeks with BoNT/E in the few examples studied (Whittaker et al. 1964; Eleopra et al. 1997; Sloop et al. 1997). BoNT/B paralysis also ranges between 2 and 4 months in humans although persistence for these serotypes appears to be somewhat less than for BoNT/A (Eleopra et al. 1997, 1998; Sloop et al. 1997; Brin et al. 1999). In rodent models, the time to recovery from muscle paralysis is consistently much less than in humans although the relative order of persistence of the different BoNT serotypes remains similar (Meunier et al. 2003; Adler et al. 1996, 2001; Sellin et al. 1983; Keller 2006; Lisk et al. 2002; Morbiato et al. 2007). BoNT/F has been tested in rats and found to have much reduced persistence compared to BoNT/A (Kauffman et al. 1985). Though recovery from BoNT intoxication is faster in rodents, neurons intoxicated by BoNT/A remain unable to recycle vesicles for a month (de Paiva et al. 1999) and observed recovery of muscle function prior to this time is hypothesized to be a consequence of nerve sprouts at the nerve endings that begin forming within days of intoxication (Brown et al. 1981; Angaut-Petit et al. 1990). The function of the original termini fully recovers during the second and third month post-intoxication and the sprouts are largely eliminated (de Paiva et al. 1999; Meunier et al. 2002). The role of sprouting in the recovery from intoxication by other BoNT serotypes has not been carefully evaluated. A careful analysis of the persistence of neurotransmitter release blockage following treatment of cultured primary neurons with each of five different BoNT serotypes demonstrated a ranking in persistence that strongly correlated to the in vivo results in humans and rodents (Foran et al. 2003). In summary, the available data to date in humans and rodents consistently ranks the persistence of intoxication for the various BoNT serotypes as $A \geq C1 > B > F > E$.

2 Mechanisms of BoNT Persistence

All BoNT serotypes intoxicate neurons by delivering the light chain (LC) protease to the cytosol which specifically cleaves one or more soluble N-ethylmaleimide-sensitive factor attachment protein receptor (SNARE) proteins. The cleaved SNARE proteins either are unable to form SNARE complexes or form inactive SNARE complexes. In

1. SNARE cleavage products of BoNT persist and maintain block to exocytosis
    a. Cleaved SNARE remains in SNARE complexes inhibiting exocytosis
    b. Cleaved SNARE remains free and inhibits new SNARE complex formation
    c. Small peptide cleaved from SNARE remains and inhibits new SNARE formation

2. BoNT protease persists and continues to cleave new SNARE proteins
    a. Protease localizes to membrane and becomes sequestered from degradation pathways
    b. Protease naturally resists native degradation pathways
    c. Protease actively recruits inhibitors of native degradation pathways
        i. Proteasome
        ii. Lysosome
    d. Protease enters a protected environment within the cytosol following translocation from the endosomes

Fig. 1 Summary of proposed mechanisms of BoNT persistence

either case, it is widely accepted that the presence of the BoNT-cleaved SNARE proteins inactivates exocytosis and release of neurotransmitters to cause the flaccid paralysis observed following exposure to BoNT (Binz 2012; Ahnet-Hilger et al. 2012). It follows then that the persistence of BoNT-induced intoxication must depend on: (1) how long the cleaved SNARE proteins remain in the cytosol and the ability of these cleaved SNARE proteins to maintain a block to exocytosis; (2) how long the BoNT protease remains in place to cleave newly synthesized SNARE proteins and/or; (3) the ability of the presynaptic terminal to remodel in a manner to overcome dysfunction. A summary of the mechanisms proposed to explain the variable persistence of the different BoNT serotypes is provided in Fig. 1. Evidence exists that all these processes may contribute to BoNT persistence in rodent models.

2.1 Role of SNARE Proteins in the Persistence of BoNT Intoxication

Studies on the mechanism of BoNT persistence have largely focused on BoNT/A and BoNT/E because these two serotypes have dramatically different persistence in mammals yet they cleave the same SNARE protein, synaptosomal associated protein of 25 kDa (SNAP-25), at positions separated by only 17 amino acids. LC/E cleavage of SNAP-25 removes a carboxyl-terminal peptide of 26 amino acids and there is evidence that this small carboxyl-terminal product is capable, on its own, of inhibiting nerve function. Specifically, several peptides corresponding to different portions of the SNAP-25 carboxyl-terminus have been shown to inhibit neurosecretion in different neuronal model systems (Ferrer-Montiel et al. 1998;

Gutierrez et al. 1995, 1997; Mehta et al. 1996). Although the carboxyl-end peptides have the ability to interfere with neurosecretion, their role in natural BoNT intoxication of motor neurons remains uncertain. For example, the carboxyl-terminal nine amino acids appear unnecessary for BoNT/A intoxication as expressing the truncated SNAP-25 corresponding only to the amino-terminal cleavage product of LC/A caused an intoxication equivalent to that from BoNT/A in insulinoma HIT-T17 cells (Huang et al. 1998). Furthermore, there is no evidence that the carboxyl-terminal peptides cleaved from SNAP-25 following LC/A or LC/E exposure play any role in the relative persistence of intoxication of these two toxins. In fact, the 9-mer peptide from cleavage by the most persistent serotype BoNT/A has an amino-terminal arginine that is predicted by the N-end rule (Varshavsky 1996) to be highly unstable while the isoleucine at the amino-terminus of the less persistent LC/E carboxyl-terminal cleavage product should be more stable. A recent study validates this prediction in neuroblastoma cells (Tsai et al. 2010) and it now seems highly unlikely that the carboxyl-terminal peptides cleaved from SNAP-25 play any role in promoting the extraordinary persistence of BoNT/A function in neurons.

The large amino-terminal SNAP-25 products resulting from the proteolytic activities of BoNT/A (SNAP-25^{A}) or BoNT/E (SNAP-25^{E}) have been proposed by several groups to play an important role in the persistence of these two toxins. If the SNAP-25 cleavage products are the primary determinant of persistence, this leads to the testable prediction that persistence of BoNT/A intoxication would be reduced to that of BoNT/E by 'super-intoxication' with BoNT/E which would truncate the SNAP-25^{A} into SNAP-25^{E}. Note that this prediction assumes that sufficient endosome recycling will remain following BoNT intoxication such that a different BoNT serotype can enter the cell. Despite the caveat, the prediction has reportedly been validated in several studies. A study by Eleopra et al. (1998) in human volunteers found that BoNT/A paralysis of the EDB foot muscle (3 IU) was much more persistent than BoNT/E paralysis of this muscle (3 IU) in the contralateral limb. When another group of volunteers were paralyzed in both EDB foot muscles with a mixture of 2 IU BoNT/A and 4 IU BoNT/E, or 4 IU BoNT/A and 2 IU BoNT/E, all muscles recovered at the same faster rate previously observed for 3 IU BoNT/E alone. Another similar study by Meunier et al. (2003) showed that intoxication by BoNT/E accelerated recovery from BoNT/A intoxication in mice while intoxication by short-lived BoNT/F, which does not cleave SNAP-25, did not affect recovery time from BoNT/A. Raciborska and Charlton (1999) employed a frog model and reported that SNAP-25^{A} was more persistent at the nerve termini than SNAP-25^{E}.

One possible explanation for BoNT/E treatment accelerating recovery from BoNT/A intoxication could be that the SNAP-25^{A} product of LC/A is much more stable than SNAP-25^{E} so cleavage of SNAP-25^{A} into SNAP-25^{E} by super-intoxication accelerates turnover of the inactive SNAP-25 and subsequent formation of functional SNARE complexes from newly synthesized proteins. Foran et al. (2003) tested this hypothesis directly by performing pulse-chase studies to compare the half-lives of the SNAP-25 forms in rat cerebellar primary neurons treated with BoNT/A or BoNT/E. The study showed that the intact or cleaved SNAP-25 proteins all have half-lives of 20–24 h and they concluded that, in their model system, the half-life of cleaved SNAP-25 "does not account for the longevity of BoNT/A-induced inhibition".

Another explanation for the accelerated recovery from BoNT/A by super-intoxication with BoNT/E has been proposed by several groups (Eleopra et al. 1998; Lisk et al. 2002; Raciborska and Charlton 1999; Bajohrs et al. 2004). These groups hypothesize that further cleavage of SNAP-25^{A} into SNAP-25^{E} allows replacement of the damaged SNAP-25 with newly synthesized, uncleaved, SNAP-25 to restore nerve function. Bajohrs et al. (2004) directly demonstrated that SNAP-25^{A} remains bound to syntaxin and localized to the plasma membrane, while cleavage to SNAP-25^{E} results in the loss of association with syntaxin and release to the cytosol. The hypothesis that cleavage to SNAP-25^{A}, but not SNAP-25^{E}, results in stable, unproductive SNARE complexes is consistent with earlier work of Otto et al. (1995) showing that cleavage of SNAP-25^{A} does not affect the formation or stability of SNARE complexes and by Hayashi et al. (1994) which showed that SNARE complexes with SNAP-25^{E} are less stable to denaturants than those with SNAP-25^{A}.

2.2 *Role of BoNT Proteases in the Persistence of BoNT Intoxication*

A number of studies report data strongly in support of the argument that the remarkable persistence BoNT/A intoxication results from retention of active BoNT/A protease within the nerve termini. An early study by Bartels et al. (1994) showed that exocytosis in bovine chromaffin cells previously intoxicated by BoNT/A could be rescued using electroporation to deliver antibodies to the cell cytosol that bind to the BoNT/A protease domain (LC). Electroporation of antibodies that bind only to the BoNT/A heavy chain domain did not restore exocytosis. The fact that exocytosis was restored to normal within a few days after antibody treatment indicates that, if the BoNT/A protease is inhibited or removed, exocytosis function is rapidly restored by endogenous synthesis of full-size SNAP-25. This implies, at least for BoNT/A, that recovery from intoxication is dependent on the loss of BoNT/A protease activity.

More recent studies, designed similar to those in the previous section, monitored recovery of neurons intoxicated by both BoNT/A and BoNT/E. A study by Keller et al. (1999) showed that intoxication of fetal mouse spinal nerve cells with BoNT/A lasted for more than 11 weeks based on SNAP-25 cleavage data, while BoNT/E intoxication persisted less than 3 weeks. When these nerve cells were first intoxicated by BoNT/A and then super-intoxicated by BoNT/E, all of the SNAP-25 became SNAP-25^{E} showing that all neurons had become intoxicated by BoNT/E. By monitoring the SNAP-25 with time post-intoxication, these authors found that SNAP-25^{E} was slowly replaced by SNAP-25^{A}. This demonstrated that the BoNT/A protease persisted and continued to cleave newly synthesized SNAP-

25 even after the BoNT/E protease was no longer active. In a similar study, Adler et al. (2001) found that persistence of intoxication by BoNT/A in a rat EDL model was not reduced by subsequent or previous intoxication with BoNT/E and the authors conclude that BoNT/A persistence in this model is a consequence of the half-life of the BoNT protease rather than that of the cleaved SNAP-25.

Another approach to understanding the factors responsible for the unusual persistence of BoNT/A intoxication was to attempt rescue of nerve function at different times post-intoxication with BoNT/A by transient expression of full-size SNAP-25 (O'Sullivan et al. 1999). In this study, cultured bovine chromaffin cells were intoxicated with BoNT/A and then transfected by plasmids promoting expression of wild-type SNAP-25 or a mutated SNAP-25 that was not cleaved by BoNT/A yet retained the ability to form functional SNARE complexes. Their results showed that BoNT/A cleaved SNAP-25^A is not a dominant block to exocytosis because function was restored by expression of the BoNT/A-insensitive mutant SNAP-25 when the chromaffin cells were transfected just a day post-intoxication. The authors then used this strategy to test whether the BoNT/A protease remained persistent in the chromaffin cells. Chromaffin cells that had been intoxicated for 16 days were transfected with plasmids to overexpress either wild-type or BoNT/A-insensitive mutant SNAP-25 and measured 5 days later for functional exocytosis. The results showed that expression of wild-type SNAP-25 remained reproducibly unable to rescue exocytosis while mutant SNAP-25 did rescue function, thus demonstrating that active BoNT/A protease persists in these cells for at least 3 weeks. The authors conclude that the "continued action of BoNT/A-LC is a major contributory factor in the prolonged inhibition of release, at least in this model system".

Ideally, studies on the role of the BoNT proteases in the persistence of intoxication would monitor the cytosolic levels of the proteases in neuronal cells at various times following toxin exposure and attempt to correlate protease levels with the cells' ability to perform exocytosis. This is not practically possible for a number of reasons. Most significant is the inability to quantify cytosolic BoNT proteases within intoxicated cells using current detection methods. Although the number of cytosolic protease molecules necessary to inactivate exocytosis is not known, it has been estimated to be no more than about a thousand (Hanig and Lamanna 1979) for BoNT proteases, and less than ten molecules for the closely related tetanus toxin protease (Erdal et al. 1995). Even these remarkably low numbers may be overestimations as for other toxins such as ricin or diphtheria toxin, a single molecule has been shown sufficient to elicit a measurable effect (cell death) (Yamaizumi et al. 1978; Eiklid et al. 1980). It is also problematic to differentiate cytosolic BoNT proteases from proteases that remain in the endosomal fraction using imprecise available immunolocalization or subcellular fractionation methods, especially if only a small percentage of BoNT proteases escape the endosomal fraction to the cytosol (Bade et al. 2004).

2.2.1 Contribution of Intraneuronal BoNT Protease Subcellular Localization to Persistence

Because of the difficulties in detecting cytosolic BoNT protease following natural intoxication, several labs have employed DNA transfection methods to promote expression of the proteases within cells to study persistence (Tsai et al. 2010; Fernandez-Salas et al. 2004, 2004; Kuo et al. 2011). Fernandez-Salas et al. overexpressed either recombinant BoNT/A or BoNT/E proteases within rat PC12 neuroblastoma cells as fusions to green fluorescent protein (GFP) and monitored protease localization by fluorescence microscopy. They found that BoNT/A protease (LC/A) colocalized with SNAP-25^{A} at the plasma membrane where intact SNAP-25 localizes. Surprisingly they found that the BoNT/E light chain protease (LC/E) was primarily found in the cytosol rather than localized at the plasma membrane. These authors suggest that the cytosolic localization of LC/E may be the reason that this protease is turned over more rapidly than the LC/A protease which they speculate remains in "a slow cycling compartment at the plasma membrane". In a recent study in a different neuroblastoma cell line (N18), Tsai et al.(2010) found that both recombinant LC/A and LC/E colocalized at the plasma membrane and argued that differential localization is not likely to be responsible for the differential persistence of these proteins. It is not possible to know whether cytosolic synthesis and overexpression of recombinant LC/A and LC/E through recombinant expression as employed in these studies produced localizations that accurately reflect the protease localization achieved following natural intoxication events in which these proteases are delivered from endosomes to the cytosol in vanishingly small amounts by a highly evolved and sophisticated translocation process (Montal 2010).

In an effort to understand the mechanistic basis for the differing subcellular localizations of LC/A and LC/E observed in their model system, some groups have studied the effects of specific protease mutations on their localization within cells. Fernandez-Salas et al. (2004) produced a variety of mutant forms of recombinant LC/A to identify signals that lead to its plasma membrane localization. They found that removal of 8 amino acids at the amino terminal end of the protease, or 22 amino acids from the carboxyl end, reduced plasma membrane association by LC/A and this indicated that signals may exist at both ends of the protease to promote plasma membrane association. These results were extended by Chen and Barbieri (2011) who used LC/A mutants to demonstrate that LC/A localization results from its binding to the SNAP-25 substrate. To reach this conclusion, the authors showed that LC/A mutants lacking the amino terminal eight amino acids, or containing lysine to alanine mutations near the amino terminus at positions 6 and 11, significantly reduced plasma membrane localization. They then showed that SNAP-25 mutants which lack their putative palmitoylation sites no longer localized to the plasma membrane and when cells were engineered to express these SNAP-25 mutants, the co-expressed LC/A also became localized to the cytosol. Since it was previously demonstrated that SNAP-25 is released to the cytosol following cleavage by LC/E but not following cleavage with LC/A (Bajohrs et al. 2004), Chen and Barbieri (2011) tested the localization of recombinant LC/A when it was co-expressed with LC/E. As expected, they found that LC/A now localized to the cytosol, presumably because the SNAP-25 to which it associates becomes localized to the cytosol following its conversion to SNAP-25^{E}. These results imply that LC/A does not naturally localize to the plasma membrane in the absence of SNAP-25, yet other studies have found plasma membrane localization for LC/A within MDCK cells which lack SNAP-25 (G. Oyler, unpublished observation). As such, other mechanisms responsible for alterations in LC localization cannot be ruled out.

Fernandez-Salas et al. (2004) hypothesized that a dileucine motif (Aiken et al. 1994) present near the carboxyl end (aa 427/428) played an important role in the plasma membrane localization of LC/A. To test this concept, they generated a LC/A mutant in which only this dileucine was altered to a dialanine (LC/A^{AA}) and compared its localization to wild-type LC/A within transfected cells. They found that this mutant displayed reduced plasma membrane localization and suggested that the dileucine in LC/A helps to promote the membrane localization of LC/A- and its characteristic persistence. The LC/A^{AA}, though, was expressed to different levels than wild-type LC/A and had reduced protease activity, and these factors may also have contributed to the altered localization.

The SNAP-25 mechanism for LC/A localization is somewhat more complicated since LC/A has been seen to associate with membranes when expressed in cell lines such as cells lacking SNAP-25 (unpublished observations, GAO). Thus, there may be additional mechanism involved in targeting LC/A to membranes.

Wang et al. (2011) attempted to directly determine whether the amino terminal amino acids of BoNT/A, or the dileucine present at the carboxyl end, contributed to the unusual persistence of this toxin. In their study, the authors engineered recombinant holotoxins in which the BoNT/A protease either lacked the seven amino terminal amino acids (residues 2–8) or had the dileucine 427/428 converted to a dialanine. They then compared the persistence of intoxication elicited by the mutant toxins compared to wild-type recombinant BoNT/A or BoNT/E holotoxins. Strikingly, the BoNT/A mutation converting the dileucine into dialanine was clearly shown to significantly reduce the persistence of intoxication from this mutant holotoxin in primary rat neuronal cells or in vivo in the mouse hind limb. To further demonstrate the role of dileucines in persistence, these authors also produced a catalytically inactive form of BoNT/A in which the LC/A protease was inactivated by mutation, with or without the dileucine mutations, and the complete LC/E protease was fused to the amino terminus. The chimeric protein that lacked the dileucine mutation, called LC_E-$BoTIM_A$, was toxic to neurons and produced intoxication with the persistence of wild-type BoNT/A. As previously shown by Bade et al. (2004) this demonstrated that a fusion of two complete BoNT proteases could be effectively delivered to the cytosol by the heavy chain translocation domain. It also demonstrated that, in this system, the mechanism responsible for the prolonged persistence of LC/A was not overcome when it was fused to LC/E, despite the intrinsic rapid turnover of LC/E alone within neurons. When the dileucine within the catalytically inactive LC/A in LC_E-$BoTIM_A$ was mutated to dialanine, the persistence of intoxication from this chimeric toxin was substantially reduced near to that of native BoNT/E. Thus, by removing the dileucine signal within the inactive LC/A fusion partner, the functional LC/E protease was no longer being stabilized within the neurons by the LC/A. Surprisingly, when these authors tested the BoNT/A holotoxin with the mutation that removed amino acids 2-7 of LC/A, this toxin elicited intoxication that had the persistence of wild-type BoNT/A. Since a similar mutation of LC/A was earlier reported to eliminate the plasma membrane localization of LC/A (Fernandez-Salas et al. 2004), Wang et al. (2011) postulated that subcellular localization probably did not play a role in BoNT/A persistence, but rather that the dileucine somehow protected the protease from proteolytic degradation.

The subcellular localizations of two additional proteases, LC/B and LC/C, have been studied following their expression by plasmid transfection. Both these proteases appeared to be dispersed throughout the cell, not localized to the plasma membrane (Fernandez-Salas et al. 2004; Land et al. 1997). These results suggest that localization to the plasma membrane is not required for protease persistence in these other long-lasting BoNT serotypes and their persistence depends on other mechanisms.

2.2.2 Contribution of the Ubiquitin-Proteasome System to BoNT Protease Persistence

Protein turnover in cells is most commonly mediated by the LAS and/or the UPS (Wong and Cuervo 2010; Hershko and Ciechanover 1998; Ciechanover 2005). The LAS is considered to be a more general system responsible for turnover of macromolecules and the UPS is utilized for more targeted and regulated protein turnover processes. While no evidence yet exists that LAS degradation plays a role in the relative persistence of the different BoNT serotypes, there is a growing body of evidence that the UPS may be an important factor. Shi et al. (2009) showed that modulation of LC/B polyubiquitination altered its protease activity. When LC/B was heavily polyubiquitinated in vitro, these authors found that the enzymatic activity was reduced by 34 %. In the same study, LC/B polyubiquitination in SHSY-5Y cells was increased about 2.5 fold when treated with phorbol 12-myristate 13-acetate (PMA), and about 3.5-fold when treated with both PMA and a proteasome inhibitor, following exposure to a very high dose of BoNT/B (10 μg/ml). The increased intracellular polyubiquitination of LC/B led to a significant reduction in intracellular vesicle associated membrane protein (VAMP) cleavage in BoNT/B treated cells although it was not possible to distinguish the contribution of reduced protease activity or accelerated LC/B turnover to this reduction.

The role of the UPS in BoNT persistence was investigated by Tsai et al. (2010) through a comparison of the polyubiquitination of the persistent LC/A versus the short-lived LC/E. These authors first showed that, in N18 cells, LC/E is much more rapidly turned over than is LC/A and that this is associated with a much heavier polyubiquitination of LC/E than of LC/A. The half-life of the LC/E was significantly increased in cells treated with proteasome inhibitors suggesting a role of the UPS in LC/E turnover. Tsai et al. (2010) then demonstrated that LC/A turnover could be accelerated by developing a specific biomolecule that promoted LC/A polyubiquitination. The biomolecule they designed was a fusion protein containing both a LC/A targeting protein and an E3-ligase domain (HECT domain from E6AP or RING domain from XIAP). As a targeting domain, the authors employed the SNAP-25 substrate for LC/A in which the protein was mutated to a form that could not be cleaved by LC/A (SNAP-25nc; R198T). These 'designer ubiquitin ligases' were co-expressed in cells with LC/A and the LC/A was found to be more heavily ubiquitinated and to have a much shorter half-life than LC/A from control cells. Kuo et al. (2011) also produced biomolecules designed to promote polyubiquitination of LC/A and LC/B, and demonstrated that these agents increased intra-neuronal cell turnover of the targeted proteases and, importantly, the agents also accelerated the recovery of cells from natural BoNT/A intoxication (see Sect. 3). These results imply that LC/A is more resistant than LC/E to polyubiquitination in intoxicated neurons and that the polyubiquitination step may be rate limiting in the regulation of LC/A turnover, and likely the protease turnover for other persistent BoNT serotypes.

Tsai et al. performed additional studies in an effort to explain why the BoNT/A protease is unusually resistant to polyubiquitination within cells in comparison to BoNT/E protease. The authors immunoprecipitated LC/A or LC/E from transfected N18 neuroblastoma cells and used proteomic tools to identify proteins that bound preferably to one or the other protease. One protein, TRAF2, was found to be specifically bound to LC/E. TRAF2 is a RING finger protein that is involved in TNF receptor signaling that has been implicated in ubiquitination (Chen 2005). To test for a possible role of TRAF2 in LC/E turnover, expression of this protein was either increased in cells by expression plasmid transfection or knocked-down by siRNA. The half-life of LC/E was shown to be reduced by overexpression of TRAF2 and increased by knockdown of TRAF2. The outcome of these studies was that TRAF2 appears to interact specifically with LC/E and promote its turnover by the UPS, and this may explain, in part, the reduced persistence of BoNT/E intoxication. In unpublished work, a number of studies have found that LC/A binds to several de-ubiquitinases (DUBs) and that knockdown of these DUBs reduces the half-life of LC/A (Oyler, personal communication), suggesting the possibility that recruitment of DUBs by LC/A may play a role in stabilizing the protease and promoting persistence of BoNT/A intoxication.

2.2.3 Contribution of the BoNT Heavy Chain to BoNT Protease Persistence

Another factor that could contribute to BoNT persistence is the uptake and protease translocation process itself. Serotype differences in the cell entry or endosomal release processes that are mediated by the BoNT translocation domains could lead to differences in the protease trafficking to its site of action or exposure of the protease to cytosolic degradation pathways. For example, differential trafficking might steer the BoNT proteases of more persistent serotypes through protected environments that result in less damage and/or more rapid access to SNARE protein substrates. Nevertheless, there is evidence, at least for BoNT/A and BoNT/E, that the receptor binding and translocation domains do not play a role in persistence. Wang et al. (2008) produced chimeras in which the 50 kDa receptor binding domain of BoNT/A was replaced by the corresponding domain from BoNT/E (AE) or in which the receptor binding domain of BoNT/E was replaced by the corresponding domain from BoNT/A (EA). For these chimeras, the persistence of intoxication appeared to be unrelated to which receptor binding domain was present on the toxin. Cells treated with the AE chimera remained intoxicated about as long as cells treated with wild-type BoNT/A, while the EA chimera persisted about as long as intoxication with BoNT/E. Similar results were obtained in a subsequent study (Wang et al. 2012) in which this laboratory produced chimeras exchanging the BoNT/A and BoNT/B receptor binding domains. Again, even though cell susceptibility to intoxication was determined by the receptor binding domain, persistence of the resulting intoxication was determined by the LC and translocation domains. Thus, for the BoNT serotypes considered the most dangerous to humans, persistence does not seem to be affected by which receptor mediates entry. However, it remains possible that the HC plays a more significant role in the persistence for other more distantly related BoNT serotypes that bind to different receptors or target different neuronal cell populations.

3 Exploiting BoNT Persistence for Therapeutic Applications

Substantial research is currently directed at enhancing the efficacy of BoNT as a treatment for a variety of conditions or to the development of new therapeutic applications, and this research includes efforts to enhance the persistence of nerve inactivation. On the flip side, there are also a number of research groups that seek to diminish the threat posed by BoNT as an agent of bioterror, and one focus is an effort to diminish the persistence of paralysis so as to promote a more rapid recovery from toxin exposure. This research has benefited greatly from the improved understanding of the mechanisms of BoNT persistence that are outlined above. As a result, research to modify the persistence of BoNT intoxication has focused on the protease domain.

3.1 Increasing BoNT Persistence for Improved Therapeutic Benefit

Wang et al. (2011) generated a chimeric toxin in an effort to combine the prolonged persistence of BoNT/A with the therapeutically beneficial features of BoNT/E such as more rapid and comprehensive intoxication. As described above (Sect. 2.2.1), this chimera, called LC_E-$BoTIM_A$, was a full-sized BoNT/A containing mutations in the light chain that eliminate LC/A protease activity in which the LC/E was appended to the amino terminus to create a protease dimer. The chimeric protein was found to intoxicate neurons with the persistence of BoNT/A and to permit the intoxication of sensory neurons not normally sensitive to BoNT/E. The authors claim that their LC_E-$BoTIM_A$ toxin is a long-acting agent with potential for pain therapy and other conditions caused by overactive cholinergic nerves.

3.2 Reversing Paralysis by Accelerated Turnover of BoNT Protease

The first report of an agent that was created with the purpose of promoting ubiquitination and accelerated turnover of a BoNT protease was by Tsai et al. (2010). In their study, described above in Sect. 2.2.2, a fusion protein was produced in which E3-ligases were joined to a mutant form of SNAP-25 (SNAP-25nc) that was not cleaved by the BoNT/A protease, LC/A. Cells expressing these 'designer ubiquitin ligases', successfully promoted increased polyubiquitination and accelerated turnover of co-expressed LC/A, presumably because the SNAP-25nc recruited the LC/A to become a substrate for E3-ligase action. This strategy is not particularly practical for therapeutic use because of difficulties commercially producing such a complex protein and delivering it to the cytosol of intoxicated neurons in patients. In addition, different non-cleavable SNARE proteins would need to be developed for each BoNT serotype and there are concerns about the off-target effects of SNARE-based E3-ligases leading to degradation of other cellular proteins.

An alternative strategy for promoting accelerated ubiquitination and turnover of BoNT proteases was proposed by Kuo et al. (2011). Like Tsai et al. (2010), the approach was to deliver a fusion protein to intoxicated neurons that would lead to polyubiquination and rapid turnover of the intoxicating BoNT protease. The new concept was to use camelid heavy chain only V_H domains (VHHs) as targeting domains that recognized BoNT proteases. These are small (14 kDa) proteins that retain their binding activity and specificity within neurons (Tremblay et al. 2010) and can be readily developed for proteases from all known BoNT serotypes. Since VHHs should be highly specific only to their LC target, side effects following their delivery to cells should be of minimal concern. In place of a large E3-ligase, Kuo et al. proposed to employ a minimal F-box domain derived from β-TrCP, a protein known to be expressed within neurons (Westbrook et al. 2008). The F-box domain is responsible for recruiting an E3-ligase to β-TrCP and promoting polyubiquitination of bound proteins. As a fusion partner with a VHH, this small F-box domain (15 kDa) was expected to be sufficient to recruit neuronal E3-ligases to the BoNT protease bound by the VHH for polyubiquitination. Kuo et al. demonstrated that expression of a VHH/F-box fusion protein, called a targeted F-box or TFB (Fig. 2), in which the VHH was specific for LC/A, led to polyubiquitination of LC/A and to a much more rapid turnover of this protease within both N2A and M17 neuroblastoma cells. Replacing the VHH with a different VHH that binds to BoNT/B protease, LC/B, resulted in a new TFB that promoted turnover of LC/B. Using the TFBs targeting LC/A or B resulted only in the accelerated turnover of the targeted protease and had no effect on the non-targeted protease, even within the same cell population. Most

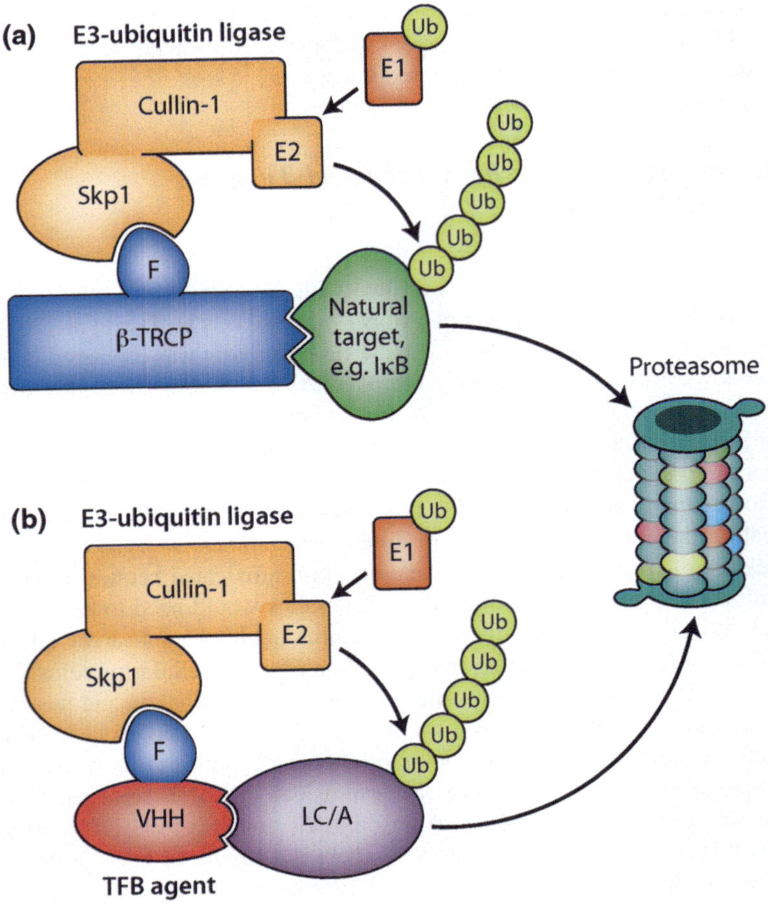

Fig. 2 Targeted F-box (*TFB*) agents exploit the natural process by which F-box proteins promote accelerated turnover of cytosolic proteins. **a**. F-box proteins participate in the proteasome-mediated turnover of natural cellular proteins. Ubiquitin (*Ub*) is activated by E1 and transferred to E2 which associates with Cullin-1 and Skp1 to form the E3 ubiquitin ligase. F-box proteins such as β-TrCP bind to natural targets, such as IκB, and recruit the E3 ubiquitin ligases through binding of the small F-box domain on β-TrCP to the Skp1 component of the ligase. The E3 ubiquitin ligase then attaches a chain of Ub molecules to the natural target and this polyubiquitinated protein becomes a substrate for proteasome-mediated degradation. **b.** TFB agents exploit natural pathways to promote accelerated turnover of selected cytosolic proteins such as BoNT proteases. The F-box domain from β-TrCP is expressed as a fusion protein to a VHH with affinity for a BoNT protease such as LC/A. This TFB agent binds to the LC in the cytosol of an intoxicated neuron and the F-box component recruits an E3-ubiquitin ligase which polyubiquitinates the LC and promotes its accelerated turnover by cytosolic proteasomes

importantly, Kuo et al. demonstrated that their TFB agents were effective in promoting the recovery of neuronal cells that were naturally intoxicated by BoNT/A. Using SNAP-25 as an indicator, M17 neuroblastoma cells expressing a LC/A-targeted TFB recovered from BoNT/A intoxication at least 2.5-fold faster than the control cells. Having validated the approach for both BoNT/A and BoNT/B, development of TFB agents targeting other BoNT serotypes should quickly follow as it requires only that VHHs binding the other LCs be identified using established methods (Tremblay et al. 2010). These results suggest it will be possible to accelerate recovery from botulism caused by any toxic serotype if the appropriate TFB agent can be successfully delivered to the cytosol of BoNT intoxicated neurons. Therapeutic delivery strategies might employ protein vehicles such as atoxic mutants of neurotoxins [e.g. BoNT (Webb et al. 2009)] or non-specific toxins that have been retargeted for neuronal specificity [e.g. *C. difficile* toxin (Krautz-Peterson et al. 2012)], or perhaps genetic vehicles such as viruses modified for motor neuron tropism.

4 Conclusions

There are seven known serotypes of BoNTs and these elicit variable levels of toxicity and persistence when different animal species become exposed. The preponderance of research on these toxins has sensibly been focused on humans and the common rodent models—so little is known regarding toxicity and persistence in most other animal species. In comparing human and rodent BoNT intoxication, there is substantial consistency with regard to the relative persistence of the different serotypes which suggests that the mechanisms of persistence that are identified in rodent models will likely predict those in humans. The persistence of paralysis occurring with BoNT intoxication is clearly dependent on the maintenance of truncated SNARE proteins following their cleavage by the BoNT proteases within motor neurons. The truncated SNARE proteins may themselves persist in the nerve endings, and/or the BoNT protease may persist and continue to cleave newly synthesized SNARE proteins. Recent literature would suggest that, for the longest lived BoNT serotype, BoNT/A, the persistence of intoxication primarily depends on the ability of the BoNT protease to resist natural turnover via the UPS. Quite possibly, with other much shorter lived BoNT proteases such as BoNT/E, it is the turnover of the truncated SNARE protein that determines the rate of recovery from intoxication. It does appear clear that efforts to develop BoNT-based therapeutics with improved persistence properties will benefit from alterations to the native toxins which increase the intracellular half-life of the protease, for example, by reducing its recognition by the UPS. Similarly, new treatments for reversing unwanted BoNT intoxication, such as following a bioterror incident, are likely to be possible through the delivery of biomolecules that accelerate the turnover rate of the BoNT protease. Such new therapeutic approaches are becoming increasingly practical as we gain new understanding of the mechanisms by which these remarkable toxin biomolecules enter and persist within their motor neuron targets.

Acknowledgments This project has been supported in part with Federal funds from the National Institute of Allergy and Infectious Diseases, National Institutes of Health, Department of Health and Human Services, under contract number N01-AI30050 and award numbers R21-AI101403, R01-AI093467, R21-AI088489. The content is solely the responsibility of the authors and does not necessarily represent the official views of the National Institute of Allergy and Infectious Diseases or the National Institutes of Health.

References

Adler M, Macdonald DA, Sellin LC, Parker GW (1996) Effect of 3, 4-diaminopyridine on rat extensor digitorum longus muscle paralyzed by local injection of botulinum neurotoxin. Toxicon: Official J Int Soc Toxinology 34:237–249

Adler M, Keller JE, Sheridan RE, Deshpande SS (2001) Persistence of botulinum neurotoxin A demonstrated by sequential administration of serotypes A and E in rat EDL muscle. Toxicon: Official J Int Soc Toxinology 39:233–243

Ahnet-Hilger G, Münster-Wandowski A, Höltje M (2012) Synaptic vesicle proteins: targets and routes for botulinum neurotoxins. doi:10.1007/978-3-642-33570-9_8

Aiken C, Konner J, Landau NR, Lenburg ME, Trono D (1994) Nef induces CD4 endocytosis: requirement for a critical dileucine motif in the membrane-proximal CD4 cytoplasmic domain. Cell 76:853–864

Angaut-Petit D, Molgo J, Comella JX, Faille L, Tabti N (1990) Terminal sprouting in mouse neuromuscular junctions poisoned with botulinum type A toxin: morphological and electrophysiological features. Neuroscience 37:799–808

Bade S, Rummel A, Reisinger C, Karnath T, Ahnert-Hilger G et al (2004) Botulinum neurotoxin type D enables cytosolic delivery of enzymatically active cargo proteins to neurones via unfolded translocation intermediates. J Neurochem 91:1461–1472

Bajohrs M, Rickman C, Binz T, Davletov B (2004) A molecular basis underlying differences in the toxicity of botulinum serotypes A and E. EMBO Rep 5:1090–1095

Bartels F, Bergel H, Bigalke H, Frevert J, Halpern J et al (1994) Specific antibodies against the Zn(2+)-binding domain of clostridial neurotoxins restore exocytosis in chromaffin cells treated with tetanus or botulinum A neurotoxin. J Biol Chem 269:8122–8127

Binz (2012) Clostridial neurotoxin light chains: devices for SNARE cleavage mediated blockade of neurotransmission. doi:10.1007/978-3-642-33570-9_7

Brin MF, Lew MF, Adler CH, Comella CL, Factor SA et al (1999) Safety and efficacy of NeuroBloc (botulinum toxin type B) in type A-resistant cervical dystonia. Neurology 53:1431–1438

Brown MC, Holland RL, Hopkins WG (1981) Motor nerve sprouting. Annu Rev Neurosci 4:17–42

Chen ZJ (2005) Ubiquitin signalling in the NF-kappaB pathway. Nat Cell Biol 7:758–765

Chen S, Barbieri JT (2011) Association of botulinum neurotoxin serotype A light chain with plasma membrane-bound SNAP-25. J Biol Chem 286:15067–15072

Ciechanover A (2005) Proteolysis: from the lysosome to ubiquitin and the proteasome. Nat Rev Mol Cell Biol 6:79–87

de Paiva A, Meunier FA, Molgo J, Aoki KR, Dolly JO (1999) Functional repair of motor endplates after botulinum neurotoxin type A poisoning: biphasic switch of synaptic activity between nerve sprouts and their parent terminals. Proc Natl Acad Sci U S A 96:3200–3205

Dolly JO, Lisk G, Foran PG, Meunier F, Mohammed N, O'Sullivan G, de Paiva A (2002) Insights into the extended duration of neuroparalysis by Botulinum Neurotoxin A relative to the other shorter-acting serotypes: differences between motor nerve terminals and cultured neurons. In: Brin MF, Hallett M, Jankovic J (ed) Scientific and therapeutic aspects of Botulinum Toxin. Lippincott Williams and Wilkins, Philadelphia, pp 91–102

Eiklid K, Olsnes S, Pihl A (1980) Entry of lethal doses of abrin, ricin and modeccin into the cytosol of HeLa cells. Exp Cell Res 126:321–326

Eleopra R, Tugnoli V, Rossetto O, Montecucco C, De Grandis D (1997) Botulinum neurotoxin serotype C: a novel effective botulinum toxin therapy in human. Neurosci Lett 224:91–94

Eleopra R, Tugnoli V, Rossetto O, De Grandis D, Montecucco C (1998) Different time courses of recovery after poisoning with botulinum neurotoxin serotypes A and E in humans. Neurosci Lett 256:135–138

Erdal E, Bartels F, Binscheck T, Erdmann G, Frevert J et al (1995) Processing of tetanus and botulinum A neurotoxins in isolated chromaffin cells. Naunyn-Schmiedeberg's Arch Pharmacol 351:67–78

Fernandez-Salas E, Steward LE, Ho H, Garay PE, Sun SW et al (2004a) Plasma membrane localization signals in the light chain of botulinum neurotoxin. Proc Natl Acad Sci U S A 101:3208–3213

Fernandez-Salas E, Ho H, Garay P, Steward LE, Aoki KR (2004b) Is the light chain subcellular localization an important factor in botulinum toxin duration of action? Mov Disord 19(Suppl 8): S23–S34

Ferrer-Montiel AV, Gutierrez LM, Apland JP, Canaves JM, Gil A et al (1998) The 26-mer peptide released from SNAP-25 cleavage by botulinum neurotoxin E inhibits vesicle docking. FEBS Lett 435:84–88

Foran PG, Mohammed N, Lisk GO, Nagwaney S, Lawrence GW et al (2003) Evaluation of the therapeutic usefulness of botulinum neurotoxin B, C1, E, and F compared with the long lasting type A. Basis for distinct durations of inhibition of exocytosis in central neurons. J Biol Chem 278:1363–1371

Gutierrez LM, Canaves JM, Ferrer-Montiel AV, Reig JA, Montal M et al (1995) A peptide that mimics the carboxy-terminal domain of SNAP-25 blocks Ca(2+)-dependent exocytosis in chromaffin cells. FEBS Lett 372:39–43

Gutierrez LM, Viniegra S, Rueda J, Ferrer-Montiel AV, Canaves JM et al (1997) A peptide that mimics the C-terminal sequence of SNAP-25 inhibits secretory vesicle docking in chromaffin cells. J Biol Chem 272:2634–2639

Hanig JP, Lamanna C (1979) Toxicity of botulinum toxin: a stoichiometric model for the locus of its extraordinary potency and persistence at the neuromuscular junction. J Theor Biol 77:107–113

Hayashi T, McMahon H, Yamasaki S, Binz T, Hata Y et al (1994) Synaptic vesicle membrane fusion complex: action of clostridial neurotoxins on assembly. EMBO J 13:5051–5061

Hershko A, Ciechanover A (1998) The ubiquitin system. Annu Rev Biochem 67:425–479

Hill KK, Smith TJ (2012) Genetic diversity within *Clostridium botulinum* serotypes, botulinum Nneurotoxin gene clusters and toxin subtypes. doi:10.1007/978-3-642-33570-9_1

Huang X, Wheeler MB, Kang YH, Sheu L, Lukacs GL et al (1998) Truncated SNAP-25 (1–197), like botulinum neurotoxin A, can inhibit insulin secretion from HIT-T15 insulinoma cells. Mol Endocrinol 12:1060–1070

Kauffman JA, Way JF Jr, Siegel LS, Sellin LC (1985) Comparison of the action of types A and F botulinum toxin at the rat neuromuscular junction. Toxicol Appl Pharmacol 79:211–217

Keller JE (2006) Recovery from botulinum neurotoxin poisoning in vivo. Neuroscience 139: 629–637

Keller JE, Neale EA, Oyler G, Adler M (1999) Persistence of botulinum neurotoxin action in cultured spinal cord cells. FEBS Lett 456:137–142

Krautz-Peterson G, Zhang Y, Chen K, Oyler GA, Feng H et al (2012) Retargeting Clostridium difficile Toxin B to Neuronal Cells as a Potential Vehicle for Cytosolic Delivery of Therapeutic Biomolecules to Treat Botulism. J Toxicol 2012:760142

Kuo CL, Oyler GA, Shoemaker CB (2011) Accelerated neuronal cell recovery from *Botulinum* neurotoxin intoxication by targeted ubiquitination. PLos ONE (in press)

Land J, Zhang H, Vaidyanathan VV, Sadoul K, Niemann H et al (1997) Transient expression of botulinum neurotoxin C1 light chain differentially inhibits calcium and glucose induced insulin secretion in clonal beta-cells. FEBS Lett 419:13–17

Mehta PP, Battenberg E, Wilson MC (1996) SNAP-25 and synaptotagmin involvement in the final Ca(2+)-dependent triggering of neurotransmitter exocytosis. Proc Natl Acad Sci U S A 93:10471–10476

Meunier FA, Schiavo G, Molgo J (2002) Botulinum neurotoxins: from paralysis to recovery of functional neuromuscular transmission. J Physiol Paris 96:105–113

Meunier FA, Lisk G, Sesardic D, Dolly JO (2003) Dynamics of motor nerve terminal remodeling unveiled using SNARE-cleaving botulinum toxins: the extent and duration are dictated by the sites of SNAP-25 truncation. Molecular and cellular neurosciences 22:454–466

Montal M (2010) Botulinum neurotoxin: a marvel of protein design. Annu Rev Biochem 79: 591–617

Morbiato L, Carli L, Johnson EA, Montecucco C, Molgo J et al (2007) Neuromuscular paralysis and recovery in mice injected with botulinum neurotoxins A and C. Eur J Neurosci 25: 2697–2704

O'Sullivan GA, Mohammed N, Foran PG, Lawrence GW, Oliver Dolly J (1999) Rescue of exocytosis in botulinum toxin A-poisoned chromaffin cells by expression of cleavage-resistant SNAP-25. Identification of the minimal essential C-terminal residues. J Biol Chem 274: 36897–36904

Otto H, Hanson PI, Chapman ER, Blasi J, Jahn R (1995) Poisoning by botulinum neurotoxin A does not inhibit formation or disassembly of the synaptosomal fusion complex. Biochem Biophys Res Commun 212:945–952

Raciborska DA, Charlton MP (1999) Retention of cleaved synaptosome-associated protein of 25 kDa (SNAP-25) in neuromuscular junctions: a new hypothesis to explain persistence of botulinum A poisoning. Can J Physiol Pharmacol 77:679–688

Sellin LC, Kauffman JA, Dasgupta BR (1983) Comparison of the effects of botulinum neurotoxin types A and E at the rat neuromuscular junction. Med biol 61:120–125

Shi X, Curran JE, Liao Z, Gordon RK (2009) The biological activity of ubiquitinated BoNT/B light chain in vitro and in human SHSY-5Y neuronal cells. J Cell Biochem 108:660–667

Sloop RR, Cole BA, Escutin RO (1997) Human response to botulinum toxin injection: type B compared with type A. Neurology 49:189–194

Tremblay JM, Kuo CL, Abeijon C, Sepulveda J, Oyler G et al (2010) Camelid single domain antibodies (VHHs) as neuronal cell intrabody binding agents and inhibitors of Clostridium botulinum neurotoxin (BoNT) proteases. Toxicon: Official J Int Soc Toxinology 56:990–998

Tsai YC, Maditz R, Kuo CL, Fishman PS, Shoemaker CB et al (2010) Targeting botulinum neurotoxin persistence by the ubiquitin-proteasome system. Proc Natl Acad Sci U S A 107:16554–16559

Varshavsky A (1996) The N-end rule: functions, mysteries, uses. Proc Natl Acad Sci U S A 93:12142–12149

Wang J, Meng J, Lawrence GW, Zurawski TH, Sasse A et al (2008) Novel chimeras of botulinum neurotoxins A and E unveil contributions from the binding, translocation, and protease domains to their functional characteristics. J Biol Chem 283:16993–17002

Wang J, Zurawski TH, Meng J, Lawrence G, Olango WM et al (2011) A dileucine in the protease of botulinum toxin A underlies its long-lived neuroparalysis: transfer of longevity to a novel potential therapeutic. J Biol Chem 286:6375–6385

Wang J, Zurawski TH, Bodeker MO, Meng J, Boddul S et al (2012) Longer-acting and highly potent chimaeric inhibitors of excessive exocytosis created with domains from botulinum neurotoxin A and B. Biochem J 444:59–67

Webb RP, Smith TJ, Wright P, Brown J, Smith LA (2009) Production of catalytically inactive BoNT/A1 holoprotein and comparison with BoNT/A1 subunit vaccines against toxin subtypes A1, A2, and A3. Vaccine 27:4490–4497

Westbrook TF, Hu G, Ang XL, Mulligan P, Pavlova NN et al (2008) SCFbeta-TRCP controls oncogenic transformation and neural differentiation through REST degradation. Nature 452:370–374

Whittaker RL, Gilbertson RB, Garrett AS Jr (1964) Botulism, type E; report of eight simultaneous cases. Ann Intern Med 61:448–454

Wong E, Cuervo AM (2010) Integration of clearance mechanisms: the proteasome and autophagy. Cold Spring Harbor perspectives in biology 2:a006734

Yamaizumi M, Mekada E, Uchida T, Okada Y (1978) One molecule of diphtheria toxin fragment A introduced into a cell can kill the cell. Cell 15:245–250

Structure-Based Drug Discovery for Botulinum Neurotoxins

Subramanyam Swaminathan

Abstract *Clostridium botulinum* neurotoxin is the most poisonous substance known to humans. It is a potential biowarfare threat and a public health hazard. The only therapeutics available is antibody treatment which will not be effective for post-exposure therapy. There are no drugs available for post-intoxication treatment. Accordingly, it is imperative to develop effective drugs to counter botulism. Available structural information on botulinum neurotoxins both alone and in complex with their substrates offers an efficient method for designing structure-based drugs to treat botulism.

Keywords Zink endoprotease · Light chain structure · Peptidic inhibitor · Pharmacophore model · Peptide analogs

Abbreviations

BoNT	Botulinum neurotoxin
SNAP-25	Synaptosomal-associated protein of 25 kDa
VAMP	Vesicle-associated membrane protein
SNARE	Soluble N-ethylmaleimide-sensitive factor attachment protein receptor
HC	Heavy chain
LC	Light chain

Contents

1 Introduction 198
2 Drug Discovery for BoNTs 201

S. Swaminathan (✉)
Biology Department, Brookhaven National Laboratory, Upton, NY 11973, USA
e-mail: swami@bnl.gov

A. Rummel and T. Binz (eds.), *Botulinum Neurotoxins*, Current Topics in Microbiology and Immunology 364, DOI: 10.1007/978-3-642-33570-9_10,

2.1 Rationale for Catalytic Domain as Drug Target for BoNT ... 202
2.2 Structure-Based Drug Discovery ... 203
3 Non-peptide Inhibitors ... 203
4 Structure-Based Drug Discovery for BoNT/A ... 203
4.1 Crystal Structures of Enzyme–Substrate Complexes and Their Use in Designing Peptide Inhibitors ... 203
4.2 Design of Structure-Based Substrate Peptide Analogs as Inhibitors ... 205
4.3 Peptide-Like Molecules as Inhibitors Based on Enzyme-Substrate Complexes ... 206
4.4 Optimum Length for Peptidic Inhibitor for BoNT/A ... 207
4.5 Pharmacophore from Co-Crystal Structures of BoNT/A and Its Use ... 207
5 Structure-Based Inhibitors for BoNT/E ... 208
6 Crystal Structure of BoNT/F-VAMP Peptide Inhibitor Complex ... 211
7 Advantages of Peptide Inhibitors ... 213
8 Concluding Remarks and Future Perspective ... 213
References ... 214

1 Introduction

Clostridium botulinum neurotoxin (BoNT) is one of the major causes of food poisoning. It affects the functioning of motoneurons leading to flaccid paralysis which is fatal, if unattended. Also, BoNT is a potential bioweapon. Accordingly, there is an urgent necessity to develop drugs to treat botulism, the disease caused by BoNT. While experimental vaccines are available to prevent botulism, their side effects are not well understood. Recently, the Centers for Disease Control and Prevention (CDC) stopped supplying this experimental vaccine. Since botulism is not a general epidemic, it is impossible to vaccinate a whole population and is not cost effective. Most importantly, the symptoms of botulism have a window of time before they can be recognized and diagnosed and this warrants development of post-exposure drugs to counter this disease. Seven serotypes of BoNT (A-G) are produced by different strains of *Clostridium botulinum* (Schiavo et al. 2000). The toxicity of BoNT works via a four-step process (Montecucco et al. 1994), viz, cell binding, internalization into neuronal cells by an energy-dependent process, translocation of the catalytic domain into the cytosol, and finally the cleavage of one of the three components of SNARE complex (Soluble N-ethylmaleimide-sensitive factor attachment protein receptor) required for membrane fusion and neurotransmitter release. Cleavage of one of the three proteins interferes with the formation of the complex thereby inhibiting neurotransmitter release leading to flaccid paralysis known as botulism. Any one of the above four steps could be a target for developing countermeasures against BoNT poisoning. While all seven serotypes (A–G) of BoNTs are poisonous, BoNT/A, B, E, and probably F affect humans, with BoNT/A being the most potent and persisting the longest within neuronal cells. Thus the focus of BoNT drug discovery has been on serotype A, though a few peptide inhibitors have been reported for serotype B, E, and F. The antibody treatments available for botulism will not be effective once the toxin has entered the neuronal cell. This makes the development of small molecule drugs that can enter the neuronal cell a top priority for post-exposure treatment.

BoNTs are produced as a single polypeptide chain (MW 150 kDa) that undergoes proteolysis, either by endogenous or exogenous protease, to provide the active dichain form of the toxin composed of a heavy chain (HC) and a light chain (LC), linked by an interchain disulfide bond. A BoNT molecule consists of three functional domains, the N-terminal half (H_N), the C-terminal half (H_C) of the HC, and the LC and all three form distinct individual structural domains (Kumaran et al. 2009; Lacy et al. 1998; Swaminathan and Eswaramoorthy 2000; Chaddock 2012) . The H_C domain, called the receptor-binding domain, further comprises two subdomains, the N-terminal H_{CN} and the C-terminal H_{CC}. H_{CC} possesses two binding sites—one for neuraminic acid containing gangliosides, GT1b, GD1b, etc., and the other for a specific protein receptor. The H_{CC} comprises a β-trefoil fold, which presents two adjacent sites for the two receptor molecules to bind. The H_{CN} domain has a lectin-binding fold but its function in BoNT is not yet clear.

The H_N domain, the translocation domain, forms a channel in the endosomal membrane and allows the catalytic domain to escape into the cytosol by a pH-dependent process. It consists of two long helices ($\sim$100 Å) forming a coiled coil and a long loop, called the belt region, that wraps around the catalytic domain (Kumaran et al. 2009; Lacy et al. 1998; Swaminathan and Eswaramoorthy 2000) and occupies a groove which would be later occupied by the substrate (Agarwal et al. 2009; Brunger et al. 2007). The mechanism of channel formation is not yet well understood (Fischer and Montal 2007; Fischer et al. 2008; Galloux et al. 2008; Koriazova and Montal 2003; Masuyer et al. 2009; Montal 2009).

The LC, which is the catalytic domain, cleaves its substrate, one of the SNARE proteins, in the cytosol. BoNTs are unique since their substrates are large polypeptides and require a large segment for optimal activity. Remarkably, each serotype cleaves a specific SNARE protein at a unique peptide bond. BoNT/A, C, and E cleave synaptosomal-associated protein of 25 kDa (SNAP-25) at specific peptide bonds, while BoNT/B, D, F, and G cleave vesicle-associated membrane protein (VAMP), each at a specific peptide bond. BoNT/C is the only serotype to cleave more than one SNARE protein, syntaxin and SNAP-25 (Schiavo et al. 2000).

The catalytic domain has an α/β fold. It is a zinc endopeptidase and contains a conserved HExxH…E zinc-binding motif in the middle of the primary sequence. The catalytic zinc occupies a deep cavity with a high negative electrostatic potential. It is coordinated by two histidines and a glutamate. The fourth coordination is from a conserved water molecule that acts as a nucleophile (Fig. 1a). The nucleophilic water molecule forms a strong hydrogen bond with the first glutamate in the zinc motif. This glutamate acts as the general base for the catalytic action. Moreover, the interactions between the conserved amino acids around the catalytic zinc (10 Å radius) are preserved among all serotypes leading to a common catalytic mechanism for all BoNTs (Agarwal et al. 2004, 2005b; Breidenbach and Brunger 2004; Kumaran et al. 2008b, 2009; Lacy et al. 1998; Silvaggi et al. 2008; Swaminathan and Eswaramoorthy 2000).

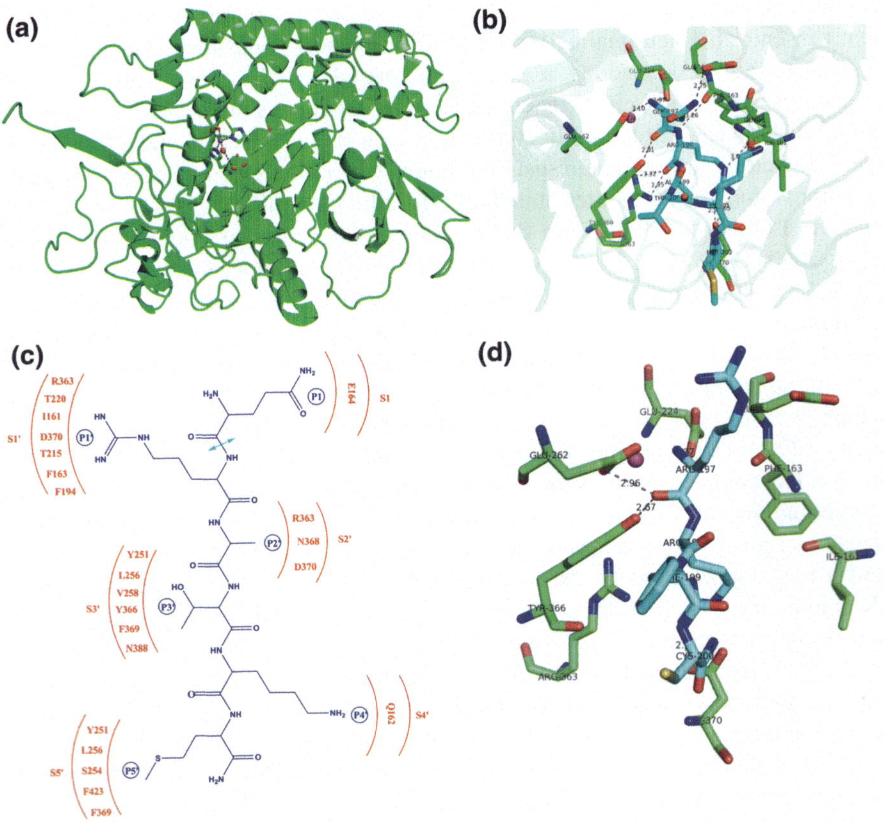

Fig. 1 **a** The catalytic domain. The catalytic domain of BoNT/B is shown in *ribbons* representation. Zinc ion (*gray*) is coordinated with the conserved His229, His233, Glu267, (shown in *stick* representation) and a nucleophilic water (*red sphere*). This coordination is the same in all BoNTs. 3EPW.pdb was used to generate this figure. **b** SNAP-25 peptide bound to BoNT/A. (197)QRATKM(202) is shown in stick model in *light blue* color and the interacting active site residues are shown in *green* in stick model. Zinc ion is shown as a sphere in *magenta*. Critical interactions between SNAP-25 peptide and the active site residues are shown in *dashed lines*. For clarity, zinc coordination is omitted. This figure was generated using 3DDA.pdb. **c** A schematic diagram representing S1 to S5′ subsites of LC/A. The co-crystal structure of (197)QRATKM(202) bound to LC/A identified the composition of subsites for P1 to P5′ residues helping in structure-based drug design for BoNT/A. (Reproduced from PLoS Pathogens, 4, e1000165, 2008). **d** Perturbation of (P1)RRGC(P3′) BoNT/A LC inhibitor peptide binding caused by changing P2′ (Gly) to Phe. The induced fit changes the conformation of P1′ and its carbonyl oxygen flips causing loss of interaction with Arg363 which is seen in Fig. 1b. This interaction is important for catalytic activity

Based on the structural and biochemical information, a model for the catalytic mechanism has been proposed (Agarwal et al. 2005a; Kumaran et al. 2008a, b; Swaminathan et al. 2004). Glu224 (BoNT/A numbering; the first Glu of the zinc-binding motif) acts as a general base by abstracting a proton from the nucleophilic water. The nucleophilic water attacks the carbonyl carbon of the scissile bond which forms the tetrahedral transition state (Kumaran et al. 2008b). The zinc ion and the conserved Arg363 and Tyr366 stabilize this intermediate tetrahedral transition state. The shuttling of protons with the help of Glu224 assists subsequent formation of the stable leaving amino group. This is consistent with the model proposed for BoNT/B and BoNT/E and is similar to that of thermolysin (Matthews 1988). The conserved Arg363 and Tyr366 help to position the substrate for Michaelis complex formation.

2 Drug Discovery for BoNTs

In the last two decades or more, the urgency of designing drugs for BoNT has led to concerted efforts in various directions toward this goal. Most of the earlier efforts were for blocking the catalytic activity after it was established that BoNT catalytic domain is a zinc endopeptidase, and a lot of effort has been invested on designing small molecule and peptidic inhibitors and it still continues. However, these efforts have not reached the stage where small molecule inhibitors could be transformed into effective drugs. After the characterization of catalytic domains of various serotypes of BoNT using biochemical, mutational, and structural tools, a number of peptide as well as nonpeptide inhibitors (NPIs) have been developed (Agarwal et al. 2004, 2005a, b, 2009; Agarwal and Swaminathan 2008; Arndt et al. 2005, 2006; Binz et al. 2002; Breidenbach and Brunger 2004; Jin et al. 2007; Kumaran et al. 2008a, b; Li et al. 1994, 2000; Sikorra et al. 2008). Most of these inhibitors have been targeted against BoNT/A.

Phosphoramidon-based inhibitors were developed for BoNT/B after it was demonstrated that it had a weak anti-BoNT activity. Based on this, a number of coumarin-based inhibitors were developed for BoNT/B (Adler et al. 1998, 1999). BABIM [bis(5-amidino-2-benzimidazoly)methane, $K_i \sim 10\ \mu M$)] was identified as an inhibitor for BoNT/B (Brewer et al. 2002; Oost et al. 2003; Sukonpan et al. 2004). Highly potent inhibitors containing β-amino thiol group were developed for BoNT/B based on the combinatorial chemistry and the structure of BoNT/B holotoxin ($K_i = 20$ nM) (Anne et al. 2003, 2005). Inhibitors to block metallopeptidase activity of tetanus toxin were discovered based on P1 and P1′ of the substrate (VAMP Gln76 and Phe77) (Martin et al. 1999). Since BoNT/A is the most potent BoNT, the focus was on developing small molecule inhibitors to block the catalytic activity of BoNT/A. Different approaches were followed. Pang has used the structural model of BoNT/A for virtual screening and has identified and designed a number of small molecule inhibitors ($K_i \sim 760$ nM) (Pang et al. 2010a, b). Schmidt worked on peptidic inhibitors based on the substrate for BoNT/A, B, and F (Schmidt and Stafford 2002, 2005; Schmidt et al. 1998). Janda et al. have used synthetic chemistry approaches (Boldt et al. 2006a, b; Capkova et al. 2007, 2008, 2009, 2010). Stevens et al. have used the crystal structures of BoNT/A to develop peptidic and peptidomimetic inhibitors (Kumaran et al. 2008b; Thompson et al. 2011; Zuniga et al. 2008). Bavari et al. have used a combined empirical and structure-based drug design (Burnett et al. 2007; Hermone et al. 2008).

2.1 Rationale for Catalytic Domain as Drug Target for BoNT

BoNT contains three well-defined domains responsible for each step of its mode of action, and thus offers multiple targets to block its toxicity. The ganglioside or the protein receptor sites (or both) in the receptor-binding domain, the channel formed by the translocation domain or the active site of the catalytic domain can be effectively used for blocking the BoNT toxicity. According to the dual-receptor model, most of the BoNTs need two receptors, a ganglioside and a protein receptor. In some cases, both the sites are occupied by gangliosides (Rummel 2012). The binding domain and the binding site of the ganglioside are well-defined for BoNT/A. However, the protein receptor site has not been mapped yet for BoNT/A though it is mapped for BoNT/B (Chai et al. 2006; Jin et al. 2006; Rummel et al. 2007; Stenmark et al. 2008). Blocking the binding sites could effectively stop the toxin from binding and entering the neuronal cell. After entering the cell, the catalytic domain has to escape from the vesicle containing the toxin into the cytosol to cleave its substrate. It is believed that the translocation domain forms a channel for the catalytic domain to escape to the cytosol. Though the mechanism of translocation is being unraveled, the shape of the channel or the composition of amino acids lining the channel is not defined structurally. However, the symptoms of botulism appear only after the toxin has entered the neuronal cell. Accordingly, blocking the binding site or the translocation channel may not be viable for post-exposure treatment.

The catalytic domain of BoNT is a zinc-dependent endopeptidase and cleaves a unique peptide bond in a specific SNARE protein. The catalytic domain has been characterized both biochemically and structurally for all serotypes. Structures of substrate–catalytic domain complex have been determined and provide information on the interactions between the substrate and the catalytic domain.

High-resolution X-ray crystal structure of the target protein is essential and a prerequisite for structure-based drug discovery (SBDD). Additionally, the architecture of the active site and the knowledge of catalytic mechanism are required. For BoNTs, the role and involvement of each amino acid at the active site is well-established structurally and also by mutagenesis. Also, all BoNTs follow a common catalytic mechanism helping in designing an inhibitor molecule for each one of them with minimal modifications. Additionally, the crystal structures of enzyme-substrate (or enzyme-substrate peptide) complex provide detailed information on the interactions between the enzyme and substrate which could be exploited to design and develop inhibitors to block the catalytic activity (Fig. 1a). For BoNTs, this is available for at least three serotypes. In view of all this, the catalytic domain is the major focus of drug discovery for BoNTs. In this chapter, the SBDD for BoNT using the catalytic domain as target will be discussed in detail.

2.2 Structure-Based Drug Discovery

SBDD makes use of knowledge of the 3-dimensional structure of the target protein and the architecture of the active site environment where the substrate binds. High-resolution X-ray crystal structures provide detailed information on both. Essentially, it depends on designing molecules which will fit this active site to block the substrate from binding to the site or will compete with the substrate for the same site. In solution, this will be a dynamic equilibrium between the inhibitor and the substrate molecule. Accordingly, the conventional drug design relies on identifying a target protein and a lead molecule which could act as an inhibitor. The lead molecule is further optimized for inhibition in an iterative manner. This lead molecule could be selected either by biochemical studies, virtual (in silico) screening, or searching for the analogs of the substrate molecule. The tight binding depends on the complementarity of the molecular forms of the active site and the inhibitor, the electrostatic charge environment of the active site, etc. In some cases, it may be possible to block an allosteric site in the target protein to achieve the same effect. However, the most effective way to block the activity is to design molecules to compete for the active site though additionally targeting allosteric sites may be more effective.

3 Non-peptide Inhibitors

Many NPIs have been discovered against various BoNT serotypes and pharmacophore models proposed for BoNT/A. Most of the recent work is based on the structural model of the catalytic domain (with or without substrate). Non-peptidic small molecules have been identified combining the structural information, molecular modeling, and combinatorial chemistry (Pang et al. 2009). Unfortunately, none of these NPIs have yielded co-crystal structures in complex with BoNT/A, so experimental proof about their binding modes and interaction with BoNT/A active site is lacking and prevents structure-based modification of the inhibitor molecules. In the following sections, we will discuss structure-based drug design starting from available crystal structures of enzyme-substrate complexes.

4 Structure-Based Drug Discovery for BoNT/A

4.1 Crystal Structures of Enzyme–Substrate Complexes and Their Use in Designing Peptide Inhibitors

Crystal structures of enzyme-substrate complex are crucial for understanding critical interactions between the enzyme and substrate and form the basis for developing substrate-based peptide inhibitors. Substrates for BoNTs are large proteins and are supposedly unstructured in their native state but become partially structured by induced fit when they bind the enzyme. However, when the substrate-enzyme complex is formed, the substrate is cleaved and the products separate and may not be in the proper bound state, and thus not helpful in understanding the interactions. Accordingly, three types of crystal structures of complexes are available for BoNT/A: (1) Crystal structure of active BoNT/A catalytic domain with uncleavable substrate peptide; (2) Crystal structure of inactive BoNT/A with cleavable substrate peptide; and (3) Crystal structure of active BoNT/A catalytic domain with a substrate-based peptide inhibitor.

Crystal structures of active BoNT/A catalytic domain with uncleavable short SNAP-25 peptides (tetra to hepta) containing the scissile bond region have been studied in detail (Kumaran et al. 2008a, b). The minimal cleavable segment of SNAP–25 by BoNT/A is (192)DEAN**QR**ATK(200), where the scissile bond (Q197–R198) is in bold (Chen and Barbieri 2006). In view of this, an uncleavable substrate peptide (197) **QR**ATKM(202) (P1 to P5′, the scissile bond being P1–P1′) was chosen for forming enzyme-substrate complex. (For the sake of clarity, single-letter amino acid codes are used for substrate peptide residues and three letter codes for enzyme residues all through this review, except in figures and figure legends. Structural studies show that the peptide tightly binds at the active site (Kumaran et al. 2008a). The crystal structure of this complex provides detailed information on the critical interactions between the enzyme and the substrate near the active site and forms the basis for designing substrate-based inhibitors (Fig. 1b). The carbonyl oxygens of P1 and P1′ are hydrogen bonded to the conserved Tyr366 and Arg363, respectively. In addition, the nucleophilic water is replaced by the free amino group of P1 which could be crucial for inhibition since it forms coordination with zinc along with the carbonyl oxygen of P1. While this structure provided invaluable information on the interactions, the inhibitory activity of the peptide was low (IC_{50} = 133 μM). However, it confirmed that the P1′ R198 is essential for cleavage by LC/A and is critical for inhibitor design as well (Schmidt and Stafford 2005). Importantly, this structure revealed the composition of the subsite associated with each substrate residue (Fig. 1c). The P1 residue (Q197) interacts with Glu164. The S1′ subsite is made up of Ile161, Phe163, Phe194, Thr215, Thr220, Arg363, and Asp370. The P1′ (R198) forms salt bridge interactions with Asp370 to help position the peptide in the right orientation. The guanidinium group has stacking interactions with Arg363 and Phe194 and these interactions help to position the peptide in proper orientation. The P2′ residue (A199) occupies subsite S2′ comprising Arg363, Asn368, and Asp370. The P3′ residue, T200 occupies a subsite formed by Tyr251, Leu256, Val258, Tyr366, Phe369, and Asn388. P4′ (K201) has a sole interaction with Gln162 and P5′ (M202) interacts with Tyr251, Ser254, Leu256, Phe369, and Phe423. The mapping of each subsite helped in designing subsequent peptide inhibitors based on the size and nature of the subsites.

The crystal structure of the complex of an inactive double mutant of BoNT/A catalytic domain with a cleavable SNAP-25 peptide (146–204) has provided the mode of substrate recognition by BoNT/A (Breidenbach and Brunger 2004). However, an inactive mutant enzyme does not allow the substrate to dock properly to form the Michaelis complex at the active site because of loss of key interactions with the enzyme. Crystal structure analysis of the double mutant of BoNT/A and SNAP-25 (146–204) revealed interactions between the enzyme and the substrate remote from the active site but the interactions at the active site were missing because of the double mutation (Tyr366Phe, Glu224Ala) (Breidenbach and Brunger 2004). Tyr366 has been shown to be important for positioning the scissile bond for cleavage. However, this seminal work unequivocally identified the exosites for BoNT/A for the first time which are now being pursued for inhibitor design.

The crystal structure of active BoNT/A catalytic domain with a relatively weak (Ki $\sim$ 2 μM) short peptide inhibitor (N-acetyl-CRATKL-amide) was determined to analyze the active site interactions (Schmidt and Stafford 2002; Silvaggi et al. 2008). In most cases, the complex of the substrate-based inhibitor with the enzyme gives a realistic picture of the interactions at the active site. However, N-acetyl-CRATKL-amide inhibitor bound structure did not exactly replicate the substrate-enzyme complex for two reasons. The N-acetyl group points to a hydrophobic pocket S1′ formed by Phe163 and Phe194 (normally occupied by the P1′ R198) and thereby forces the side chain of C197 to face the catalytic zinc or vice versa. The Sγ of C197 is oxidized to sulfenic acid and coordinates with zinc. This forces the peptide to slip down by one residue and the carbonyl oxygen of P1 forms a hydrogen bond with Arg363 instead of with the usual Tyr366 (see below). Altogether, these three crystal structures have helped in designing substrate-based peptide or peptidomimetic inhibitors for BoNT/A.

4.2 *Design of Structure-Based Substrate Peptide Analogs as Inhibitors*

Since the subsite of each amino acid of the substrate peptide, QRATKM, is identified in BoNT/A LC (Fig. 1c), modification of one residue at a time has been followed (Kumar et al. 2012b). P1 (Q197) of the peptide is in a region with negative electrostatic potential though P1 itself interacts with only one enzyme residue, Glu164. The negative electrostatic potential suggests that a positively charged residue will have better complementarity. When Q197 is replaced with R197, a positively charged residue, the interactions improve. The (197)RRATKM(202) peptide showed better interaction, especially with an additional salt bridge between R197 and Glu164 but does not improve the inhibitory activity significantly over QRATKM (IC_{50} = 95 vs 133 μM). To reduce the number of possibilities and also since K and M had slightly larger thermal factors indicating disorder, the inhibitor design is restricted to tetrapeptides comprising P1 to P3′. The P1′ residue (R198) occupies a negatively charged deep pocket and fits well, making salt bridge and stacking interactions. Since biochemical evidence on activity has shown that arginine is absolutely essential at P1′ position for inhibitory activity, P1′ was left unchanged (Chen and Barbieri 2006). The next residue A199 (P2′) is a small hydrophobic residue whose side chain cannot make any hydrogen bond or ionic contact. To allow some flexibility in this region it was changed to glycine. T200 (P3′) is in a hydrophobic environment. This was changed to C200 since it is capable of interacting with charged or hydrophilic groups. Thus, the peptide RRGC was designed and the structure showed good interactions between the enzyme and the peptide, and the K_i improved to 157 nM ($IC_{50} = 1.5\ \mu M$) (Kumaran et al. 2008b and Kumar et al. 2012b). Interestingly, these changes also induce a better fit for P1′ (R198) at S1′ site (Fig. 1b). Also, the stacking interaction among P1′, Phe194, and Arg363 is better than that in QRATKM or RRATKM complexes. Comparison of the electrostatic potential of these two also shows that the charge complementarity is better in RRGC than in QRATKM, as expected. Keeping RRG constant, the P3′ residue was changed to larger hydrophobic residues successively as RRGL, RRGM, RRGI, and RRGF. But the first three were not better than RRGC. When P3′ was changed to phenylalanine, it had better hydrophobic and aromatic interactions in the S3′ site resulting in better inhibitory activity ($IC_{50} = 0.9\ \mu M$) (Kumar et al. 2012b).

Though P2′ is alanine in the natural substrate and is glycine in peptide inhibitor RRGC, a bigger hydrophobic residue could be accommodated at the S2′ site. When P2′ is changed to phenylalanine, the carbonyl oxygen of P1′ flips over and because of this the critical interactions between the carbonyl oxygen of P1′ and Arg363, and the salt bridge between P1′ (R198) and Asp370 are lost and the stacking interaction is disturbed. This causes the side chain of P1′ to take a different rotamer position though it occupies the same cavity (Fig. 1d). The negatively charged S1′ site is big enough to accommodate different orientations of the guanidinium group. The volume of the side chain of arginine is $\sim 135\ \text{Å}^3$ compared to the subsite S1′ which is $\sim 370\ \text{Å}^3$. This also points to the fact that a bigger peptidomimetic could fit in that cavity allowing tighter binding but might perturb the structure. However, the inhibitory potency of RRFC was very similar to RRGC (1.8 vs 1.5 μM) (Kumar et al. 2012b). The flipping of carbonyl bond of P1′ was also observed with Ac-CRATKML structure where the peptide has slipped by one residue because of the coordination of Sγ of C197 with Zn (Schmidt et al. 1998; Silvaggi et al. 2008).

Changing one residue at a time may be advantageous but may not always give the desired result since the conformation of the peptide changes to accommodate to the site resulting in perturbation in the structure and loss of some critical interactions. This may both be the strength and weakness of peptide inhibitors since they have large flexibility. As described later, a pharmacophore model based on experimental structures is useful in designing effective inhibitors.

4.3 Peptide-Like Molecules as Inhibitors Based on Enzyme-Substrate Complexes

A peptide-like molecule (PLM) based on the inhibitor CRATKML called I1 (*DNP-DAB*-RWT-*DAB*-ML where *DNP-DAB* is 4-(2,4-dinitrophenylamino)-2-amino-butanoic acid and *DAB* is 2,4-diaminobutanoic acid) has been designed:

DNP-DAB-RWT-*DAB*-ML (Zuniga et al. 2008, 2010). Only C197 and K201 were modified giving a K_i of 41 nM (Zuniga et al. 2008), the best so far for a substrate-based inhibitor for BoNT/A. Interestingly, this modified peptide had a helical conformation unlike the extended conformation that has been found in other peptide inhibitors. Later, using an iterative structure-based approach, a number of peptide mimics were designed and synthesized combining the features from the most potent PLM I1 ($K_i = 41$ nM) and RRGC ($K_i = 157$ nM) resulting in JTH-NB72-35, -38, and -39. Of the new peptidomimetics, JTH-NB72-35 (RR-Nph-Aib-AML, where Aib is aminoisobutyric acid and Nph is naphthylene) was found to be the most potent ($K_i = 314.5$ nM) though these compounds are worse than I1 or RRGC in their inhibitory potency against BoNT/A (Zuniga et al. 2010). However, only JTH-NB72-39 could be co-crystallized. This peptide showed a similar effect observed with RRFC where the carbonyl bond of P1′ flipped causing loss of interactions probably leading to higher K_i value. Incidentally, P2′ residue is phenylalanine in both inhibitors (Fig. 1d).

4.4 Optimum Length for Peptidic Inhibitor for BoNT/A

In order to find the optimum length of the peptide inhibitor, the peptide length was increased or decreased by one or two residues at a time either at the C- or N-terminus and no improvement in inhibitory activity was observed. A single (R) or dipeptide (RR) was also tried. 2,4-dichlorocinnamic hydroxamate, an analog of arginine, has been tried earlier but K_i of 300 nM was worse than that of RRGC (Hale et al. 2010). Based on this, it was concluded that a tetrapeptide is the optimum length for a substrate-based inhibitor. Crystal structures of various tetrapeptides in complex with BoNT/A catalytic domain were determined and used to analyze structure activity relation (Hale et al. 2010; Kumar et al. 2012b).

4.5 Pharmacophore from Co-Crystal Structures of BoNT/A and Its Use

SBDD uses the complementarity of the interface between the substrate and the enzyme, especially near the active site. An inhibitor is designed to complement the molecular features, both in shape and chemical properties, near the active site. A pharmacophore describing the molecular features necessary to recognize the substrate at the active site is modeled based on the interactions between the enzyme and the substrate and helps in designing powerful inhibitors. Structural information from BoNT/A LC-SNAP-25 (146–204) complex structure was used to arrive at a pharmacophore model. Since some of the loops near the active site are flexible, molecular dynamics was used to generate different dynamic conformers to model 2-mercapto-3-phenylpropionyl RATKML, an inhibitor with $K_i = 330$ nM (Burnett et al. 2007).

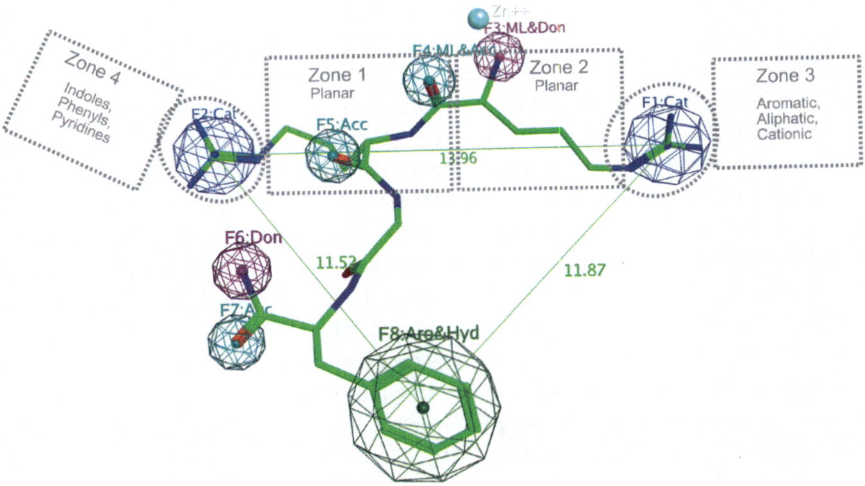

Fig. 2 Pharmacophore model for BoNT/A inhibitors. This pharmacophore model was arrived at using a number of SNAP-25 peptide-BoNT/A light chain complexes. The pharmacophore model has: *1* two positive ion features, out of which one also has π-stacking interaction, *2* metal ligands, *3* a hydrogen bond acceptor, *4* a hydrogen bond donor/acceptor pair and *5* an aromatic/hydrophobic feature. Reproduced from Acta Crystallograph. D, 68, 511, 2012

This pharmacophore model was used in virtual screening to identify lead inhibitor molecules with the National Cancer Institute, USA (NCI) open repository. Later, this pharmacophore was further modified based on more docking trials. The refined pharmacophore described by Burnett et al. (2007) was arrived at by combining the structural information and docking of small molecules and comprises three zones and was used for further virtual screening of small molecule libraries. The pharmacophore model based on a number of co-crystal structures of BoNT/A LC and tetrapetide inhibitors in addition to empirical results from small molecule docking is also available (Kumar et al. 2012b). From the inhibition data and the co-crystal structures, certain features emerged as major contributors in the inhibitory activity. These features were combined together in a pharmacophore model. The pharmacophore model has: (1) two positive ion features, out of which one also has π-stacking interaction, (2) metal ligands, (3) a hydrogen bond acceptor, (4) a hydrogen bond donor/acceptor pair, and (5) an aromatic/hydrophobic feature. This pharmacophore is based on experimental structural data and will form the basis for future modifications of the tetrapeptide to a suitable peptidomimetic drug candidate (Fig. 2).

5 Structure-Based Inhibitors for BoNT/E

For BoNT/E, the minimal substrate required for cleavage is (178)ID**RI**ME(183) but adding (184)KAD(186) just enhances the substrate cleavage (Chen and Barbieri 2006). The crystal structure of BoNT/E in complex with uncleavable SNAP-

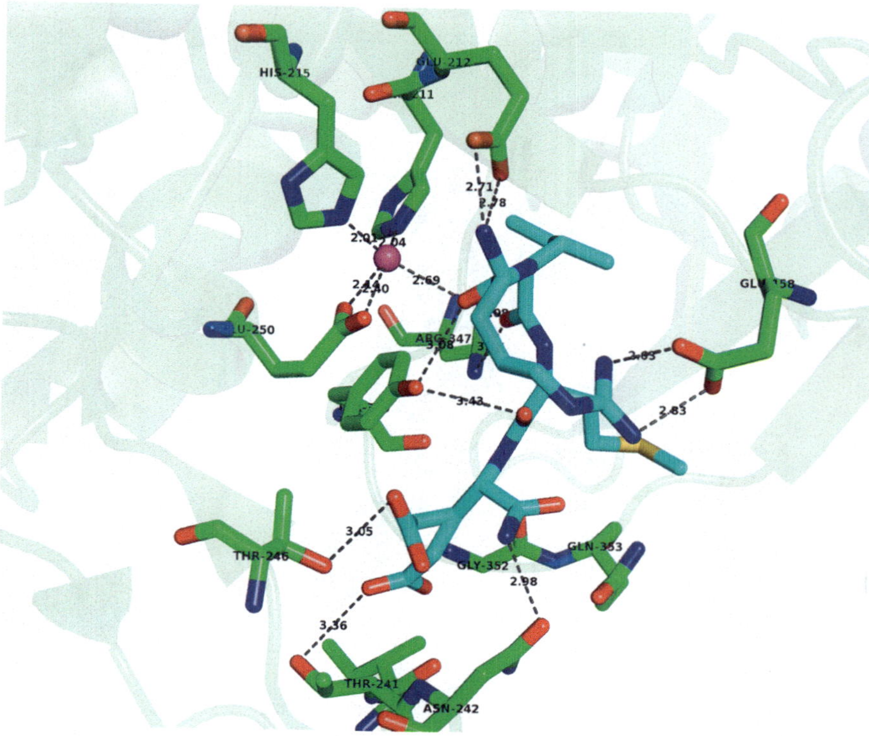

Fig. 3 Peptide inhibitor for BoNT/E. SNAP-25 peptide RIME (180–183, P1–P3′) bound to BoNT/E is shown in stick model in *cyan*. BoNT/E residues interacting with the SNAP-25 peptide are shown as *green* stick model. Zinc is shown as a sphere (*magenta*). BoNT/E is in *ribbons* representation in light *green*. This figure was generated using 3D3X.pdb

25 peptide RIME (180–183, P1–P3′; 3D3X.pdb) provides detailed information regarding the interaction between the enzyme and the substrate peptide (Agarwal and Swaminathan 2008; Chen and Barbieri 2007) (Fig. 3). Though the RIME peptide competes with SNAP-25 substrate peptide it was a poor inhibitor with $K_i \sim 69$ μM. However, the structure of the complex brings out many important and interesting features and defines the subsites for P1 to P3′. The hydrogen bonds between the carbonyl oxygens of P1 and P1′ with the residues Arg347 and Tyr350 homologous to Arg363 and Tyr366 of LC/A, respectively, confirmed that these two conserved residues play a critical role in presenting the scissile bond to the catalytic center in proper orientation in all BoNT serotypes. P1 establishes many hydrogen bonds with S1 site. Most importantly, the hydrophobic P1′ residue (I181) is in a hydrophobic pocket formed by Thr159, Phe191, and Thr208 of LC/E. Biochemical evidence has also shown that Phe191 is crucial since changing it to Asn decreases the activity of the BoNT/E LC 100 fold. Additionally, P3′ (M182) also faces a hydrophobic pocket formed by Phe191, Tyr354, and Tyr356 confirming the importance of Phe191. Thus, the complex structure explains the

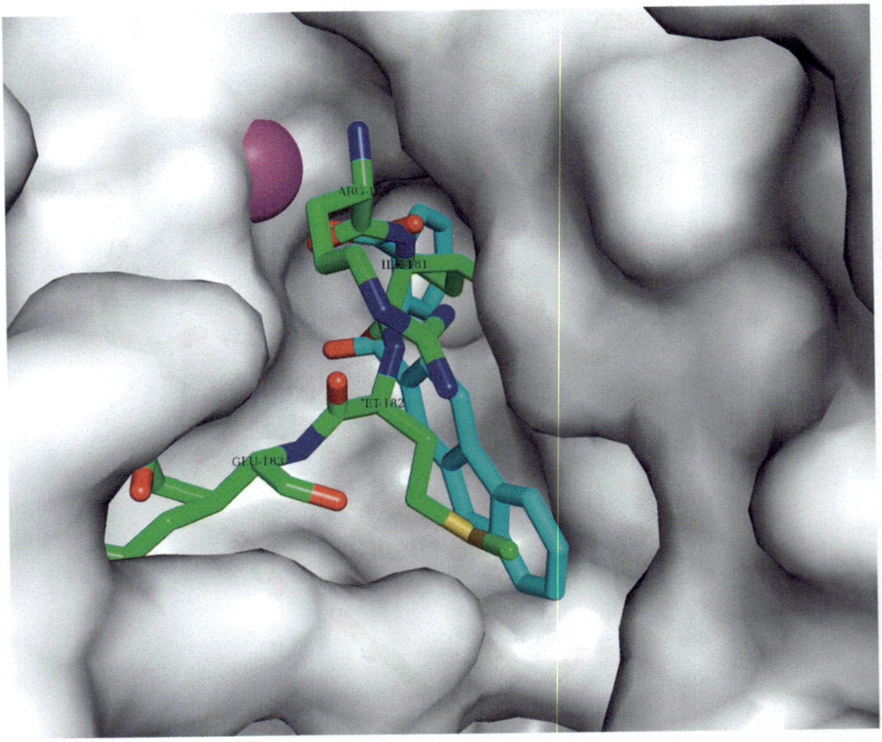

Fig. 4 A small molecule inhibitor bound to BoNT/E. NCI77053 (2-(9H-fluoren-2-ylcarbonyl)benzoic acid) was identified ($K_i \sim 1.2$ μM) by virtual screening using BoNT/E-RIME complex structure. The small molecule inhibitor is superposed on RIME and shows that P1′ and P2′ residues are in the same orientation as benzene and fluorene rings

hydrophobic nature of S1′ and S2′ sites. Also, a charged residue (D179) at P2 may be required for proper docking as seen by modeling and supported by mutagenesis (Agarwal and Swaminathan 2008; Chen and Barbieri 2007). Additionally, even though there are three regions in SNAP-25 with the amino acid sequence RI, BoNT/E cleaves only RI in the unique region (178)IDRIME(183) and neither in (57)LERIEE(62) nor in (189)KTRIDE(194). The combined requirement of a charged P2 residue and a hydrophobic P2′ residue in the scissile bond region prevents the other two RI peptides being cleaved. Unfortunately, this structure does not provide any information on the exosites which would help in designing longer peptidomimetics. Systematic variation of this tetrapeptide can now be undertaken to find a suitable peptide or peptidomimetic inhibitor for BoNT/E LC.

On the basis of this model, the virtual screening using the crystal structure of the catalytic domain from the complex structure (devoid of the substrate peptide; 3D3X.pdb) has helped in identifying a number of small molecule inhibitors. One such molecule NCI77053 (2-(9H-fluoren-2-ylcarbonyl)benzoic acid) shows reasonably good inhibition with $K_i = 1.2$ μM and provides a lead molecule to work with. The orientation of the docked NCI77053 in the BoNT/E LC shows that the interactions discussed above are mimicked and may be the reason for good inhibitory activity (Fig. 4). This discovery has opened up new avenues to identify more analogs of 77053 for better inhibition based on structure–activity relationship (Kumar et al. 2012a). However, this molecule has neither been tested in cell-based assays nor co-crystallized with LC/E yet.

6 Crystal Structure of BoNT/F-VAMP Peptide Inhibitor Complex

In the case of defining the BoNT/F substrate minimum requirements two different approaches were followed (Schmidt and Stafford 2005; Sikorra et al. 2008). Both groups have identified the optimal length of the substrate required for cleavage by BoNT/F, A21-D64, or P22-D65. Schmidt's group has systematically truncated the protein both at the N and C terminals to arrive at the optimum length while Binz's group has used mutagenic studies to arrive at the same result. In addition to this, sulfhydryl group containing cysteine or D-cysteine was substituted for P1 (Q58) or P1′ (K59) and it was found that cysteine substitution at P1 works better as inhibitor. The same studies also found that while P2, P1′, and P2′ are crucial for catalytic activity, P1 is not and is tolerant to substitution with other amino acids. Similar results have also been shown for other BoNT serotypes. Of the various polypeptide lengths tested, P22-Q58 seemed to be the optimum since increasing the length up to residue S75 significantly decreased binding affinity. Based on this study, two inhibitors, VAMP-2-P22-Q58-Cys(D) and VAMP-2-T27-Q58-Cys(D), were tested. Both showed excellent inhibition property ($K_i = 1.0$ and 1.9 nM, respectively) (Schmidt and Stafford 2005)..

Crystal structures of both these inhibitors in complex with BoNT/F catalytic domain revealed many critical interactions and allowed a few key exosites to be defined (Agarwal et al. 2009) (Fig. 5). It also identified and confirmed the crucial role of amino acids of both the substrate and the enzyme. Additionally, Arg133 and Arg171 of LC/F were identified to be critical for catalytic activity. However, the inhibitors are large polypeptides and may not follow Lipinski's rule of five which describes properties important for a drug including their absorption, distribution, metabolism, and excretion (ADME). Based on the crystal structure, an optimum peptide inhibitor can be designed similar to those for BoNT/A. The co-crystal structure established the importance of P2 (D57), P3 (R56), and P4 (E55). P2 forms a salt bridge with Arg240 and Arg263 of BoNT/F, P3 forms a salt bridge with Glu164, and P4 does the same with Arg171. These three substrate residues are crucial for the substrate binding and recognition and are important for inhibitor design also. This suggests that a smaller polypeptide (less than 7 aa) could be designed and suitably modified for better inhibition. Such efforts are under way. Taken together, these results from biochemical analysis and the co-crystal

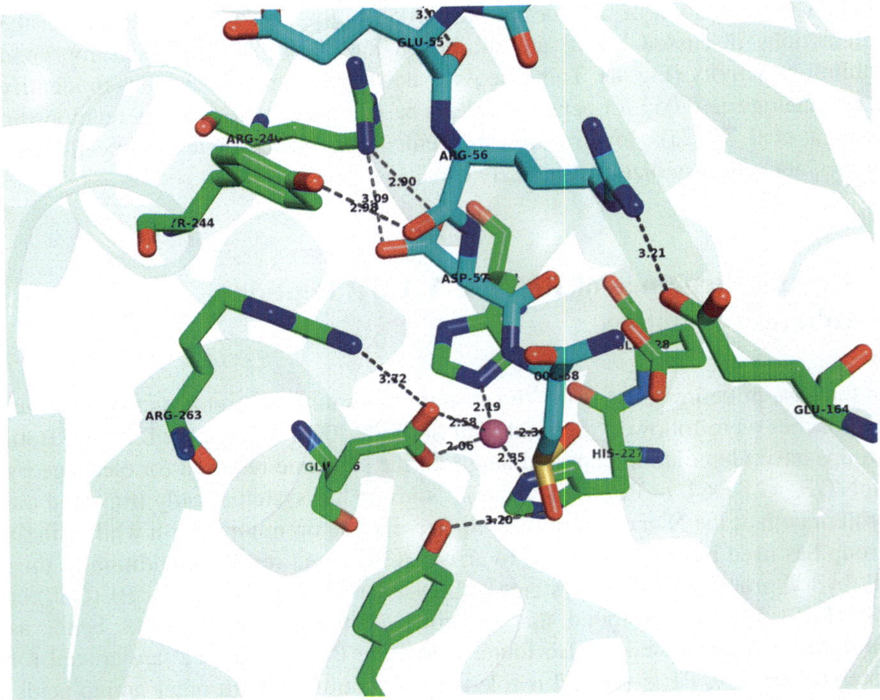

Fig. 5 VAMP peptide inhibitor bound to BoNT/F. A close-up view of VAMP-27-58/Gln58-Cys(D) inhibitor (K_i = 1.9 nM) bound to BoNT/F light chain is shown in close-up view near the active site. The inhibitor and BoNT/F are shown in stick models in *cyan* and *green*, respectively. Zinc is shown as a sphere in *magenta*. For clarity, only a limited number of interactions are shown. BoNT/F is also shown in ribbon model in light *green*. Coordinates from 3FII.pdb are used for generating this figure

structure will be useful for designing structure-based inhibitors with better efficiency. Incidentally, the substrate-based peptide inhibitors for BoNT/F terminate at P1, while those for BoNT/A begin at P1.

While co-crystal structures of enzyme-substrate peptides are available for BoNT/A, E, and F, such data are lacking for other serotypes. If such structures are determined in future they will greatly help in designing substrate-based peptide inhibitors. Though BoNT/F and BoNT/D cleave adjacent peptide bonds in VAMP-2 (Gln58-Lys59 vs Lys59-Leu60), the optimal length of substrate required for cleavage is different for these two as also the exosites required for the substrate to bind (Sikorra et al. 2008). Extensive biochemical work has been done analyzing the residues of the substrate playing important role in substrate cleavage. Likewise BoNT/A and BoNT/C cleave adjacent peptide bonds in SNAP-25 but again have different requirement for substrate cleavage (Sikorra et al. 2008; Vaidyanathan et al. 1999). However, for both BoNT/C and BoNT/D there is no published report on SBDD. SBDD has been pursued but there is no co-crystal structure of BoNT/B with inhibitor to analyze structure activity relationship for modifying the inhibitor for better efficacy (Anne et al. 2003, 2005).

7 Advantages of Peptide Inhibitors

Significant work has been done in the recent past on the discovery of peptide-based inhibitors of BoNTs. Various aspects of peptide-based inhibitors that have been explored are discussed below. Peptide-based inhibitors provide an excellent opportunity for the discovery of drug candidates targeting protease activity of BoNTs. The natural complementarity of the substrate peptide with the enzyme active site can be easily mimicked in the peptide-based inhibitors and there is scope for introducing rigidity, decreasing rotational freedom, and specificity. The natural substrate for BoNT/A is SNAP-25 that wraps around the protease like the belt region of holotoxin. The active site of BoNT is shallow except for the S1′ pocket, which accommodates the P1′ amino acid of the substrate. So, in principle, competitive inhibitors as large as the whole substrate are possible. To conform to Lipinski's rule of five, and design an inhibitor with molecular weight <500 Da, it may be necessary to focus around the S1′ site and the cofactor zinc that is at the catalytic center. A number of tetrapeptides that meet Lipinski's criteria for molecular weight have been tested on BoNT/A. Apart from exhibiting inhibition of the purified BoNT/A in nanomolar concentrations, these tetrapeptides also inhibited BoNT/A cleavage of the endogenous SNAP-25 in mouse brain lysates. These peptides entered the neuronal cell lines, Neuro-2A and BE(2)-M17, without any adverse effect on metabolic functions as measured by ATP production and P-38 phosphorylation. Also, the peptides showed biological activity in cultured chick motor neurons, and rat and mouse cerebellar neurons for more than 40 h and inhibited BoNT/A protease action inside the neurons in a dose- and time-dependent fashion. All these properties make tetrapeptide inhibitors excellent candidates for drug development (Hale et al. 2010). However, the pharmacokinetics like ADME has not been tested in vivo experiments.

8 Concluding Remarks and Future Perspective

Research in the field of drug discovery is growing fast with many different approaches. Many factors need to be taken into consideration with the peptide- and structure-based drug design. The flexibility of the loops in the enzyme molecule may pose a problem during docking, especially since the target molecule is treated as a rigid molecule during docking. This is avoided by considering multiple conformations of the loops near the active site or conducting molecular dynamics simulations to achieve a minimum energy conformation prior to docking. The candidate peptides are also flexible and the results from docking simulations may not represent the binding mode in solution. Biochemical assays and the co-crystal structures provide confirmatory evidence for inhibition and the mode of binding. Another problem is the lifetime of peptide inside the cells. This may be avoided by blocking the terminal regions or by designing peptidomimetics. One interesting aspect is that the substrate peptide forms a beta strand near the active site and forms an antiparallel beta sheet with a beta strand of the enzyme. Hydrogen bonded synthetic mimics of the secondary structure of the substrate peptide may be tried to avoid flexibility of the peptide and to disrupt the protein–protein interactions (Adler et al. 2011). All the peptide inhibitors discussed in this chapter are peptides blocking the active site. Exosites are crucial for substrate recognition and this fact could be taken into account and used to enhance the inhibitory effect. However, this increases the peptide length and may not give an optimum molecular mass to obey Lipinski's rule. Could two small molecules, one blocking the active site and the other the exosite, be used synergistically? Research in this direction is yet to be explored. Another aspect that has not been explored is to identify pan-active inhibitors which will act on more than one serotype, though a crossover inhibitor, chicoric acid, has been identified for BoNT/A (Silhár et al. 2010) and BoNT/B (Salzameda et al. 2011). The commonality of the active site and the common catalytic mechanism (Binz 2012) suggest that it may be possible. In conclusion, there are challenges in designing peptide-based inhibitors, but they have the potential to provide drug candidates for botulism. Small molecule inhibitors have been identified and tested in vitro for their effect in blocking the catalytic activity. Recently, these are being tested in cell-based assays and efforts to deliver them into the cell are also being undertaken. Their pharmacological effects must also be studied before they can become a drug approved by Food and Drug Administration. In summary, there are challenges to overcome, but the pace at which the research is moving gives an optimistic outlook.

Acknowledgments The author is thankful to his colleagues and collaborators for contributing to this research. Special thanks are due to Drs. S. Eswaramoorthy, D. Kumaran, R. Agarwal, S.A. Ahmed, and G. Kumar. Research was supported by an award from DTRA BO742081 under DOE prime contract No. DEAC02-98CH10886 with Brookhaven National Laboratory.

References

Adler M, Nicholson JD, Hackley BE (1998) Efficacy of a novel metalloprotease inhibitor on botulinum neurotoxin B activity. FEBS Lett 429:234–238

Adler M, Nicholson JD, Starks DF, Kane CT, Cornille F, Hackley BE (1999) Evaluation of phosphoramidon and three synthetic phosphonates for inhibition of botulinum neurotoxin B catalytic activity. J Appl Toxicol 19:S5–S11

Adler MJ, Jamieson AG, Hamilton AD (2011) Hydrogen-bonded synthetic mimics of protein secondary structure as disruptors of protein–protein interactions. Curr Top Microbiol Immunol 348:1–23

Agarwal R, Binz T, Swaminathan S (2005a) Analysis of active site residues of botulinum neurotoxin E by mutational, functional and structural studies: Glu335Gln is an apoenzyme. Biochemistry 44:8291–8302

Agarwal R, Binz T, Swaminathan S (2005b) Structural analysis of botulinum neurotoxin serotype f light chain: implications on substrate binding and inhibitor design. Biochemistry 44: 11758–11765

Agarwal R, Eswaramoorthy S, Kumaran D, Binz T, Swaminathan S (2004) Structural analysis of botulinum neurotoxin type E catalytic domain and its mutant Glu212->Gln reveals the pivotal role of the Glu212 carboxylate in the catalytic pathway. Biochemistry 43:6637–6644

Agarwal R, Schmidt JJ, Stafford RG, Swaminathan S (2009) Mode of VAMP substrate recognition and inhibition of *Clostridium botulinum* neurotoxin F. Nat Struct Mol Biol 16:789–794

Agarwal R, Swaminathan S (2008) SNAP-25 substrate peptide (residues 180–183) binds to but bypasses cleavage by catalytically active Clostridium botulinum neurotoxin E. J Biol Chem 283:25944–25951

Anne C, Turcaud S, Blommaert AG, Darchen F, Johnson EA, Roques BP (2005) Partial protection against Botulinum B neurotoxin-induced blocking of exocytosis by a potent inhibitor of its metallopeptidase activity. Chem Bio Chem 8:1375–1380

Anne C, Turcaud S, Quancard J, Teffo F, Meudal H, Fournie-Zaluski M-C, Roques BP (2003) Development of potent inhibitors of botulinum neurotoxin type B. J Med Chem 46:4648–4656

Arndt JW, Chai Q, Christian T, Stevens RC (2006) Structure of botulinum neurotoxin type D light chain at 1.65 A resolution: repercussions for VAMP-2 substrate specificity. Biochem Cell Biol 45:3255–3262

Arndt JW, Yu W, Bi F, Stevens RC (2005) Crystal structure of botulinum neurotoxin type G light chain: serotype divergence in substrate recognition. Biochemistry 44:9574–9580

Binz T (2012) Neurotoxin light chains: devices for snare cleavage mediated blockade of neurotransmission. doi:10.1007/978-3-642-33570-9_7

Binz T, Bade S, Rummel A, Kollewe A, Alves J (2002) Arg^{362} and Tyr^{365} of the botulinum neurotoxin type A light chain are involved in transition state stabilization. Biochemistry 41:1717–1723

Boldt GE, Eubanks LM, Janda KD (2006a) Identification of a botulinum neurotoxin A protease inhibitor displaying efficacy in a cellular model. Chem Commun 7:3063–3065

Boldt GE, Kennedy JP, Hixon MS, McAllister LA, Barbieri JT, Tzipori S, Janda KD (2006b) Synthesis, characterization and development of a high-throughput methodology for the discovery of botulinum neurotoxin A inhibitors. J Comb Chem 8:513–521

Breidenbach MA, Brunger A (2004) Substrate recognition strategy for botulinum neurotoxin serotype A. Nature 432:925–929

Brewer M, Oost T, Sukonpan C, Pereckas M, Rich DH (2002) Sequencing hydroxyethylamine-containing peptides via Edman degradation. Org Lett 4:3469–3472

Brunger AT, Breidenbach MA, Jin R, Fischer A, Santos JS, Montal M (2007) Botulinum neurotoxin heavy chain belt as an intramolecular chaperone for the light chain. PLoS Pathog 3:1191–1194

Burnett JC, Ruthel G, Stegmann CM, Panchal RG, Nguyen TL, Hermone AR, Stafford RG, Lane DJ, Kenny TA, McGrath CF, Wipf P, Stahl AM, Schmidt JJ, Gussio R, Brunger AT, Bavari S (2007) Inhibition of metalloprotease botulinum serotype A from a pseudo-peptide binding mode to a small molecule that is active in primary neurons. J Biol Chem 282:5004–5014

Capkova K, Hixon MS, McAllister LA, Janda KD (2008) Toward the discovery of potent inhibitors of botulinum neurotoxin A: development of a robust LC MS based assay operational from low to subnanomolar enzyme concentrations. Chem Commun (Camb) 30:3525–3527

Capkova K, Hixon MS, Pellett S, Barbieri JT, Johnson EA, Janda KD (2010) Benzylidene cyclopentenediones: First irreversible inhibitors against botulinum neurotoxin A's zinc endopeptidase. Bioorg Med Chem Lett 20:206–208

Capkova K, Salzameda NT, Janda KD (2009) Investigations into small molecule non-peptidic inhibitors of the botulinum neurotoxins. Toxicon 54:575–582

Capkova K, Yoneda Y, Dickerson TJ, Janda KD (2007) Synthesis and structure-activity relationships of second-generation hydroxamate botulinum neurotoxin A protease inhibitors. Bioorg Med Chem Lett 17:6463–6466

Chaddock J (2012) Transforming the domain structure of botulinum neurotoxins into novel therapeutics. doi:10.1007/978-3-642-33570-9_13

Chai Q, Arndt JW, Dong M, Tepp WH, Johnson EA, Chapmann ER, Stevens RC (2006) Structural basis of cell surface receptor recognition by botulinum neurotoxin B. Nature (London), New Biology 444:1019–1020

Chen S, Barbieri JT (2006) Unique substrate recognition by botulinum neurotoxins serotypes A and E. J Biol Chem 281:10906–10911

Chen S, Barbieri JT (2007) Multiple pocket recognition of SNAP25 by botulinum neurotoxin serotype E. J Biol Chem 282:25540–25547

Fischer A, Montal M (2007) Crucial role of the disulfide bridge between botulinum neurotoxin light and heavy chains in protease translocation across membranes. J Biol Chem 282: 29604–29611

Fischer A, Mushrush DJ, Lacy DB, Montal M (2008) Botulinum neurotoxin devoid of receptor binding domain translocates active protease. PLoS Pathog 4:e1000245

Galloux M, Vitrac H, Montagner C, Raffestin S, Popoff MR, Chenal A, Forge V, Gillet D (2008) Membrane Interaction of botulinum neurotoxin A translocation (T) domain. The belt region is a regulatory loop for membrane interaction. J Biol Chem 283:27668–27676

Hale M, Oyler G, Swaminathan S, Ahmed SA (2010) Basic tetrapeptides as potent intracellular inhibitors of type A botulinum neurotoxin protease activity. J Biol Chem 286:1802–1811

Hermone AR, Burnett JC, Nuss JE, Tressler LE, Nguyen TL, Solaja BA, Vennerstorm JL, Schmidt JJ, Wipf P, Bavari S, Gussio R (2008) Three-Dimensional Database Mining Identifies a Unique Chemotype that Unites Structurally Diverse Botulinum Neurotoxin Serotype A Inhibitors in a Three-Zone Pharmacophore. Chem Med Chem 3:1905–1912

Jin R, Rummel A, Binz T, Brunger AT (2006) Botulinum neurotoxin B recognizes its protein receptor with high affinity and specificity. Nature (London) New Biology 444:1092–1095

Jin R, Sikorra S, Stegmann CM, Pich A, Binz T, Brunger AT (2007) Structural and biochemical studies of botulinum neurotoxin serotype C1 light chain protease: implications for dual substrate specificity. Biochemistry 46:10685–10693

Koriazova L, Montal M (2003) Translocation of botulinum neurotoxin light chain protease through the heavy chain channel. Nat Struct Biol 10:13–18

Kumar G, Agarwal R, Swaminathan S (2012a) Discovery of a fluorene class of compounds as inhibitors of botulinum neurotoxin serotype E by virtual screening. Chem Commun (Camb) 48:2412–2414

Kumar G, Kumaran D, Ahmed SA, Swaminathan S (2012b) Peptide inhibitors of botulinum neurotoxin serotype A: design, inhibition, co crystal structures, structure-activity relationship and pharmacophore modeling. Acta Crystallogr D Biol Crystallogr 68:511–520

Kumaran D, Eswaramoorthy S, Furey W, Navaza J, Sax M, Swaminathan S (2009) Domain organization in *Clostridium botulinum* neurotoxin type E is unique: Its implication in faster translocation. J Mol Biol 386:233–245

Kumaran D, Rawat R, Ahmed SA, Swaminathan S (2008a) Substrate binding mode and its implication on drug design for botulinum neurotoxin A. PLoS Pathog 4:e1000165

Kumaran D, Rawat R, Ludivico ML, Ahmed SA, Swaminathan S (2008b) Structure and substrate based inhibitor design for clostridium botulinum neurotoxin serotype A. J Biol Chem 283:18883–18891

Lacy DB, Tepp W, Cohen AC, DasGupta BR, Stevens RC (1998) Crystal structure of botulinum neurotoxin type A and implications for toxicity. Nat Struct Biol 5:898–902

Li L, Binz T, Niemann H, Singh BR (2000) Probing the mechanistic role of glutamate residues in the zinc-binding motif of type A botulinum neurotoxin light chain. Biochemistry 39: 2399–2405

Li Y, Foran P, Fairweather NF, Paiva Ad, Weller U, Dougan G, Dolly JO (1994) A single mutation in the recombinant light chain of tetanus toxin abolishes its proteolytic activity and removes the toxicity seen after reconstitution with native heavy chain. Biochemistry 33: 7014–7020

Martin L, Cornille F, Turcaud S, Meudal H, Roques BP, Fourine-Zaluski MC (1999) Metallopeptidase inhibitors of tetanus toxin: a combinatorial approach. J Med Chem 42:515–525

Masuyer G, Thiyagarajan N, James PL, Marks PHH, Chaddock J, Acharya KR (2009) Crystal structure of a catalytically active, non-toxic endopeptidase derivative of Clostridium botulinum toxin A. Biochem Biophys Res Commun 381:50–53

Matthews BW (1988) Structural basis of the action of thermolysin and related zinc peptides. Acc Chem Res 21:333–340

Montal M (2009) Translocation of botulinum neurotoxin light chain protease by the heavy chain protein-conducting channel. Toxicon 54:565–569

Montecucco C, Papini E, Schiavo G (1994) Bacterial protein toxins penetrate cells via a four-step mechanism. FEBS Lett 346:92–98

Oost T, Sukonpan C, Brewer M, Goodnough M, Tepp W, Johnson EA, Rich DH (2003) Design and synthesis of substrate-based inhibitors of botulinum neurotoxin type B metalloprotease. Biopolymers 71:602–619

Pang YP, Davis J, Wang S, Park JG, Nambiar MP, Schmidt JJ, Millard CB (2010a) Small molecules showing significant protection of mice against botulinum neurotoxin serotype A. PLoS ONE 5:e10129

Pang YP, Vummenthala A, Mishra RK, Park JG, Wang S, Davis J, Millard CB, Schmidt JJ (2009) Potent new small-molecule inhibitor of botulinum neurotoxin serotype A endopeptidase developed by synthesis-based computer-aided molecular design. PLoS ONE 4:e7730

Pang YP, Vummenthala A, Mishra RK, Park JG, Wang S, Davis J, Millard CB, Schmidt JJ (2010b) Potent new small-molecule inhibitor of botulinum neurotoxin serotype A endopeptidase developed by synthesis-based computer-aided molecular design. PLoS ONE 4:e7730

Rummel A (2012) Double receptor anchorage of botulinum neurotoxins accounts for their exquisite neurospecificity. doi:10.1007/978-3-642-33570-9_4

Rummel A, Eichner T, Weil T, Karnath T, Gutcaits A, Mahrhold S, Sandhoff K, Proia RL, Acharya KR, Bigalke H, Binz T (2007) Identification of the protein receptor binding site of botulinum neurotoxins B and G proves the double-receptor concept. Proc Natl Acad Sci USA 104:359–364

Salzameda NT, Eubanks ME, Zakhari JS, Tsuchikama K, DeNunzio NJ, Allen KA, Hixon SS, Janda KD (2011) A cross-over inhibitor of the botulinum neurotoxin light chain B: a natural product implicating an exosite mechanism of action. Chem Commun 47:1713–1715

Schiavo G, Matteoli M, Montecucco C (2000) Neurotoxins affecting neuroexocytosis. Physiol Rev 80:717–766

Schmidt JJ, Stafford RG (2002) A high-affinity competitive inhibitor of type A botulinum neurotoxin protease activity. FEBS Lett 532:423–426

Schmidt JJ, Stafford RG (2005) Botulinum neurotoxin serotype F: identification of substrate recognition requirements and development of inhibitors with low nanomolar affinity. Biochemistry 44:4067–4073

Schmidt JJ, Stafford RG, Bostian KA (1998) Type A botulinum neurotoxin proteolytic activity: development of competitive inhibitors and implications for substrate specificity at the S1' binding subsite. FEBS Lett 435:61–64

Sikorra S, Henke T, Galli T, Binz T (2008) Substrate recognition mechanism of VAMP/synaptobrevin-cleaving clostridial neurotoxins. J Biol Chem 283:21145–21152

Silhár P, Capková K, Salzameda NT, Barbieri JT, Hixon MS, Janda KD (2010) Botulinum neurotoxin A protease: discovery of natural product exosite inhibitors. J Am Chem Soc 132:2868–2869

Silvaggi NR, Wilson D, Tzipori S, Allen KN (2008) Catalytic features of the botulinum neurotoxin a light chain revealed by high resolution structure of an inhibitory peptide complex. Biochemistry 47:5736–5745

Stenmark P, Dupuy J, Imamura A, Kiso M, Stevens RC (2008) Crystal structure of botulinum neurotoxin type A in complex with the cell surface co-receptor GT1b-insight into the toxin-neuron interaction. PLoS Pathog 4:e1000129

Sukonpan C, Oost T, Goodnough M, Tepp W, Johnson EA, Rich DH (2004) Synthesis of substrates and inhibitors of botulinum neurotoxin type A metalloprotease. J Pept Res 63: 181–193

Swaminathan S, Eswaramoorthy S (2000) Structural analysis of the catalytic and binding sites of *Clostridium botulinum* neurotoxin B. Nat Struct Biol 7:693–699

Swaminathan S, Eswaramoorthy S, Kumaran D (2004) Structure and enzymatic activity of botulinum neurotoxins. Movement Disord 19(8):S17–S22

Thompson AA, Jiao GS, Kim S, Thai A, Cregar-Hernandez L, Margosiak SA, Johnson AT, Han GW, O'Malley S, Stevens RC (2011) Structural characterization of three novel hydroxamate-based zinc chelating inhibitors of the Clostridium botulinum serotype A neurotoxin light chain metalloprotease reveals a compact binding site resulting from 60/70 loop flexibility. Biochemistry 50:4019–4028

Vaidyanathan VV, Yoshino K, Jahnz M, Dorries C, Bade S, Nauenburg S, Niemann H, Binz T (1999) Proteolysis of SNAP-25 isoforms by botulinum neurotoxin types A, C, and E: Domains and amino acid residues controlling the formation of enzyme-substrate complexes and cleavage. J Neurochem 72:327–337

Zuniga JE, Hammill JT, Drory O, Nuss JE, Burnett JC, Gussio R, Wipf P, Bavari S, Brunger AT (2010) Iterative structure-based peptide-like inhibitor design against the botulinum neurotoxin serotype A. PLoS ONE 5:e11378

Zuniga JE, Schmidt JJ, Fenn T, Burnett JC, Arac D, Gussio R, Stafford RG, Badie SS, Bavari S, Brunger AT (2008) A potent peptidomimetic inhibitor of botulinum neurotoxin serotype A has a very different conformation than SNAP-25 substrate. Structure (Camb) 16:1588–1597

Complexity of Botulinum Neurotoxins: Challenges for Detection Technology

Martin B. Dorner, K. Melanie Schulz, Skadi Kull and Brigitte G. Dorner

Abstract The detection of botulinum neurotoxins (BoNT) is extremely challenging due to their high toxicity and the multiple BoNT variants. To date, seven serotypes with more than 30 subtypes have been described, and even more subtypes are expected to be discovered. The fact that the BoNT molecules are released as large complexes of different size and composition adds further complexity to the issue. Currently, in the diagnostics of botulism, the mouse bioassay (MBA) is still considered as gold standard for the detection of BoNT in complex sample materials. Over the years, different functional, immunological, and spectrometric assays or combinations thereof have been developed, supplemented by DNA-based assays for the detection of the organism. In this review, advantages and limitations of the current technologies will be discussed, highlighting some of the intricacies of real sample analysis.

Keywords Botulinum neurotoxin · Detection · Activity · ELISA · Endopeptidase assay · Mouse bioassay · Mass spectrometry

M. B. Dorner · K. M. Schulz · S. Kull · B. G. Dorner (✉)
Centre for Biological Security 3, Microbial Toxins, Robert Koch-Institut,
Nordufer 20 13353 Berlin, Germany
e-mail: DornerB@rki.de

M. B. Dorner
e-mail: DornerM@rki.de

K. M. Schulz
e-mail: km.schulz@yahoo.de

S. Kull
e-mail: skull@fz-borstel.de

A. Rummel and T. Binz (eds.), *Botulinum Neurotoxins*, Current Topics in Microbiology and Immunology 364, DOI: 10.1007/978-3-642-33570-9_11,

Abbreviations

ALISSA	Assay using a large immuno-sorbent surface area
BoNT	Botulinum neurotoxin
ELISA	Enzyme-linked immunosorbent assay
ESI	Electrospray ionization
FRET	Förster resonance energy transfer
GE	Genome equivalents
HA	Hemagglutinin
HC	Heavy chain
LC	Light chain
LC	Liquid chromatography
LFA	Lateral flow assay
mAb	Monoclonal antibody
MALDI	Matrix-assisted laser desorption/ionization
MBA	Mouse bioassay
MPN ASSAY	Mouse phrenic nerve hemidiaphragm assay
MS	Mass spectrometry
NTNHA	Non-toxic nonhemagglutinin
orf	Open reading frame
pAb	Polyclonal antibody
PCR	Polymerase chain reaction
PTC	Progenitor toxin complex
SNAP-25	Synaptosome-associated protein of 25 kDa
SNARE	Soluble *N*-ethylmaleimide-sensitive factor attachment protein receptor
TeNT	Tetanus neurotoxin
TOF	Time-of-flight
VAMP	Vesicle-associated membrane protein

Contents

1 Complexity of Botulinum Neurotoxins 221
2 Mode of Action 223
3 Challenges in BoNT Detection 224
4 Activity-Based Detection Assays 225
  4.1 In Vivo and Ex Vivo Tests 225
  4.2 Endopeptidase Assays 227
  4.3 Endopeptidase-Mass Spectrometry (Endopep-MS) Assay 229
5 Immunological Assays 231
  5.1 Classical Sandwich ELISA 231
  5.2 Selected Examples of Different ELISA Formats 232
  5.3 Rapid Detection Tools Based on ELISA Formats 235
6 BoNT Detection by Mass Spectrometry 236
7 DNA-Based Detection of BoNT-Producing Bacteria 237
8 Laboratory Diagnostics of Botulism: Conclusion and Perspectives 239
References 244

1 Complexity of Botulinum Neurotoxins

Botulism, the disease caused by botulinum neurotoxins (BoNTs), has first been described in 1822 by Kerner as sausage poisoning (*botulus* = Latin: sausage). At the turn of the nineteenth century the causative agent, the anaerobic bacterium *Bacillus botulinus* (since 1923 *Clostridium botulinum*), or more precisely a heat-labile toxin in the culture supernatant, was identified by van Ermengem during his investigation of a botulism outbreak 1895 in Ellezelles, Belgium (van Ermengem 1897). The idea that botulism is caused by a single toxin produced by a single species (*Clostridium botulinum*) was shaken already some years later when in 1904 another strain of *C. botulinum* was isolated from bean salad in Darmstadt, Germany (Landmann 1904). Landmann was the first one to notice differences in optimal growth temperature and culture characteristics between the very first strains isolated. These differences were confirmed by Leuchs who also showed that the toxin produced by the Ellezelles strain was of a different serotype than the one produced by the Darmstadt strain (Leuchs 1910). Unfortunately, both strains were lost, but it seems likely that the Ellezelle strain belonged to a nonproteolytic strain of serotype B, whereas the Darmstadt strain was probably a proteolytic strain of serotype A (Meyer and Gunnison 1929). Today, the species *C. botulinum* is recognized as being inhomogeneous and is divided into four groups (I–IV) with distinct physiological characteristics (growth temperature, spore heat-resistance, salt tolerance etc.). The analysis of 16S rRNA sequences separates the four groups and places them together with other nonneurotoxic clostridia (Collins and East 1998; Peck 2009). Apart from *C. botulinum*, some strains of *C. baratii* and *C. butyricum* harbor the botulinum neurotoxin (*bont*) gene; thus, botulism can be caused by six distinct neurotoxin-producing species: *C. botulinum* group I–IV, *C. baratii*, and *C. butyricum*.

Since the beginning of the twentieth century, it has become clear that not only the BoNT-producing clostridia represents a heterogeneous family, but also that the neurotoxin is more divergent than originally anticipated. Some years after the discovery of BoNT/B and A the serotype C was identified in the 1920s (Bengtson 1922; Seddon 1922; Theiler and Robinson 1927). The group of serotypes was extended by D, E, F, and finally G between 1935 and 1969 (Gunnison et al. 1936; Hazen 1937; Møller and Scheibel 1960; Giménez and Cicarelli 1970), (Hill and Smith 2012). The serotypes show between 32 and 65 % identity at the amino acid level. It became apparent that certain serotypes can be produced by different *C. botulinum* groups and other clostridia species. For instance, serotype F can be produced by *C. botulinum* of group I, group II, and *C. baratii* (Peck 2009). In particular for serotype C, some of the controversies observed could be resolved by the identification of the toxins C2 and C3, which add to toxicity of the producing strains without being neurotoxins but belonging to the ADP-ribosyltransferase superfamily (Aktories et al. 2011; Just et al. 2011).

Many riddles remained unsolved before the era of modern molecular biology allowed for a more detailed analysis and interpretation of the *bont* gene locus. The comparison of *bont* sequences revealed a great degree of heterogeneity at the nucleotide and amino acid levels (Hill and Smith 2012). This heterogeneity led to the introduction of different subtypes for a given serotype, which vary up to 36 % at the amino acid level. Until today, six subtypes of serotype A have been described (A1–A6), seven subtypes of serotype B (B1–B7), eight subtypes of serotype E (E1–E8), and seven subtypes of serotype F (F1–F7), and more are still to be expected (Hill et al. 2007; Lúquez et al. 2009; Umeda et al. 2009; Raphael et al. 2010a; Macdonald et al. 2011; Kalb et al. 2012a). The differences within the subtypes of a given serotype are greatest in serotype A (16 %) and serotype F (36 %) as compared to serotype B (7 %) and serotype E (6 %) at the amino acid level. For serotypes C and D, mosaic toxins named C/D and D/C, respectively, have been described (Moriishi et al. 1996), shedding some light on cross-inhibition of these two particular serotypes by certain antisera. No subtypes have been identified yet for serotype G, which to date has not been linked with natural disease.

The differences observed among the subtypes at the nucleotide and amino acid levels have major implications for detection methods relying on either protein epitopes (e.g., enzyme-linked immunosorbent assay [ELISA]-based detection) or sequence recognition (e.g., polymerase chain reaction [PCR]-based assays). The issue is made even more complex by the fact that in bacterial supernatants the neurotoxins are not found as holotoxins, but are associated with a nontoxic non-hemagglutinin (NTNHA), and additionally—depending on serotype and subtype—with up to three different hemagglutinins (HA1 of 33 kDa, HA2 of 17 kDa, HA3 of 70 kDa (East and Collins 1994; Fujinaga et al. 1994; Inoue et al. 1996)). The composition of the progenitor toxin complex (PTC) out of BoNT, NTNHA, and other accessory proteins depends on the genes located within the neurotoxin gene cluster and other yet unidentified factors. Two different neurotoxin clusters have been described. One is named the *ha⁺orfX⁻* cluster and consists of the bicistronically expressed *bont* and *ntnha* genes, and three genes coding for the hem-agglutinins in reverse orientation, separated by *botR*, an alternative sigma factor as regulatory element. Alternatively, a second cluster is known which is called *ha⁻orfX⁺* cluster in which the *ha* genes are replaced by three open reading frames called *orfX1*, *orfX2* and *orfX3*, and *p21* that codes for a positive regulatory protein analogous to *botR* (Gu and Jin 2012) of this issue of CTMI. In many cases, the type of neurotoxin cluster is unique for a given subtype; however, some BoNT/A1-producing strains which occur mostly in an *ha⁺orfX⁻* cluster can also be found associated with an *ha⁻orfX⁺* cluster (Raphael et al. 2008; Lúquez et al. 2009). When present in an *ha⁺orfX⁻* cluster, the complex consisting of BoNT and NTNHA, also called M-PTC (~300 kDa; 12S), can associate with hemagglutinins leading to higher molecular weight complexes (L-PTC, ~600 kDa; 16S); the subtype A1 is able to form even larger complexes of about 900 kDa (LL-PTC; 19S).

While the crystal structure of the M-PTC has just been resolved ((Gu et al. 2012); (Gu and Jin 2012) of this issue of CTMI), the exact stoichiometry and structure of L- and LL-PTC are still under investigation (Inoue et al. 1996; Hasegawa et al. 2007). When BoNT and NTNHA are expressed within an *ha⁻orfX⁺* cluster, only the M-PTC has been purified. Eventually, minor amounts of OrfX proteins have been identified in association with the M-PTC or have been observed in crude toxin preparations (Lin et al. 2010). So far the function of the OrfX proteins remains unclear.

2 Mode of Action

In terms of function, the seminal work of Burgen, Jahn, Montecucco, and others made it clear that from the different components of the neurotoxin complex the exquisite specificity and neurotoxicity is mediated by the 150 kDa BoNT molecule (Burgen et al. 1949; Jahn and Niemann 1994; Montecucco and Schiavo 1994). BoNTs act as endopeptidases at the neuromuscular junction, cleaving components involved in neurotransmitter release, thus leading to neuromuscular paralysis.

Active BoNTs are dichain molecules consisting of a 50 kDa *N*-terminal light chain (LC) that is responsible for enzymatic activity and a 100 kDa *C*-terminal heavy chain (HC) that is involved in receptor binding and cellular uptake (Rummel 2012; Bercseny et al. 2012; Fischer 2012; Binz 2012 and Ahnert-Hilger et al. 2012 of this issue of CTMI). After oral uptake into the body, BoNT molecules are protected from the harsh conditions in the gastrointestinal tract by NTNHA which shields the toxin in a pH-dependent manner (Gu et al. 2012). Within the small intestine, BoNT complexes bind to microvilli mediated by the hemagglutinins (Fujinaga 2006, 2010). The mechanism by which the BoNT complex crosses the epithelial barrier is still under debate. Two mechanisms have been suggested: receptor-mediated endocytosis with subsequent transcytosis and ternary HA complex mediated destruction of the intercellular junctions, followed by paracellular influx (Maksymowych and Simpson 1998, 2004; Fujinaga et al. 2009; Fujinaga 2010), (Fujinaga et al. 2012). Directly after absorption of the BoNT complex, it dissociates before it reaches the blood circulation that takes it to its neuronal target cells (Sakaguchi 1982). At the neuromuscular junctions, the BoNT binds to serotype-specific ganglioside and glycoprotein receptors on the presynaptic membrane of neurons and is internalized through endocytosis (Montecucco and Schiavo 1994; Brunger and Rummel 2009). In a pH-dependent process, the BoNT HC confers translocation of the BoNT LC into the cytoplasm (Koriazova and Montal 2003). After the disulfide bond that links LC und HC is reduced, the released LC acts as a zinc-dependent endopeptidase. It selectively cleaves proteins of the soluble *N*-ethylmaleimide-sensitive factor attachment protein receptor (SNARE) complex which normally mediates the release of acetylcholine from synaptic vesicles (Montecucco and Schiavo 1994). The SNARE complex is formed by the assembly of the proteins synaptosome-associated protein (SNAP)-25, syntaxin and vesicle-associated membrane protein (VAMP)/synaptobrevin. While BoNT/A, C, and E cleave at different sites of SNAP-25, BoNT/C also targets syntaxin. BoNT/B, D, F, G, and the closely related tetanus neurotoxin (TeNT) cleave at distinct sites of VAMP (Montecucco and Schiavo 1994). After cleavage of any of the above-mentioned SNARE proteins, the formation of the SNARE complex is inhibited, resulting in the blockage of neurotransmitter release. This leads to the classical paralytic symptoms of botulism. The estimated lethal dose of complexed crystalline serotype A for humans is 1 μg/kg, 10–13, and 1–2 ng/kg for oral, inhalational, and intravenous exposure, respectively (Arnon et al. 2001).

3 Challenges in BoNT Detection

Generally, there are two different fields in BoNT research which require highly sensitive detection of BoNT molecules:

(i) Analysis of suspected cases of botulism in humans and animals (the focus of this review);
(ii) Potency testing of highly purified pharmacological products used for medical or cosmetic (Pellett 2012 of this issue of CTMI)

The two fields have quite contrary requirements for diagnostic approaches:

In the case of botulism diagnostics, the focus lies on the detection of all serotypes and subtypes, including known and unknown subtypes. The detection methods have to be compatible with different clinical, food, and environmental matrices (see below). It is critical to get a timely response, since therapeutic measures have to be taken in due time. On the other hand, it is sufficient to get a rough estimate of the toxin's functional activity.

In the case of potency testing, however, the correct recognition of the serotype is second-tier since it is already defined through the industrial production process. Pharmacological products are composed of highly pure toxins or toxin complexes plus additives and stabilizing proteins suspended in physiological buffer; therefore, matrix effects are usually not critical for their detection. With respect to the length of the whole production process, the time to result is not important for the detection assay. The main focus, however, is on the highly precise, statistically valid potency determination of the neurotoxin, which has to reflect all four steps of the BoNT action: binding to specific surface receptors, internalization into neurons, translocation of the LC into the cytoplasm, and finally proteolytic cleavage of SNARE proteins.

Botulism occurs in three major forms: (i) Food-borne botulism is caused by ingestion of food contaminated with BoNT; (ii) Wound botulism occurs after uptake and growth of *C. botulinum* in wounds with parallel production of BoNT; (iii) Infant botulism is caused by colonization of the intestinal tract and toxin production (Johnson and Montecucco 2008). Depending on the form of botulism, different sample materials are usually analyzed: in the case of food-borne botulism serum, feces, and suspected food; in wound botulism wound swabs, pus, tissue, and serum; and in infant botulism feces, serum, intestinal contents, suspected food, and environmental samples (Lindström and Korkeala 2006). With respect to diagnostics of suspicious botulism samples, attention has to be paid to the fact that the toxin occurs in different forms in different sample matrices. As discussed above, botulinum toxin is not a single protein but in fact a heterogeneous family of neurotoxins, comprising seven serotypes and more than 30 subtypes, thus challenging modern detection methods. In food matrices and bacterial culture supernatants, the toxin is usually present in one of its complexed forms. While the complex is stable at acidic pH, it dissociates spontaneously at physiological pH and high ionic strength (Sakaguchi 1982; Simpson 2004); this is of relevance for the analysis of food samples. In serum samples, however, free BoNT is found and the exact fate of the associated complex proteins is not clear to date. Due to the high toxicity of BoNT, their detection methods should be (i) highly sensitive down to the low pg/mL (fM–pM) range; (ii) able to detect all serotypes and subtypes including both the free neurotoxins and the high-molecular weight complexes; and (iii) compatible with a range of complex matrices. Additional points of interest are the simplicity of the assay, the potential for automation, and the robust assay performance including a high precision and accuracy.

4 Activity-Based Detection Assays

4.1 *In Vivo and Ex Vivo Tests*

Already in the 1920s, the mouse bioassay (MBA) for toxin detection was proposed: a BoNT-containing solution (e.g., patient serum, culture supernatant, or food extract) is injected intraperitoneally into mice and symptoms are observed for several hours up to 4 days. Characteristically, mice sequentially show ruffled fur, labored but not rapid breathing, a characteristic wasp-like abdomen with narrowed waist caused by increased respiratory effort due to paralysis of the diaphragm, weakness of limbs that progresses to total paralysis, and gasping for breath followed by death as a result of respiratory failure (Bengtson 1921). Mostly, the symptoms can be distinguished from symptoms caused by other toxins, e.g., TeNT which causes spastic paralysis. However, it should be noted that large doses of TeNT have been shown to initially mimic botulism symptoms (Matsuda et al. 1982). This phenomenon has been observed in a human patient, too ((König et al. 2007); TeNT and BoNT/B cleave VAMP at the very same position, but in different anatomic locations: while TeNT is retrogradely transported to the central nervous system, BoNT acts at the neuromuscular junction. One possible explanation for a sequential change from botulism to tetanus symptoms is that in the presence of high amounts of TeNT not all toxin is retrogradely transported from the neuromuscular junction to the central nervous system, thus inducing VAMP-cleavage at the neuromuscular junction similar to BoNT serotype B; Bercsenyi et al. 2012). Death of mice in the absence of neurological symptoms is not an acceptable indication of botulism (or tetanus), because it may be nonspecifically caused by other microorganisms, chemicals present in the test fluids, or injection trauma (Kautter and Solomon 1977). It is important to perform a number of specificity controls. Since the BoNT molecules are heat labile, a heat treatment of the sample material (15 min at 95 °C) should render it nontoxic, thus failing to induce any clinical symptoms. Furthermore, a trypsin activation step may be required for the detection of toxins of weak or nonproteolytic strains. Confirmation and neurotoxin typing is performed by mouse-protection tests using polyvalent or, even better, monovalent neutralizing antibodies: on simultaneous application of toxin-containing material and the respective neutralizing antibodies, the mice are rescued and no symptoms occur. To estimate the quantity of BoNT in a sample, different dilutions are injected into mice and symptoms are followed as described above. The quantity of toxin in the sample is then estimated by relating the maximum dilution killing the mice to the known mouse lethal dose: 10 pg for serotype A (Ferreira 2001).

The MBA is currently the only widely accepted method, the "gold standard", for confirmation of active BoNT molecules and is included in official methods and national guidelines (e.g., AOAC Official Method 977.26 or the German Standard DIN 10102). While the performance of the test is of serious ethical concern, it still has several advantages over other methods:

(i) It has an exquisite sensitivity of between 10 and 100 pg/mL, depending on the serotype and subtype analyzed (Ohishi and Sakaguchi 1980; Sugiyama 1980; Sharma et al. 2006);
(ii) The MBA displays all four steps of BoNT action and the physiological outcome: binding, uptake, translocation, and target cleavage resulting in inhibition of neurotransmitter release and muscle paralysis;
(iii) All serotypes and subtypes can be detected in their free and complexed form; and
(iv) The assay is compatible with the use of complex matrices like serum, feces, gastric content, wound samples, food samples, and bacterial cultures (after clarification and pH adjustment).

Apart from the ethical concern, however, the MBA has a number of technical disadvantages. Depending on the amount of toxin present in the sample, the assay takes 1–4 days to yield a result, and a precise quantitation of BoNT in a sample requires many animals. Inter-laboratory comparisons have shown that MBA results and precise quantitation may be variable depending on the age and strain of mice and other factors (McLellan et al. 1996; Sesardic et al. 2003). Additionally, recent work has shown that the potency of, e.g., BoNT/B cannot be directly transferred from mice to man due to differences in protein receptor amino acid sequences in both species (Strotmeier et al. 2012).

The ethical concerns of the MBA encouraged the development of different alternative assays. They are still refined in vivo assays, but nonlethal and with greatly reduced suffering of animals. These assays measure the local paralysis induced by BoNT, e.g., flaccid paralysis (Sesardic et al. 1996; Jones et al. 2006), abdominal ptosis (Takahashi et al. 1990), hind limp paralysis (Sugiyama et al. 1975; Pearce et al. 1994; Aoki 2001), grip strength (Meyer et al. 1979; Torii et al. 2011), and toe-spread reflex (Wilder-Kofie et al. 2011). Other assays use an electromyographic measurement of the compound action muscle potential to quantify BoNT activity or anti-BoNT antibodies (Sakamoto et al. 2009; Torii et al.

2010a, b). Generally, these in vivo assays have not been used frequently for the detection of BoNT out of complex food, environmental, or clinical samples. Rather they have been used to quantify BoNT from pharmaceutical-grade toxin preparations (Huber et al. 2008) or for the detection of antibodies against BoNT (Sesardic et al. 2004; Jones et al. 2006). However, these assays still require several days to perform and are in vivo tests with more or less objective readouts.

To avoid suffering of animals, replacement methods for the MBA have been described. Most widely used is an ex vivo test, the rat or mouse phrenic nerve hemidiaphragm (MPN) assay (Burgen et al. 1949). In this test, the phrenic nerve connected with the hemidiaphragm muscle is prepared from sacrificed mice or rats and transferred to a culture bath. The phrenic nerve is electrically stimulated and the resulting muscle twitches are measured. Upon addition of BoNT, the time required to decrease the amplitude to 50 % of the starting value, the paralytic half-time, is measured as a function of the dose of BoNT applied (Simpson and Tapp 1967; Simpson 1973, 1974; Habermann et al. 1980). Serotyping is performed in a way similar to the MBA, using monovalent antibodies. Apart from quantitation of BoNT activity, the hemidiaphragm assay has also been used to detect antibodies against BoNT (Dressler et al. 2005; Rasetti-Escargueil et al. 2009, 2011). Like the MBA, the MPN assay has the advantage to measure all four steps of BoNT action and the physiological endpoint (muscle paralysis). The duration of the assay is much shorter (<4 h) while its sensitivity is similar to the one of the MBA. However, less animals are needed and their suffering is greatly reduced; nevertheless, animals are still required. A disadvantage of the method is that it is sensitive to matrix interference with components of real samples.

As pure in vitro tests, cell culture-based assays have been developed which are also able to display four steps in BoNT action. These assays are mainly suited to quantify the activity of purified BoNT preparations (Pellett 2012 of this issue of CTMI).

4.2 Endopeptidase Assays

In vitro, activity assays focusing on the endopeptidase activity of the LC of BoNT have been developed and improved since the identification of their substrates. Basically, endopeptidase assays display the serotype-specific proteolytic cleavage of SNARE proteins in conjunction with technically different readouts.

One of the most straightforward ways to detect cleavage of SNARE proteins is by immunoblotting. Very soon after the elucidation of the endopeptidase activity of BoNT toward SNARE proteins, immuno blots of toxin-treated synaptosomes or neuronal cells were probed with anti-SNAP-25 or anti-VAMP antibodies to visualize the substrate cleavage (Poulain et al. 1993; Schiavo et al. 1993). Twenty years later, this is still a useful technique in the field of basic research, e.g., to study cellular uptake kinetics (Pier et al. 2011). This type of endopeptidase assay is particularly useful for deducing BoNT activity by analyzing cell lysates for

SNARE cleavage, e.g., to show the persistence of BoNT activity ex vivo or to demonstrate the anterograde axonal transport and transcytosis of catalytically active BoNT/A (Keller et al. 1999a; Restani et al. 2011).

In 1996, Shone and colleagues started to use immobilized peptides derived from SNAP-25 and VAMP which are cleaved by serotypes A and B, respectively. The newly generated amino acid terminus, the neoepitope, was then recognized specifically by enzyme-labeled antibodies, thus allowing the quantitation of the enzymatic activity in conjunction with a BoNT standard curve (Hallis et al. 1996). Antibody-based detection of the neoepitope was also used by the group of Sesardic to detect and quantify substrate cleavage by serotypes A, E, and C in purified toxin solutions (Jones et al. 2008, 2009). In a converse approach, the uncleaved SNARE substrate was detected by an antibody, which does not recognize the cleaved products: after coating of the VAMP substrate to microtiter plates, BoNT activity was correlated with the loss of detectable substrate (Keller et al. 1999b).

Another widely used endopeptidase technology applies the use of Förster resonance energy transfer (FRET, (Förster 1948)): here a SNARE peptide harboring the specific BoNT cleavage site is labeled with a fluorescence donor and a fluorescence acceptor. As long as the fluorescence donor–acceptor pair is located in close vicinity in the uncleaved substrate molecule, the fluorescence of the excited donor is absorbed (quenched) by the fluorescence acceptor molecule. Upon substrate cleavage, the two fluorophores are separated, so that the fluorescence of the donor is no longer quenched and can be measured. This principle was used by different groups to detect the activity of different BoNT serotypes with sensitivities between 35 pg/mL and 150 ng/mL depending on serotype, FRET substrate, and assay time used (Anne et al. 2001; Dong et al. 2004; Rasooly and Do 2008; Rasooly et al. 2008; Pires-Alves et al. 2009; Poras et al. 2009; Gilmore et al. 2011; Ruge et al. 2011). The principle has also been implemented into portable devices with sensitivities in the ng/mL range (Sapsford et al. 2008; Kostov et al. 2009; Sun et al. 2010; Balsam et al. 2011) and is the basis for commercial substrates like SNAPtide® (Shine et al. 2002).

However, many of the assays mentioned have not been tested on complex matrices yet and are indeed aiming at inhibitor or potency testing rather than detection of BoNT dissolved in complex matrices. As enzymatic assays, endopeptidase assays are relatively sensitive toward changes in reaction conditions, and testing of complex matrices can dramatically affect assay performance (Rasooly and Do 2008; Rasooly et al. 2008). To reduce matrix interference, an immunoaffinity enrichment step has been introduced, where the toxin is captured from the matrix using antibody-coated magnetic microbeads prior to performing the endopeptidase reaction, resulting in assay sensitivities similar to those of the MBA (Wictome et al. 1999a, b; Rasooly and Do 2008; Rasooly et al. 2008; Piazza et al. 2011). Immunoaffinity enrichment is advantageous since it separates the toxin from other proteases which might cleave the SNARE peptide unspecifically, preventing false-positive results. Of particular concern is the protease trypsin which is present in the gastrointestinal tract of vertebrates and in feces. Trypsin cleaves SNAP-25 at exactly the same position as BoNT/C. The problem of unspecific cleavage by other proteases is more substantial in endopeptidase assays recognizing proteolytic cleavage via FRET or other readouts as compared to assays which detect the precise cleavage product by a neoepitope-specific antibody or by mass spectrometry (see below). As a control for unspecific cleavage, Tucker and co-workers used a mutated SNAP-25 that could not be cleaved by serotypes A, C, and E (Piazza et al. 2011). Nevertheless, immunoaffinity enrichment combined with an FRET-based endopeptidase assay can be highly sensitive. The group of Kalkum developed an assay using a large immunosorbent surface area (ALISSA) for extraction of the toxins, resulting in excellent sensitivities of about 0.5 fg/mL for the detection of BoNT/A and E; matrices like serum, juice, and milk have been successfully analyzed (Bagramyan et al. 2008; Bagramyan and Kalkum 2011). Depending on the antibody used for immunoaffinity enrichment, the endopeptidase assay detects the LC activity only. Therefore, it can be advantageous to combine HC-specific antibodies for extraction with neoepitope-specific antibodies for the cleaved SNAP-25 in a single reaction (Liu et al. 2012).

To include cell binding in the in vitro activity assay, Shone and colleagues included a ganglioside-binding step in their endopeptidase method. To capture BoNT, they used synaptosomes, which contain gangliosides, followed by an endopeptidase assay, resulting in assay sensitivities in the range of the MBA (Evans et al. 2009).

4.3 Endopeptidase-Mass Spectrometry (Endopep-MS) Assay

Another variation of the endopeptidase format was developed by Barr and colleagues who coupled the endopeptidase format with a precise mass spectrometric detection and identification of the cleavage products, a method known as Endopeptidase-Mass Spectrometry (Endopep-MS) (Barr et al. 2005; Boyer et al. 2005). The detection and identification of the cleavage products can be performed either by matrix-assisted laser desorption/ionization time-of-flight (MALDI-TOF) MS or by liquid chromatography (LC) electrospray ionization (ESI) tandem MS (MS/MS). After implementing an immunoaffinity enrichment step, the Endopep-MS approach turned out to be very useful for the detection of BoNT/A, B, E and F activity out of serum, feces and organ homogenates. This was done either by using serotype-specific antibodies (Kalb et al. 2005, 2006, 2008; Gaunt et al. 2007; Parks et al. 2011; Wang et al. 2011) or by using a pan-reactive antibody directed against a common epitope present in the H_N of all four toxins (Kalb et al. 2010). In addition, BoNT/C, D, and G can also be detected by the Endopep-MS assay (Moura et al. 2011; Terilli et al. 2011). In the case of BoNT/C and D, the Endopep-MS method is even more precise in discriminating the two serotypes than the MBA, since the latter often suffers from a substantial cross-reactivity of antisera against serotypes C and D used for functional blockade due to the occurrence of BoNT/CD and DC mosaics (Hedeland et al. 2011).

Using an internal isotope-labeled standard with identical composition as one of the product peptides, but a different mass, Endopep-MS is able to precisely quantify the activity of toxin present in a sample. Depending on the serotype and complexity of the matrix analyzed, sensitivities between 0.05 and 50 pg/mL can be reached, similar to or even better than those reached by using the MBA (Boyer et al. 2005; Kalb et al. 2006, 2010; Moura et al. 2011). In contrast to endopeptidase assays using FRET or neoepitope recognition, the combination of immunoaffinity enrichment, endopeptidase reaction, and MS-based detection allows for simultaneous measurement of activity plus unambiguous identification of the corresponding immunocaptured toxin (Kalb et al. 2005, 2006, 2011a; Wang et al. 2011).

While the MBA is sensitive to all known and unknown BoNT serotypes and subtypes, in vitro endopeptidase assays have to include specific substrates for the BoNT serotypes aimed at. Until recently, all subtypes within a given serotype have been reported to share the same cleavage site on their respective SNARE protein. However, this principle was recently shaken by the finding that BoNT/F5 cleaves VAMP-2 at a different site from that of all other BoNT/F subtypes known. The cleavage site is located four amino acids upstream and has been identified by Endopep-MS (Kalb et al. 2012b). Also, not all subtypes of a serotype must recognize a given substrate equally well, as the detailed comparison of BoNT/F subtypes elucidated (Kalb et al. 2011a). It is also notable that the catalytic activity depends on the nature of the substrate and on the assay conditions used. For example, it has been shown that the catalytic activity of the BoNT subtypes A1–A4 tested on an SNAP-25 peptide was different from their activity on a longer recombinant SNAP-25 fragment. In particular, BoNT/A3, which showed 50 % of BoNT/A1 activity on a longer recombinant SNAP-25 fragment (aa 141–206), cleaved the peptide substrate at a much faster rate than subtype A1 (Henkel et al. 2009). Some of the discrepancies can be explained by the fact that SNAP-25 is not only bound around the active site of BoNT, but also by so-called α- and β-exosites upstream and downstream of the active site (Breidenbach and Brunger 2004; Brunger et al. 2008; Henkel et al. 2009). With respect to assay conditions, it has been shown that certain buffer components such as NaCl can reduce or even abolish cleavage of SNAP-25 or VAMP (Ferracci et al. 2011; Jones et al. 2011; Piazza et al. 2011).

In contrast to other in vitro methods, e.g., immunological assays which detect the presence of the protein, endopeptidase assays detect the functional activity of BoNT molecules in vitro. This is a clear advantage and is relevant, e.g., in the field of food safety. Endopeptidase assays inherently amplify the signal intensity by the catalytic reaction, since one BoNT molecule is able to cleave several substrate molecules. This results in excellent sensitivities, even beyond the sensitivity of the MBA. However, like most enzymatic assays, endopeptidase assays are quite sensitive to interference with matrix components. Therefore, immunoaffinity enrichment is performed to analyze real samples. Depending on the capture antibody or reagent used, the assay usually detects the activity of the LC plus the presence of the HC. Endopep-MS has been proven to be very useful to simultaneously measure the activity of BoNT molecules in parallel with an unambiguous identification of the immunocaptured toxin (Fig. 1). The only disadvantage is that

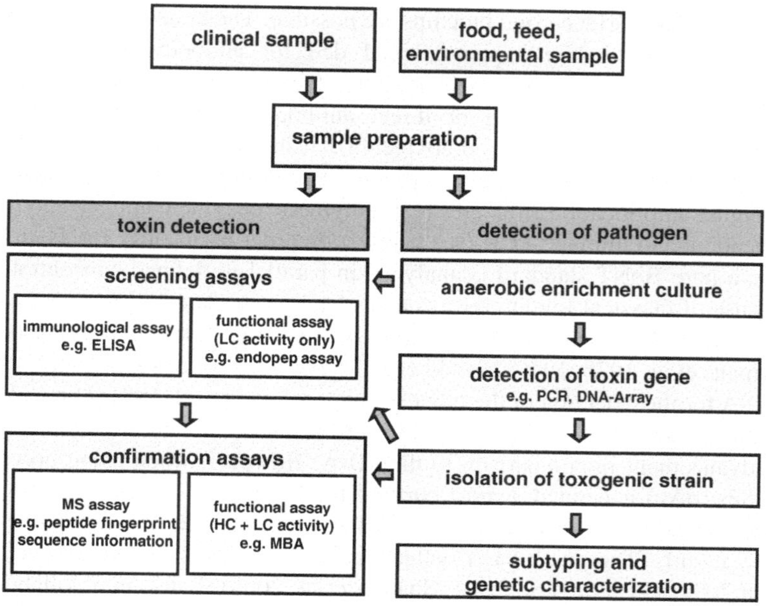

Fig. 1 Laboratory diagnostics of botulism. Depending on the type of botulism, clinical, food or environmental samples are analyzed for BoNT (*left*) and for the BoNT-producing pathogen (*right*), either directly or after anaerobic enrichment culture. A successful strategy combines fast and easy screening methods with confirmation assays providing information on the activity and/or identity of the holotoxin. The detection, isolation and genetic characterization of the toxin-producing strain deliver important additional information in an epidemiological investigation

this technology requires expensive equipment and a highly specialized technical expertise, and is therefore less common in clinical routine laboratories.

5 Immunological Assays

5.1 Classical Sandwich ELISA

ELISA-based technologies are by far the in vitro methods most commonly employed for BoNT detection. This is due to their generally high sensitivity, the simplicity of the assay which can be easily done in clinical routine laboratories, short assay times, and a robust assay performance including high precision and accuracy. After thorough validation, ELISA-based techniques can be applied to detect BoNT in a range of complex matrices. However, quality and validity of obtained data strictly depend on the quality and combination of the antibodies used.

In sandwich ELISA-based assays, a capture antibody is immobilized onto a solid support to capture BoNT. In the classical assay format these are plastic microtiter plates; other supports like silica microbeads, affinity columns, magnetic microbeads, glass surfaces, and biochips are possible. The bound analyte is usually detected via a secondary hapten-coupled detector antibody plus an enzyme conjugate to develop a measurable readout. Different assay variations can be performed, e.g., using indirect or direct antibody–antigen interactions, using different haptens (e.g., biotin, digoxigenin), using different read-out systems (e.g., UV/VIS absorption, fluorescence, and chemiluminescence), and using different signal amplification strategies (e.g. polymeric enzyme conjugates, tyramide amplification, and immuno-PCR (see below)). In order to quantify BoNT in a real sample, a pure BoNT standard is analyzed in parallel in defined concentrations.

The use of classical immunoassays dates back to the late 1970s when the first sandwich ELISA systems for the detection of BoNT/A, B, and E were introduced (Notermans et al. 1978, 1979; Kozaki et al. 1979). They were soon complemented by ELISA for the detection of the other serotypes (Lewis et al. 1981; Lee and Yang 1982; Notermans et al. 1982b). Results were routinely obtained within 4–6 h, a clear advancement in comparison to the MBA. However, when using polyclonal antibodies (pAb) generated against purified BoNT or BoNT complexes, the ELISAs were clearly less sensitive than the MBA with detection limits usually down to a few ng/mL (Thomas 1991; Doellgast et al. 1993, 1994; Szílagyi et al. 2000; Ferreira 2001; Ferreira et al. 2004; Sharma et al. 2006). Later, in a collaborative effort, sandwich ELISAs based on pAb against BoNT/A, B, E, and F have been validated and compared to the MBA, resulting in assay sensitivities of 0.1–1 ng/mL. The method was approved as AOAC Official Method 2002.08 by The Scientific Association Dedicated to Analytical Excellence (Ferreira 2001; Ferreira et al. 2003). Sharma and colleagues developed an amplified immunoassay based on digoxigenin-labeled pAb to detect the same serotypes in food matrices with assay sensitivities of 60–176 pg/mL (Sharma et al. 2006). Even more sensitive and specific immunoassays have been developed with the accessibility of monoclonal antibodies (mAb) (Köhler and Milstein 1975): sandwich ELISAs incorporating mAb resulted in assay sensitivities of around 1–100 pg/mL from culture supernatants, beverages, and complex food matrices (Shone et al. 1985; Ferreira et al. 1987, 1990; Gibson et al. 1987, 1988; Ekong et al. 1995; Varnum et al. 2006; Chiao et al. 2008b; Stanker et al. 2008; Volland et al. 2008; Brooks et al. 2010; Scotcher et al. 2010; Weingart et al. 2010).

5.2 *Selected Examples of Different ELISA Formats*

Albeit the sensitivity and specificity of BoNT-specific sandwich ELISA systems have been improved, the detection from complex matrices is still challenging. In *bead-based immunoassays* the capture antibody is immobilized onto microbeads, allowing to separate the toxin from the matrix, a step which is not possible in conventional plate-bound ELISA formats. Especially, magnetic microbeads have proved to be useful for extracting BoNT from complex and even colloidal matrices (Gessler et al. 2006; Pauly et al. 2009; Garber et al. 2010). This immunoaffinity enrichment step is not only useful for ELISAs, but is also often used in combination with other techniques (e.g., endopeptidase assays, mass spectrometry, cp. 12.4.2) (Bagramyan et al. 2008; Parks et al. 2011; Piazza et al. 2011; Liu et al. 2012).

In this context, flow cytometry has been shown to be a useful technique to detect and to quantify bead-bound BoNT molecules through a fluorescent readout (Anderson and Taitt 2008; Ozanich et al. 2009; Warner et al. 2009). A further development is *suspension array technologies* (e.g., the Luminex® xMAP technology). Fluorescent suspension arrays use either polystyrene or magnetic microspheres which are embedded with precise ratios of either two or three different fluorescent dyes, thus yielding arrays of 100 or 500 bead sets, respectively. The different bead sets are spectrally unique and are individually addressed by the measurement instrument. When immobilizing different antibodies to different bead sets, simultaneous and miniaturized immunoassays can be performed out of a single sample, thus reducing materials as well as cost and effort. Pauly and colleagues used color coded, magnetic Luminex beads to set up a pentaplex immunoassay for different biological toxins, among them BoNT/A and B. The detection limits were 21 and 73 pg/mL, respectively, out of 50 μL sample volume and could be further improved by magnetic immunoaffinity enrichment. Additionally, this method worked well to detect the toxins from different food matrices (Pauly et al. 2009). A similar approach was followed by Garber and colleagues for the detection of BoNT/A and five additional toxins with sensitivities of 1 ng/mL in spiked food samples (Garber et al. 2010). They took advantage of the multiplex approach by using different antibody pairs for each toxin, some recognizing different epitopes or displaying different binding kinetics, and implemented up to eight control assays. By doing so, they were able to reduce the likelihood of false-positive and false-negative results when testing toxins in complex matrices. In practice, fluorescent suspension arrays with up to 20–30 analytes in parallel are possible using a sandwich-ELISA principle.

Electrochemiluminescence-based immunoassays are usually performed on magnetic beads, too. They use a secondary reporter antibody, which is covalently coupled to a ruthenium (II) tris (bipyridyl) complex that becomes luminescent in the presence of an electric potential. The method has been used to detect BoNT/A, B, E, and F from clinical samples and food matrices with sensitivities of 50 pg/mL to 5 ng/mL, depending on the assay conditions used (Guglielmo-Viret et al. 2005; Rivera et al. 2006; Phillips and Abbott 2008). An advantage of this technique is its high signal-to-noise ratio due to the absence of optical background signals. However, the sensitivity boost compared to other ELISA readouts is limited and the technology is quite expensive, thus a widespread application is less likely.

Immuno-PCR is a further modified ELISA format using DNA-labeled detection antibodies (Wu et al. 2001; Chao et al. 2004; Adler et al. 2008; Rajkovic et al. 2012). Upon binding of the detection antibody to its target molecule, the oligonucleotide tag is amplified by PCR, resulting in assay sensitivities of 1 pg/mL for BoNT/A in buffer (Chao et al. 2004). In a different approach, Mason and colleagues described an ultrasensitive immunoassay based on liposomes with encapsulated DNA reporters and gangliosides embedded in the lipid bilayer as detection reagent (Mason et al. 2006). After binding of a BoNT to a specific immobilized capture antibody and subsequent binding of the ganglioside to the toxin, the liposomes were ruptured to release the reporter DNA for amplification by real-time-PCR. The assay combined the detection of the toxin with detection of ganglioside binding as part of the functional HC activity. Thereby the assay reached a sensitivity so far unmatched of 0.02 fg/mL for BoNT/A in water (Mason et al. 2006). However, this method has not been tested on complex matrices yet, so its wider applicability will have to be worked out in the future.

While the above-mentioned examples showed that current ELISA-based technologies can technically reach the sensitivity of the MBA and beyond, the main problem when analyzing unknown samples is still to discriminate true-positive from false-positive signals and to exclude false-negative results. It is per se difficult to anticipate the degree of cross-reactivity between the antibodies used and the matrix components in an unknown sample. Therefore, a thorough validation of the ELISA is necessary that includes determination of recovery rates from the most important matrices. Generally, ELISA systems based on pAb show a higher tendency for cross-reactivity compared to ELISAs using mAb. Especially, if the pAb has been generated against BoNT complexes or bacterial supernatants rather than the pure holotoxins, high titers against accessory proteins or unrelated proteins have been observed which can result in substantial cross-reactivity (Sakaguchi et al. 1974; Sugiyama et al. 1974; Betley and Sugiyama 1979; Notermans et al. 1982b; Sakaguchi 1982; Dezfulian et al. 1984; Kumar et al. 1994; Ferreira 2001). It has also been noticed that culture supernatants of *C. botulinum* shared antigens found in supernatants of other clostridia (Poxton 1984; Poxton and Byrne 1984) and gave rise to cross-reactivity with nontoxogenic clostridia (Lewis et al. 1981; Thomas 1991).

Certain matrices have been recognized as being difficult to analyze for the presence of BoNT. Especially, fecal specimens were shown to be problematic (Dezfulian et al. 1984; Viscidi et al. 1984). By diluting infant botulism stool samples in 40 % fetal bovine serum, Dezfulian and colleagues were able to decrease interference with fecal specimens and obtained a good correlation between MBA and their ELISA format (Dezfulian et al. 1984). Certain food matrices are rich in avidin (e.g., egg white); therefore, they might cause problems when employing ELISA formats using biotinylated detection antibodies. In this case, detection via digoxigenin/anti-digoxigenin amplification offers an alternative with low background signals due to the absence of endogenous digoxigenin in all prokaryotic and eukaryotic cells (except for *Digitalis purpurea* (Dorner et al. 2003)).

MAb are generally more specific and less sensitive to cross-reactivity than pAb, but their supreme specificity can be a pitfall in the light of the different subtypes of BoNT serotypes described. Optimally, ELISAs based on mAb have to be tested against all the different subtypes of a given serotype. Indeed, a variation or lack in recognition of a certain subtype has been observed for some mAb and mAb-based ELISAs (Gibson et al. 1987, 1988; Smith et al. 2005; Kalb et al. 2009, 2011b; Brunt et al. 2010).

Failure of a mAb to recognize individual subtypes of a serotype has implications for all of those assays in which this mAb is used, e.g., for immunoaffinity enrichment strategies. This problem is usually less pronounced with pAb recognizing numerous epitopes on a target protein, but also pAb have been shown to neutralize BoNT subtypes differently (Kozaki et al. 1977) or in extreme fail to recognize a certain subtype: Brunt and colleagues showed that a particular pAb directed against BoNT/F1 failed to recognize culture supernatants of a nonproteolytic F strain (Brunt et al. 2010). Notably, the highest divergence among the subtypes is found in BoNT/F (36 %) (Raphael et al. 2010a).

With respect to the detection of active versus inactive BoNT, it should be noted that only very few ELISA formats based on mAb have been shown to be able to discriminate the active toxin from its denatured form (Weingart et al. 2010). This might be relevant in thermal inactivation studies when comparing ELISA results with activity-based results.

In the light of the different sources of error connected with ELISA-based technologies, it is recommended to use this method for screening purposes to detect BoNT in complex matrices. Results should be confirmed by MBA or by other independent technologies having a high confirmatory power like mass spectrometry (Fig. 1; (Ferreira 2001; Ferreira et al. 2003)). Additionally, the detection, isolation, and characterization of the toxin-producing organism delivers important information.

5.3 *Rapid Detection Tools Based on ELISA Formats*

In a clinical case of botulism, first ELISA results can be obtained within several hours. Together with the characteristic clinical picture of acute botulism, this is timely enough in most cases to start medical treatment. However, in certain situations it is desirable to obtain results within 1 h, e.g., in a suspected case of intentional food poisoning in a bioterrorism scenario. Due to the fact that BoNTs are ranked among the category A agents of highest priority in the field of bioterrorism by the *Centers for Disease Control and Prevention* (Atlanta, USA), a number of on-site detection technologies have been developed over the last decade.

Among them, *lateral flow assays* (*LFA*) have been developed for commercial use, best known from pregnancy test kits (Posthuma-Trumpie et al. 2009). LFAs are hand-held devices based on immunochromatography on paper strips such as nitrocellulose. The sample is applied to one end of the strip and migrates by capillary action to the opposite end of the strip. While migrating along the strip the sample molecules first bind to an immobilized detection antibody conjugated to gold nanoparticles or dyes. The antibody–antigen complex further continues to migrate along the strip and is captured in the detection zone by a capture antibody, resulting in a visible change of color. Several LFAs for the detection of different BoNT serotypes have been developed with detection limits usually between 0.3 and 250 ng/mL (Chiao et al. 2004, 2008a; Klewitz et al. 2006; Attrée et al. 2007; Han et al. 2007). Sharma and co-workers evaluated two commercial products on spiked food samples and found detection limits of above 20 ng/mL for BoNT complexes of serotypes A, B, and E (Sharma et al. 2005). Others reported that some commercial tests were unable to recognize the purified holotoxin but detected the toxin complexes only (Gessler et al. 2007).

Although the sensitivity of LFA is clearly lower than laboratory-based ELISA technologies, they offer several advantages: they are inexpensive, easy to use without any sophisticated equipment, and have a rapid read-out time (usually 20 min). This makes them a good tool for field use by untrained personnel. It is often argued that in a case of a bioterrorism incident high levels of toxin are likely to be found; hence LFAs with their limited sensitivity might nevertheless be useful.

As an alternative, *column-based immunochromatography* tests requiring only few handling steps have been developed (ABICAP® technology). In this technique, the capture antibody is immobilized on a 3-dimensional immunofiltration column made of sintered material providing a large inner surface. For a colorimetric readout the technology uses the increased sensitivity of polymeric enzyme conjugates covalently attached to a streptavidin conjugate. Using this rather straightforward technology, BoNT/C and D have been detected from culture supernatants within 40 min with sensitivities close to that of the MBA (Gessler et al. 2005). Peck and colleagues recently expanded the method to detect BoNT/B, E, and F (Brunt et al. 2010). BoNT/A has been detected from different clinical samples, food, or powder materials with a detection limit in the low pg/mL-range, similar to that of the MBA (Attrée et al. 2007).

Apart from these technically simple devices, a number of sophisticated biosensor technologies based on different principles have been established. Most of them detect BoNT in the ng/mL-range (Ogert et al. 1992; Shriver-Lake et al. 1993; Kumar et al. 1994; O'Brien et al. 2000; Varnum et al. 2006; Grate et al. 2009; Ren and Pearton 2012). Only few of them have been tested with complex matrices like food, clinical, or environmental samples (Ganapathy et al. 2008; Weingart et al. 2012).

6 BoNT Detection by Mass Spectrometry

Mass spectrometry (MS) is a powerful tool to detect and to unambiguously identify analytes.

The principle is that charged ions are generated by an ion source, separated on the basis of their mass-to-charge (*m*/z) ratio, and finally recorded on a detector (Boyer et al. 2011). Different types of ionization methods are commonly used for biological substances, e.g., MALDI and electrospray ionization (ESI). In tandem (MS/MS) mass spectrometry, multiple cycles of MS analysis are performed, usually in conjunction with a fragmentation or dissociation process. This allows for protein sequencing of the analyte.

Per se the molecular mass of a protein is not a unique characteristic; therefore, large proteins are usually enzymatically fragmented by proteases (e.g., trypsin, chymotrypsin), delivering a characteristic peptide fingerprint. By searching in protein databases, the peptide fragments are then assigned to an individual protein.

The first characterization of BoNT serotypes A–F by enzymatic digest followed by MALDI-TOF MS and ESI–MS/MS has been described by van Baar and colleagues (van Baar et al. 2002, 2004). The approach was extended to detect all BoNT serotypes together with their nontoxic accessory proteins from the high molecular weight complexes with nano-Liquid Chromatography (LC)-ESI–MS/MS (Hines et al. 2005). For this approach, μg-amounts of pure BoNT or BoNT complex were necessary.

Compared to other in vitro methods described so far, the significance of data obtained is higher for MS-based methods, since they allow for unambiguous identification of the toxins by a unique peptide fingerprint pattern or a protein sequence. However, MS methods are usually not as sensitive as other in vitro methods and require pre-enrichment or purification steps to be compliant with complex matrices.

To this end, Klaubert and colleagues presented a method to detect and identify complexes of BoNT/A, B, E, and F out of bacterial culture supernatants (Klaubert et al. 2009): starting with a culture volume of 1 mL, they used a peptic sample pretreatment strategy combined with 2D-nano-LC–ESI–MS/MS to identify the toxins; around 30 fmol toxin could be detected.

In a different approach, a multiplex immunoaffinity enrichment strategy for BoNT/A and B, their respective complexes, and other biological toxins has been used, followed by tryptic digest and MALDI-TOF MS-based detection of characteristic peptide fingerprints (Kull et al. 2010). The approach turned out to be successful at detecting the toxins out of beverages with a detection limit of 300–500 fmol. Starting from an anaerobic enrichment culture of a suspected specimen in a real case of botulism, the multiplex approach correctly identified the BoNT serotype and subtype involved.

An interesting proteomics approach was recently published by Barr and coworkers: in order to identify new BoNT subtypes, they created an amino acid substitution database in which every position of the BoNT protein sequence was substituted against all other possible amino acids. The database allowed for the precise identification of multiple BoNT/B subtypes including the novel subtype BoNT/B7 with no DNA required (Kalb et al. 2012a).

As described above, Endopep-MS has been proven to be very useful for simultaneously measuring the activity of BoNT molecules in parallel to an unambiguous identification of the immunocaptured toxin. The method has been successfully applied to a number of clinical and food matrices (Kalb et al. 2005, 2006, 2011a; Hedeland et al. 2011; Parks et al. 2011; Wang et al. 2011).

7 DNA-Based Detection of BoNT-Producing Bacteria

In parallel to methods aiming at detection of the neurotoxins, most laboratories engaged in botulism diagnostics use technically independent DNA-based methods to screen suspect samples for BoNT-producing organisms and to perform an epidemiological investigation.

By far, the most commonly employed methods are PCR-based techniques (Mullis et al. 1986; Saiki et al. 1988), many of which aim at detecting *bont* genes by conventional or quantitative amplification reactions (Szabo et al. 1992, 1993; Franciosa et al. 1994, 1996; Fach et al. 1995, 2009; Takeshi et al. 1996; Aranda et al. 1997; Braconnier et al. 2001; Kimura et al. 2001; Craven et al. 2002; Popoff and Walker 2003; Akbulut et al. 2004; Takeda et al. 2005; Yoon et al. 2005; Lindström and Korkeala 2006; Artin et al. 2007; Fenicia et al. 2007; Heffron and Poxton 2007; Prévot et al. 2007; Sánchez-Hernández et al. 2008; Sakuma et al. 2009; Hill et al. 2010; Lindberg et al. 2010; Takahashi et al. 2010). Since conventional PCR is difficult to quantify and requires a post-PCR step to visualize and to verify the PCR product, many modern approaches use quantitative PCR (qPCR) formats. Inclusion of a fluorogenic probe in qPCR assays was shown to increase specificity and to allow simultaneous detection of a number of genes via differently labeled probes. Using qPCR, 10–100 genome equivalents (GE) can be readily detected in about 2 h. Including DNA purification and dilution steps, this has led to detection limits of 10^3–10^5 GE/mL (Fach et al. 2009; Kirchner et al. 2010).

Since in botulism diagnosis more than one serotype is of concern, multiplex reactions covering several serotypes simultaneously have been reported in recent years. In particular, assays able to detect *bont/a*, */b*, */e*, and */f*, which are known to be pathogenic to humans, in a single reaction have been developed (Lindström et al. 2001; Shin et al. 2007; Kirchner et al. 2010; Satterfield et al. 2010; Fach et al. 2011). Assays covering *bont/c*, */d*, and their mosaic forms have been developed to meet the needs of veterinary medicine (Anniballi et al. 2012; Woudstra et al. 2012).

From a diagnostic point of view, assays including an internal amplification control allow for a more accurate evaluation of results, which is mandatory under certain quality control schemes, and procedures have thus been implemented accordingly (Braconnier et al. 2001; Akbulut et al. 2004; Messelhäusser et al. 2007; De Medici et al. 2009; Kirchner et al. 2010; Fach et al. 2011; Fenicia et al. 2011; Anniballi et al. 2012).

The differences observed among the subtypes at the genetic level have, of course, major implications on PCR-based assays. It has been noted that some PCR assays fail to detect certain subtypes due to sequence variations (De Medici et al. 2009). Thus, whenever new subtypes are reported it is important to re-analyze the capacity of the assay used, and, if necessary, to amend primer and probe sequences. Conversely, differences between subtypes have also been used to specifically differentiate them (Umeda et al. 2009, 2010).

In addition to assays focusing on the detection of the *bont* genes, PCR approaches have been described that amplify the *ntnha* gene located directly upstream of *bont* within the toxin gene cluster (Raphael and Andreadis 2007; Hill et al. 2010). The *ntnha* gene is present in all *bont*-containing gene clusters but is less divergent than the *bont* itself (Peck 2009). Thus, it has been used as a valuable surrogate marker for *bont*-positive clostridia (Raphael and Andreadis 2007; Hill et al. 2010).

With respect to botulism cases, a caveat is that the detection of *bont* (i) cannot account for the amount of toxin produced; and (ii) does not necessarily discriminate between intact and silent genes (Franciosa et al. 1994).

As an alternative to a deeper analysis of the genome of toxin-producing clostridia, DNA-microarrays have been used to differentiate serotypes (Gauthier et al. 2005) and to reveal information on the different neurotoxin clusters (Raphael et al. 2008, 2010b). These and other methods described (e.g. amplified fragment length polymorphism, pulsed-field gel electrophoresis, multilocus sequence typing, and whole-genome sequencing) are valuable for characterizing the genome of BoNT-producing clostridia and help to complement epidemiological investigations, but—except for whole genome approaches—do not necessarily deduce the BoNT sero- or subtype (Lindström and Korkeala 2006).

8 Laboratory Diagnostics of Botulism: Conclusion and Perspectives

As botulism is a potentially life-threatening illness, a rapid diagnosis is important to start medical treatment in due time. In this context, reliable laboratory diagnostics is essential to support and to confirm the suspected diagnosis. Starting from clinical samples (serum, feces) and, if appropriate, also food, feed, or environmental samples, the detection of the toxin itself remains the standard method. Supporting information is obtained from the detection of the toxin-producing pathogen, either directly out of the sample material or after anaerobic enrichment culture. Both pathways together provide important and technically independent pieces of information in an epidemiological investigation (Fig. 1).

While the MBA is still seen as the "gold standard" in BoNT detection, numerous in vitro methods have been established based on different functional, immunological, and spectrometric principles or combinations thereof. Technically challenging is the fact that BoNT is not a single molecule but occurs in different serotypes and subtypes. Ideally, a BoNT detection method is able to detect them all, providing a similar or better sensitivity than the MBA. Furthermore, it should be fast and easy to perform with a high precision and accuracy. Finally, an ideal BoNT detection method should be compatible with the analysis of complex matrices.

Immunological techniques offer the advantage of being highly sensitive and specific, and are easy to perform in routine laboratories. A number of assay formats have been tested with complex matrices and reached sensitivities close to the MBA and beyond, also in a multiplex format (Table 1). However, the main difficulty remains the discrimination of true-positive from false-positive signals and the exclusion of false-negative results when analyzing unknown samples. Thorough validation of an ELISA for BoNT detection in different matrices is indispensable, because the extent of cross-reactivity between the antibodies used and the matrix components in an unknown sample is difficult to predict. Also when

Table 1 An overview of common BoNT detection assays

Method[a]	Serotype detected	Detection limit	Analysis time	Matrices tested[b]	References
In vivo/ex vivo activity assays					
MBA	A, B, C, D, E, F, G	10–100 pg/mL	0.5–4 days	cs, se, fo, fe	Bengtson (1921), Kautter and Solomon (1977), Hatheway and McCroskey (1987), Sesardic et al. (2003)
Phrenic nerve hemidiaphragm assay	A, B, E	20–600 pg/mL	2–4 h	cs	Simpson (1973, 1974), Rasetti-Escargueil et al. (2009)
Endopeptidase assays					
Neoepitope detection	A, B, C, E	0.04–200 pg/mL	4–24 h	cs, se, fo	Hallis et al. (1996), Wictome et al. (1999a, b), Jones et al. (2008, 2009), Evans et al. (2009), Liu et al. (2012)
Fluorescence/FRET detection	A, B, D, E, F, G	0.035–150 ng/mL	15 min–20 h	se	Anne et al. (2001), Schmidt et al. (2001), Dong et al. (2004), Poras et al. (2009), Ruge et al. (2011), Joshi (2012)
Immunocapture, FRET detection	A, B, E	0.5–500 fg/mL 0.5–38 ng/mL	3–6 h	cs, se, fo	Bagramyan et al. (2008), Bagramyan and Kalkum (2011), Rasooly and Do (2008, 2010), Rasooly et al. (2008), Piazza et al. (2011)
Portable endopeptidase assay	A	25–62.5 ng/mL	2–3 h	–	Sapsford et al. (2005, 2008), Kostov et al. (2009), Sun et al. (2009, 2010), Balsam et al. (2011)
Endopep-MS	A, B, C, D, E, F, G	0.05–50 pg/mL	4–17 h	cs, se, fo, fe	Barr et al. (2005), Boyer et al. (2005), Kalb et al. (2005, 2006, 2010, 2011a), Hedeland et al. (2011), Moura et al. (2011), Terilli et al. (2011), Wang et al. (2011)

(continued)

Table 1 (continued)

Method[a]	Serotype detected	Detection limit	Analysis time	Matrices tested[b]	References
Immunoassays					
ELISA (pAb)	A, B, C, D, E, F, G	0.1–100 ng/mL	5 h-2 days	cs, se, fo, fe	Notermans et al. (1978), Kozaki et al. (1979), Notermans et al. (1979), Lewis et al. (1981), Lee and Yang (1982), Notermans et al. (1982a, b), Dezfulian et al. (1984), Michalik et al. (1986), Thomas (1991), Doellgast et al. (1993), Potter et al. (1993), Doellgast et al. (1994), Szílagyi et al. (2000), Ferreira (2001), Ferreira et al. (2001, 2004), Poli et al. (2002), Sharma et al. (2006), Rajkovic et al. (2012)
ELISA (mAb)	A, B, C, D, E, F	1–1000 pg/mL	5–7 h	cs, se, fo	Shone et al. (1985), Ferreira et al. (1987, 1990), Gibson et al. (1987, 1988), Ekong et al. (1995), Guglielmo-Viret et al. (2005), Gessler et al. (2006), Varnum et al. (2006), Stanker et al. (2008), Volland et al. (2008), Brooks et al. (2010), Scotcher et al. (2010)
Electrochemiluminescence immunoassay	A, B, E, F	50–5000 pg/mL	2–3 h	se, fo	Guglielmo-Viret et al. (2005), Rivera et al. (2006), Phillips and Abbott (2008)
Immuno-PCR	A	0.02–4000 fg/mL	3–10 h	fo	Wu et al. (2001), Chao et al. (2004), Mason et al. (2006), Warner et al. (2009)
Flow cytometric assay/suspension arrays	A, B	10–5000 pg/mL	15 min–4 h	cs, fo	Varnum et al. (2006), Grate et al. (2009), Ozanich et al. (2009), Pauly et al. (2009), Warner et al. (2009), Garber et al. (2010)
LFA	A, B, D, E	0.3–250 ng/mL	10–30 min 2–4 h	cs, se, fo, fe	Chiao et al. (2004, 2008a), Sharma et al. (2005), Klewitz et al. (2006), Attrée et al. (2007), Gessler et al. (2007), Han et al. (2007), Yamashiro et al. (2007)

(continued)

Table 1 (continued)

Method[a]	Serotype detected	Detection limit	Analysis time	Matrices tested[b]	References
Immunochromatography columns	A, B, C, D, E, F	0.01–50 ng/ml	40 min	cs, se, fo	Gessler et al. (2005), Attrée et al. (2007), Brunt et al. (2010)
Different biosensor technologies	A, B, F	1–5 ng/ml	1–25 min	–	Ogert et al. (1992), Shriver-Lake et al. (1993, Kumar et al. (1994), O'Brien et al. (2000), Ladd et al. (2008), Ren and Pearton (2012)
Protein identification					
Mass spectrometry	A, B, C, D, E, F, G	49–375 ng/ml	8–14 h	cs, se, fo, fe	van Baar et al. (2002, 2004), Hines et al. (2005), Kalb et al. (2005, 2008, 2010, 2011a, 2012a), Gaunt et al. (2007), Klaubert et al. (2009), Kull et al. (2010), Hedeland et al. (2011), Moura et al. (2011), Parks et al. (2011), Terilli et al. (2011), Wang et al. (2011)
DNA-based detection					
PCR	A, B, C, D, E, F, G	10^3–10^5 GE/mL 10–100 GE	0.5–4 h	cs, se, fo, fe	Szabo et al. (1994), Braconnier et al. (2001), Lindström et al. (2001), Akbulut et al. (2004), Messelhäusser et al. (2007), De Medici et al. (2009), Kirchner et al. (2010), Fach et al. (2011), Fenicia et al. 2011), Anniballi et al. (2012)

[a] *ELISA* enzyme-linked immunosorbent assay; *PCR* polymerase chain reaction; *LFA* lateral flow assay; *FRET* Förster resonance energy transfer; *ALISSA* assay with a large immuno-sorbent surface area

[b] Complex matrices tested—buffer only; *cs* culture supernatant; *se* serum; *fo* food/feed; *fe* feces

using validated ELISA, it is still recommended to confirm results by either functional or mass spectrometric methods with a high confirmative power.

Compared to immunological methods detecting the presence of BoNT, functional in vitro assays like endopeptidase assays or the MBA offer the advantage of detecting the activity of BoNT. Endopeptidase assays focus on the activity of the BoNT LC only, but can include the presence of the HC depending on the antibodies employed for immunocapture. Since endopeptidase assays inherently amplify the signal intensity by the catalytic reaction, they reach very good sensitivities even beyond the MBA (Table 1). However, as enzymatic assays they can be quite sensitive to interference with matrix components, in particular other proteases. Hence, results obtained should be confirmed by MBA or by MS-based methods.

Generally, MS-based methods have the advantage to deliver unambiguous results, although they are still somewhat less sensitive and more time consuming than other methods. However, in combination with immunoaffinity enrichment, Endopep-MS and the parallel identification of the immunocaptured BoNT by peptide fingerprint or protein sequencing has proven to be very sensitive and enabled analysis of BoNT in clinical and food matrices (Table 1). In summary, a suitable combination of modern BoNT detection methods based on different technical approaches—functional, immunological, or spectrometric—is necessary and able to deliver confirmed results in a reasonable amount of time.

Tremendous progress has been made in the development of in vitro BoNT detection assays. However, more needs to be done. Highly specific and affine BoNT antibodies as central tools for enrichment strategies are not easily available, and only few of them have been commercialized. Along the same line, there is a lack of commercially available in vitro tests (e.g., ELISA, endopeptidase assays) which have been thoroughly validated on a broad range of complex matrices. Another problem is that currently there is no certified BoNT reference material available which can be used to compare different analytical approaches in expert laboratories. In the future, it will be important to strengthen quality assurance for the detection of BoNT and BoNT-producing clostridia by organizing regular proficiency tests.

On a technical level, it is anticipated that in vitro BoNT detection methods will be further multiplexed and miniaturized to detect and to differentiate the growing number of BoNT subtypes. Array-based platforms for protein detection and genetic characterization will become more important and replace singleplex detection methods. With the rising sequencing capabilities and the associated drop in cost it is expected that whole-genome sequencing will more and more replace classical genetic typing methods currently in use to characterize strains. At the same time, the gain of knowledge will allow a deeper understanding of physiological processes within BoNT-producing clostridia.

Acknowledgments We would like to thank Diana Pauly and Ursula Erikli for critically reading the manuscript. This work was supported by grants from the Federal Ministry of Education and Research (BiGRUDI project, 13N9601; zoonosis project, 01KI1021).

References

Adler M, Wacker R, Niemeyer CM (2008) Sensitivity by combination: immuno-PCR and related technologies. Analyst 133:702–718

Akbulut D, Grant KA, McLauchlin J (2004) Development and application of real-time PCR assays to detect fragments of the *Clostridium botulinum* types A, B, and E neurotoxin genes for investigation of human foodborne and infant botulism. Foodborne Pathog Dis 1:247–257

Aktories K, Lang AE, Schwan C et al (2011) Actin as target for modification by bacterial protein toxins. FEBS J 278:4526–4543

Anderson GP, Taitt CR (2008) Amplification of microsphere-based microarrays using catalyzed reporter deposition. Biosens Bioelectron 24:324–328

Anne C, Cornille F, Lenoir C et al (2001) High-throughput fluorogenic assay for determination of botulinum type B neurotoxin protease activity. Anal Biochem 291:253–261

Anniballi F, Auricchio B, Delibato E et al (2012) Multiplex real-time PCR SYBR green for detection and typing of group III *Clostridium botulinum*. Vet Microbiol 154:332–338

Ahnert-Hilger G, Münster-Wandowski A, Höltje M (2012) Synaptic vesicle proteins: Targets and routes for Botulinum nurotoxins. doi:10.1007/978-3-642-33570-9-8

Aoki KR (2001) A comparison of the safety margins of botulinum neurotoxin serotypes A, B, and F in mice. Toxicon 39:1815–1820

Aranda E, Rodriguez MM, Asensio MA et al (1997) Detection of *Clostridium botulinum* types A, B, E and F in foods by PCR and DNA probe. Lett Appl Microbiol 25:186–190

Arnon SS, Schechter R, Inglesby TV et al (2001) Botulinum toxin as a biological weapon: medical and public health management. J Am Med Assoc 285:1059–1070

Artin I, Bjorkman P, Cronqvist J et al (2007) First case of type E wound botulism diagnosed using real-time PCR. J Clin Microbiol 45:3589–3594

Attrée O, Guglielmo-Viret V, Gros V et al (2007) Development and comparison of two immunoassay formats for rapid detection of botulinum neurotoxin type A. J Immunol Methods 325:78–87

Bagramyan K, Kalkum M (2011) Ultrasensitive detection of botulinum neurotoxins and anthrax lethal factor in biological samples by ALISSA. Methods Mol Biol 739:23–36

Bagramyan K, Barash JR, Arnon SS et al (2008) Attomolar detection of botulinum toxin type A in complex biological matrices. PLoS ONE 3:e2041

Balsam J, Ossandon M, Kostov Y et al (2011) Lensless CCD-based fluorometer using a micromachined optical Söller collimator. Lab Chip 11:941–949

Barr JR, Moura H, Boyer AE et al (2005) Botulinum neurotoxin detection and differentiation by mass spectrometry. Emerg Infect Dis 11:1578–1583

Bengtson I (1921) Direct inoculation test for *B. botulinus* toxin. Public Health Rep 36:1665–1671

Bengtson I (1922) Preliminary note on a toxin-producing anaerobe isolated from the larvae of *Lucilia caesar*. Public Health Rep 37:164–170

Bercsenyi k, Giribaldi F, Schiavo G (2012) The elusive compass of clostridial Neurotoxins: Deciding when and where to go. doi:10.1007/978-3-642-33570-9-5

Betley MJ, Sugiyama H (1979) Noncorrelation between mouse toxicity and serologically assayed toxin in *Clostridium botulinum* type A culture fluids. Appl Environ Microbiol 38:297–300

Binz T (2012) Clostridial Neurotoxin light chains:Devices for SNARE cleavage Mediated blockde of neurotransmission. doi:10.1007/978-3-642-33570-9-7

Boyer AE, Moura H, Woolfitt AR et al (2005) From the mouse to the mass spectrometer: detection and differentiation of the endoproteinase activities of botulinum neurotoxins A–G by mass spectrometry. Anal Chem 77:3916–3924

Boyer AE, Gallegos-Candela M, Lins RC et al (2011) Quantitative mass spectrometry for bacterial protein toxins—a sensitive, specific, high-throughput tool for detection and diagnosis. Molecules 16:2391–2413

Braconnier A, Broussolle V, Perelle S et al (2001) Screening for *Clostridium botulinum* type A, B, and E in cooked chilled foods containing vegetables and raw material using polymerase chain reaction and molecular probes. J Food Prot 64:201–207
Breidenbach MA, Brunger AT (2004) Substrate recognition strategy for botulinum neurotoxin serotype A. Nature 432:925–929
Brooks CE, Clarke HJ, Finlay DA et al (2010) Culture enrichment assists the diagnosis of cattle botulism by a monoclonal antibody based sandwich ELISA. Vet Microbiol 144:226–230
Brunger AT, Rummel A (2009) Receptor and substrate interactions of clostridial neurotoxins. Toxicon 54:550–560
Brunger AT, Jin R, Breidenbach MA (2008) Highly specific interactions between botulinum neurotoxins and synaptic vesicle proteins. Cell Mol Life Sci 65:2296–2306
Brunt J, Webb MD, Peck MW (2010) Rapid affinity immunochromatography column-based tests for sensitive detection of *Clostridium botulinum* neurotoxins and *Escherichia coli* O157. Appl Environ Microbiol 76:4143–4150
Burgen ASV, Dickens F, Zatman LJ (1949) The action of botulinum toxin on the neuro-muscular junction. J Physiol 109:10–24
Chao H-Y, Wang Y-C, Tang S-S et al (2004) A highly sensitive immuno-polymerase chain reaction assay for *Clostridium botulinum* neurotoxin type A. Toxicon 43:27–34
Chiao DJ, Shyu RH, Hu CS et al (2004) Colloidal gold-based immunochromatographic assay for detection of botulinum neurotoxin type B. J Chromatogr B Analyt Technol Biomed Life Sci 809:37–41
Chiao DJ, Wey JJ, Shyu RH et al (2008a) Monoclonal antibody-based lateral flow assay for detection of botulinum neurotoxin type A. Hybridoma (Larchmt) 27:31–35
Chiao DJ, Wey JJ, Tang SS (2008b) Monoclonal antibody-based enzyme immunoassay for detection of botulinum neurotoxin type A. Hybridoma (Larchmt) 27:43–47
Collins MD, East AK (1998) Phylogeny and taxonomy of the food-borne pathogen *Clostridium botulinum* and its neurotoxins. J Appl Microbiol 84:5–17
Craven KE, Ferreira JL, Harrison MA et al (2002) Specific detection of *Clostridium botulinum* types A, B, E, and F using the polymerase chain reaction. J AOAC Int 85:1025–1028
De Medici D, Anniballi F, Wyatt GM et al (2009) Multiplex PCR to detect botulinum neurotoxin-producing clostridia in clinical, food and environmental samples. Appl Environ Microbiol 75:6457–6461
Dezfulian M, Hatheway CL, Yolken RH et al (1984) Enzyme-linked immunosorbent assay for detection of *Clostridium botulinum* type A and type B toxins in stool samples of infants with botulism. J Clin Microbiol 20:379–383
Doellgast GJ, Triscott MX, Beard GA et al (1993) Sensitive enzyme-linked immunosorbent assay for detection of *Clostridium botulinum* neurotoxins A, B, and E using signal amplification via enzyme-linked coagulation assay. J Clin Microbiol 31:2402–2409
Doellgast GJ, Beard GA, Bottoms JD et al (1994) Enzyme-linked immunosorbent assay and enzyme-linked coagulation assay for detection of *Clostridium botulinum* neurotoxins A, B, and E and solution-phase complexes with dual-label antibodies. J Clin Microbiol 32:105–111
Dong M, Tepp WH, Johnson EA et al (2004) Using fluorescent sensors to detect botulinum neurotoxin activity in vitro and in living cells. Proc Natl Acad Sci U S A 101:14701–14706
Dorner BG, Steinbach S, Hüser MB et al (2003) Single-cell analysis of the murine chemokines MIP-1alpha, MIP-1beta, RANTES and ATAC/lymphotactin by flow cytometry. J Immunol Methods 274:83–91
Dressler D, Lange M, Bigalke H (2005) Mouse diaphragm assay for detection of antibodies against botulinum toxin type B. Mov Disord 20:1617–1619
East AK, Collins MD (1994) Conserved structure of genes encoding components of botulinum neurotoxin complex M and the sequence of the gene coding for the nontoxic component in nonproteolytic *Clostridium botulinum* type F. Curr Microbiol 29:69–77
Ekong TAN, McLellan K, Sesardic D (1995) Immunological detection of *Clostridium botulinum* toxin type A in therapeutic preparations. J Immunol Methods 180:181–191

Evans ER, Skipper PJ, Shone CC (2009) An assay for botulinum toxin types A, B and F that requires both functional binding and catalytic activities within the neurotoxin. J Appl Microbiol 107:1384–1391

Fach P, Gibert M, Griffais R et al (1995) PCR and gene probe identification of botulinum neurotoxin A-, B-, E-, F-, and G-producing *Clostridium* spp. and evaluation in food samples. Appl Environ Microbiol 61:389–392

Fach P, Micheau P, Mazuet C et al (2009) Development of real-time PCR tests for detecting botulinum neurotoxins A, B, E, F producing *Clostridium botulinum, Clostridium baratii* and *Clostridium butyricum.* J Appl Microbiol 107:465–473

Fach P, Fenicia L, Knutsson R et al (2011) An innovative molecular detection tool for tracking and tracing *Clostridium botulinum* types A, B, E, F and other botulinum neurotoxin producing Clostridia based on the GeneDisc cycler. Int J Food Microbiol 145:S145–S151

Fenicia L, Anniballi F, De MD et al (2007) SYBR green real-time PCR method to detect *Clostridium botulinum* type A. Appl Environ Microbiol 73:2891–2896

Fenicia L, Fach P, van Rotterdam BJ et al (2011) Towards an international standard for detection and typing botulinum neurotoxin-producing Clostridia types A, B, E and F in food, feed and environmental samples: a European ring trial study to evaluate a real-time PCR assay. Int J Food Microbiol 145:S152–S157

Ferracci G, Marconi S, Mazuet C et al (2011) A label-free biosensor assay for botulinum neurotoxin B in food and human serum. Anal Biochem 410:281–288

Ferreira JL (2001) Comparison of amplified ELISA and mouse bioassay procedures for determination of botulinal toxins A, B, E, and F. J AOAC Int 84:85–88

Ferreira JL, Hamdy MK, Herd ZL et al (1987) Monoclonal antibody for the detection of *Clostridium botulinum* type A toxin. Mol Cell Probes 1:337–345

Ferreira JL, Hamdy MK, McCay SG et al (1990) Monoclonal antibody to type F *Clostridium botulinum* toxin. Appl Environ Microbiol 56:808–811

Ferreira JL, Eliasberg SJ, Harrison MA et al (2001) Detection of preformed type A botulinal toxin in hash brown potatoes by using the mouse bioasssay and a modified ELISA test. J AOAC Int 84:1460–1464

Ferreira JL, Maslanka S, Johnson E et al (2003) Detection of botulinal neurotoxins A, B, E, and F by amplified enzyme-linked immunosorbent assay: collaborative study. J AOAC Int 86:314–331

Ferreira JL, Eliasberg SJ, Edmonds P et al (2004) Comparison of the mouse bioassay and enzyme-linked immunosorbent assay procedures for the detection of type A botulinal toxin in food. J Food Prot 67:203–206

Fischer A (2012) Synchronized Chapterone function of Botulinum neurotoxin domain Mediates light chain translocation into neutrons.doi:10.1007/978-3-642-33570-9-6

Förster T (1948) Zwischenmolekulare Energiewanderung und Fluoreszenz. Ann Phys 437:55–75

Franciosa G, Ferreira JL, Hatheway CL (1994) Detection of type A, B, and E botulism neurotoxin genes in *Clostridium botulinum* and other *Clostridium* species by PCR: evidence of unexpressed type B toxin genes in type A toxigenic organisms. J Clin Microbiol 32:1911–1917

Franciosa G, Fenicia L, Caldiani C et al (1996) PCR for detection of *Clostridium botulinum* type C in avian and environmental samples. J Clin Microbiol 34:882–885

Fujinaga Y (2006) Transport of bacterial toxins into target cells: pathways followed by cholera toxin and botulinum progenitor toxin. J Biochem 140:155–160

Fujinaga Y (2010) Interaction of botulinum toxin with the epithelial barrier. J Biomed Biotechnol 2010:974943

Fujinaga Y, Inoue K, Shimazaki S et al (1994) Molecular construction of *Clostridium botulinum* type C progenitor toxin and its gene organization. Biochem Biophys Res Commun 205: 1291–1298

Fujinaga Y, Matsumura T, Jin Y et al (2009) A novel function of botulinum toxin-associated proteins: HA proteins disrupt intestinal epithelial barrier to increase toxin absorption. Toxicon 54:583–586

Fujinaga Y, Sugawara Y,Mtsumura T (2012) Uptake of Botulinum Neurotoxin in the intestine.doi: 10.1007/978-3-642-33570-9-3

Ganapathy R, Padmanabhan S, Eric YP et al (2008) Rapid detection of botulinum neurotoxins A, B, E, and F by optical immunoassay. Front Biosci 13:5432–5440
Garber EA, Venkateswaran KV, O'Brien TW (2010) Simultaneous multiplex detection and confirmation of the proteinaceous toxins abrin, ricin, botulinum toxins, and Staphylococcus enterotoxins A, B, and C in food. J Agric Food Chem 58:6600–6607
Gaunt PS, Kalb SR, Barr JR (2007) Detection of botulinum type E toxin in channel catfish with visceral toxicosis syndrome using catfish bioassay and endopep mass spectrometry. J Vet Diagn Invest 19:349–354
Gauthier M, Cadieux B, Austin JW et al (2005) Cloth-based hybridization array system for the detection of *Clostridium botulinum* type A, B, E, and F neurotoxin genes. J Food Prot 68:1477–1483
Gessler F, Hampe K, Böhnel H (2005) Sensitive detection of botulinum neurotoxin types C and D with an immunoaffinity chromatographic column test. Appl Environ Microbiol 71:7897–7903
Gessler F, Hampe K, Schmidt M et al (2006) Immunomagnetic beads assay for the detection of botulinum neurotoxin types C and D. Diagn Microbiol Infect Dis 56:225–232
Gessler F, Pagel-Wieder S, Avondet MA et al (2007) Evaluation of lateral flow assays for the detection of botulinum neurotoxin type A and their application in laboratory diagnosis of botulism. Diagn Microbiol Infect Dis 57:243–249
Gibson AM, Modi NK, Roberts TA et al (1987) Evaluation of a monoclonal antibody-based immunoassay for detecting type A *Clostridium botulinum* toxin produced in pure culture and an inoculated model cured meat system. J Appl Bacteriol 63:217–226
Gibson AM, Modi NK, Roberts TA et al (1988) Evaluation of a monoclonal antibody-based immunoassay for detecting type B *Clostridium botulinum* toxin produced in pure culture and an inoculated model cured meat system. J Appl Bacteriol 64:285–291
Gilmore MA, Williams D, Okawa Y et al (2011) Depolarization after resonance energy transfer (DARET): a sensitive fluorescence-based assay for botulinum neurotoxin protease activity. Anal Biochem 413:36–42
Giménez DF, Cicarelli AS (1970) Studies on strain 84 of *Clostridium botulinum*. Zentralbl Bakteriol Parasitenkd Infektionskr Abt 1 Orig Reihe A 215:212–220
Grate JW, Warner MG, Ozanich RM Jr et al (2009) Renewable surface fluorescence sandwich immunoassay biosensor for rapid sensitive botulinum toxin detection in an automated fluidic format. Analyst 134:987–996
Gu S, Rumpel S, Zhou J et al (2012) Botulinum neurotoxin is shielded by NTNHA in an interlocked complex. Science 335:977–981
Gu S, Jin R (2012) Assembly and function of the Botulinum neurotoxin progenitor complex. doi: 10.1007/978-3-642-33570-9-2
Guglielmo-Viret V, Attree O, Blanco-Gros V et al (2005) Comparison of electrochemiluminescence assay and ELISA for the detection of *Clostridium botulinum* type B neurotoxin. J Immunol Methods 301:164–172
Gunnison JB, Cummings JR, Meyer KF (1936) *Clostridium botulinum* Type E. Proc Soc Exp Biol Med 35:278–280
Habermann E, Dreyer F, Bigalke H (1980) Tetanus toxin blocks the neuromuscular transmission in vitro like botulinum A toxin. Naunyn Schmiedebergs Arch Pharmacol 311:33–40
Hallis B, James BA, Shone CC (1996) Development of novel assays for botulinum type A and B neurotoxins based on their endopeptidase activities. J Clin Microbiol 34:1934–1938
Han SM, Cho JH, Cho IH et al (2007) Plastic enzyme-linked immunosorbent assays (ELISA)-on-a-chip biosensor for botulinum neurotoxin A. Anal Chim Acta 587:1–8
Hasegawa K, Watanabe T, Suzuki T et al (2007) A novel subunit structure of *Clostridium botulinum* serotype D toxin complex with three extended arms. J Biol Chem 282:24777–24783
Hatheway CL, McCroskey LM (1987) Examination of feces and serum for diagnosis of infant botulism in 336 patients. J Clin Microbiol 25:2334–2338
Hazen EL (1937) A strain of *B. botulinus* not classified as type A, B, or C. J Infect Dis 60:260–264
Hedeland M, Moura H, Baverud V et al (2011) Confirmation of botulism in birds and cattle by the mouse bioassay and Endopep-MS. J Med Microbiol 60:1299–1305

Heffron A, Poxton IR (2007) A PCR approach to determine the distribution of toxin genes in closely related *Clostridium* species: *Clostridium botulinum* type C and D neurotoxins and C2 toxin, and *Clostridium novyi* α toxin. J Med Microbiol 56:196–201

Henkel JS, Jacobson M, Tepp W et al (2009) Catalytic properties of botulinum neurotoxin subtypes A3 and A4. Biochemistry 48:2522–2528

Hill KK, Smith TJ, Helma CH et al (2007) Genetic diversity among botulinum neurotoxin-producing clostridial strains. J Bacteriol 189:818–832

Hill KK, Smith TJ (2012) Genetic diversity within Clostridium botulinum serotypes, botulinum neurotoxin Gene. Cluster and toxin subtypes. doi: 10.1007/978-3-642-33570-9-1

Hill BJ, Skerry JC, Smith TJ et al (2010) Universal and specific quantitative detection of botulinum neurotoxin genes. BMC Microbiol 10:267

Hines HB, Lebeda F, Hale M et al (2005) Characterization of botulinum progenitor toxins by mass spectrometry. Appl Environ Microbiol 71:4478–4486

Huber A, France RM, Riccalton-Banks L et al (2008) The intercostal NMJ assay: a new alternative to the conventional LD50 assay for the determination of the therapeutic potency of botulinum toxin preparations. Altern Lab Anim 36:141–152

Inoue K, Fujinaga Y, Watanabe T et al (1996) Molecular composition of *Clostridium botulinum* type A progenitor toxins. Infect Immun 64:1589–1594

Jahn R, Niemann H (1994) Molecular mechanisms of clostridial neurotoxins. Ann N Y Acad Sci 733:245–255

Johnson EA, Montecucco C (2008) Botulism. Handb Clin Neurol 91:333–368

Jones RG, Alsop TA, Hull R et al (2006) Botulinum type A toxin neutralisation by specific IgG and its fragments: a comparison of mouse systemic toxicity and local flaccid paralysis assays. Toxicon 48:246–254

Jones RG, Ochiai M, Liu Y et al (2008) Development of improved SNAP25 endopeptidase immuno-assays for botulinum type A and E toxins. J Immunol Methods 329:92–101

Jones RG, Liu Y, Sesardic D (2009) New highly specific botulinum type C1 endopeptidase immunoassays utilising SNAP25 or syntaxin substrates. J Immunol Methods 343:21–27

Jones RG, Liu Y, Halls C et al (2011) Release of proteolytic activity following reduction in therapeutic human serum albumin containing products: detection with a new neoepitope endopeptidase immunoassay. J Pharm Biomed Anal 54:74–80

Joshi SG (2012) Detection of biologically active botulinum neurotoxin-A in serum using high-throughput FRET-assay. J Pharmacol Toxicol Methods 65:8–12

Just I, Rohrbeck A, Hülsenbeck SC et al (2011) Therapeutic effects of *Clostridium botulinum* C3 exoenzyme. Naunyn Schmiedebergs Arch Pharmacol 383:247–252

Kalb SR, Goodnough MC, Malizio CJ et al (2005) Detection of botulinum neurotoxin A in a spiked milk sample with subtype identification through toxin proteomics. Anal Chem 77:6140–6146

Kalb SR, Moura H, Boyer AE et al (2006) The use of endopep-MS for the detection of botulinum toxins A, B, E, and F in serum and stool samples. Anal Biochem 351:84–92

Kalb SR, Smith TJ, Moura H et al (2008) The use of endopep-MS to detect multiple subtypes of botulinum neurotoxins A, B, E, and F. Int J Mass Spectrom 278:101–108

Kalb SR, Lou J, Garcia-Rodriguez C et al (2009) Extraction and inhibition of enzymatic activity of botulinum neurotoxins/A1,/A2, and/A3 by a panel of monoclonal anti-BoNT/A antibodies. PLoS ONE 4:e5355

Kalb SR, Garcia-Rodriguez C, Lou J et al (2010) Extraction of BoNT/A,/B,/E, and/F with a single, high affinity monoclonal antibody for detection of botulinum neurotoxin by endopep-MS. PLoS ONE 5:e12237

Kalb SR, Pirkle JL, Barr JR (2011a) Mass spectrometric detection of botulinum neurotoxin by measuring its activity in serum and milk. In: Banoub J (ed) Detection of biological agents for the prevention of bioterrorism. Springer, Dordrecht, pp 115–129

Kalb SR, Santana WI, Geren IN et al (2011b) Extraction and inhibition of enzymatic activity of botulinum neurotoxins/B1,/B2,/B3,/B4, and/B5 by a panel of monoclonal anti-BoNT/B antibodies. BMC Biochem 12:58

Kalb SR, Baudys J, Rees JC et al (2012a) De novo subtype and strain identification of botulinum neurotoxin type B through toxin proteomics. Anal Bioanal Chem 403:215–226
Kalb SR, Baudys J, Webb RP et al (2012b) Discovery of a novel enzymatic cleavage site for botulinum neurotoxin F5. FEBS Lett 586:109–115
Kautter DA, Solomon HM (1977) Collaborative study of a method for the detection of *Clostridium botulinum* and its toxins in foods. J Assoc Off Anal Chem 60:541–545
Keller JE, Neale EA, Oyler G et al (1999a) Persistence of botulinum neurotoxin action in cultured spinal cord cells. FEBS Lett 456:137–142
Keller JE, Nowakowski JL, Filbert MG et al (1999b) Rapid microplate assay for monitoring botulinum neurotoxin B catalytic activity. J Appl Toxicol 19:S13–S17
Kimura B, Kawasaki S, Nakano H et al (2001) Rapid, quantitative PCR monitoring of growth of *Clostridium botulinum* type E in modified-atmosphere-packaged fish. Appl Environ Microbiol 67:206–216
Kirchner S, Krämer KM, Schulze M et al (2010) Pentaplexed quantitative real-time PCR assay for the simultaneous detection and quantification of botulinum neurotoxin-producing clostridia in food and clinical samples. Appl Environ Microbiol 76:4387–4395
Klaubert B, Vujtovic-Ockenga N, Wermter R et al (2009) Determination of botulinum toxins after peptic sample pre-treatment by multidimensional nanoscale liquid chromatography and nano-electrospray ion-trap mass spectrometry. J Chromatogr B Analyt Technol Biomed Life Sci 877:1084–1092
Klewitz T, Gessler F, Beer H et al (2006) Immunochromatographic assay for determination of botulinum neurotoxin type D. Sens Actuators B Chem 113:582–589
Köhler G, Milstein C (1975) Continuous cultures of fused cells secreting antibody of predefined specificity. Nature 256:495–497
König K, Ringe H, Dorner BG et al (2007) Atypical tetanus in a completely immunized 14-year-old boy. Pediatrics 120:e1355–e1358
Koriazova LK, Montal M (2003) Translocation of botulinum neurotoxin light chain protease through the heavy chain channel. Nat Struct Biol 10:13–18
Kostov Y, Sergeev N, Wilson S et al (2009) A simple portable electroluminescence illumination-based CCD detector. Methods Mol Biol 503:259–272
Kozaki S, Miyazaki S, Sakaguchi G (1977) Development of antitoxin with each of two complementary fragments of *Clostridium botulinum* type B derivative toxin. Infect Immun 18:761–766
Kozaki S, Dufrenne J, Hagenaars AM et al (1979) Enzyme linked immunosorbent assay (ELISA) for detection of *Clostridium botulinum* type B toxin. Jpn J Med Sci Biol 32:199–205
Kull S, Pauly D, Störmann B et al (2010) Multiplex detection of microbial and plant toxins by immunoaffinity enrichment and matrix-assisted laser desorption/ionization mass spectrometry. Anal Chem 82:2916–2924
Kumar P, Colston JT, Chambers JP et al (1994) Detection of botulinum toxin using an evanescent wave immunosensor. Biosens Bioelectron 9:57–63
Ladd J, Taylor AD, Homola J et al (2008) Detection of botulinum neurotoxins in buffer and honey using a surface plasmon resonance (SPR) sensor. Sens Actuators B Chem 130:129–134
Landmann G (1904) Ueber die Ursache der Darmstädter Bohnenvergiftung. Hyg Rundsch 14:449–452
Lee JK, Yang KH (1982) Enzyme-linked immunosorbent assay (ELISA) for detection of *Clostridium* type F toxin. Kor J Appl Microbiol Bioeng 10:205–209
Leuchs J (1910) Beiträge zur Kenntnis des toxins und antitoxins des *Bacillus botulinus*. Z Hyg 65:55–84
Lewis GE Jr, Kulinski SS, Reichard DW et al (1981) Detection of *Clostridium botulinum* type G toxin by enzyme-linked immunosorbent assay. Appl Environ Microbiol 42:1018–1022
Lin G, Tepp WH, Pier CL et al (2010) Expression of the *Clostridium botulinum* A2 neurotoxin gene cluster proteins and characterization of the A2 complex. Appl Environ Microbiol 76:40–47

Lindberg A, Skarin H, Knutsson R et al (2010) Real-time PCR for *Clostridium botulinum* type C neurotoxin (BoNTC) gene, also covering a chimeric C/D sequence-application on outbreaks of botulism in poultry. Vet Microbiol 146:118–123

Lindström M, Korkeala H (2006) Laboratory diagnostics of botulism. Clin Microbiol Rev 19:298–314

Lindström M, Keto R, Markkula A et al (2001) Multiplex PCR assay for detection and identification of *Clostridium botulinum* types A, B, E, and F in food and fecal material. Appl Environ Microbiol 67:5694–5699

Liu YY, Rigsby P, Sesardic D et al (2012) A functional dual-coated (FDC) microtiter plate method to replace the botulinum toxin LD(50) test. Anal Biochem 425:28–35

Lúquez C, Raphael BH, Maslanka SE (2009) Neurotoxin gene clusters in *Clostridium botulinum* type Ab strains. Appl Environ Microbiol 75:6094–6101

Macdonald TE, Helma CH, Shou Y et al (2011) Analysis of *Clostridium botulinum* serotype E strains by using multilocus sequence typing, amplified fragment length polymorphism, variable-number tandem-repeat analysis, and botulinum neurotoxin gene sequencing. Appl Environ Microbiol 77:8625–8634

Maksymowych AB, Simpson LL (1998) Binding and transcytosis of botulinum neurotoxin by polarized human colon carcinoma cells. J Biol Chem 273:21950–21957

Maksymowych AB, Simpson LL (2004) Structural features of the botulinum neurotoxin molecule that govern binding and transcytosis across polarized human intestinal epithelial cells. J Pharmacol Exp Ther 310:633–641

Mason JT, Xu L, Sheng ZM et al (2006) Liposome polymerase chain reaction assay for the sub-attomolar detection of cholera toxin and botulinum neurotoxin type A. Nat Protoc 1:2003–2011

Matsuda M, Sugimoto N, Ozutsumi K et al (1982) Acute botulinum-like intoxication by tetanus neurotoxin in mice. Biochem Biophys Res Commun 104:799–805

McLellan K, Das RE, Ekong TA et al (1996) Therapeutic botulinum type A toxin: factors affecting potency. Toxicon 34:975–985

Messelhäusser U, Zucker R, Ziegler H et al (2007) Nachweis von *Clostridium botulinum* Typ A, B, E und F mittels real-time-PCR. J Verbr Lebensm 2:198–201

Meyer KF, Gunnison JB (1929) European strains of *Cl. botulinum.* XXXVI. J Infect Dis 45:96–105

Meyer OA, Tilson HA, Byrd WC et al (1979) A method for the routine assessment of fore- and hindlimb grip strength of rats and mice. Neurobehav Toxicol 1:233–236

Michalik M, Grzybowski J, Ligięza J et al (1986) Enzyme-linked immunosorbent assay (ELISA) for the detection and differentiation of *Clostridium botulinum* toxins type A and B. J Immunol Methods 93:225–230

Møller V, Scheibel I (1960) Preliminary report on the isolation of an apparently new type of *Cl. botulinum.* Acta Pathol Microbiol Scand 48:80

Montecucco C, Schiavo G (1994) Mechanism of action of tetanus and botulinum neurotoxins. Mol Microbiol 13:1–8

Moriishi K, Koura M, Abe N et al (1996) Mosaic structures of neurotoxins produced from *Clostridium botulinum* types C and D organisms. Biochim Biophys Acta 1307:123–126

Moura H, Terilli RR, Woolfitt AR et al (2011) Studies on botulinum neurotoxins type/C1 and mosaic/DC using endopep-MS and proteomics. FEMS Immunol Med Microbiol 61:288–300

Mullis K, Faloona F, Scharf S et al (1986) Specific enzymatic amplification of DNA in vitro: the polymerase chain reaction. Cold Spring Harb Symp Quant Biol 51:263–273

Notermans S, Dufrenne J, Schothorst M (1978) Enzyme-linked immunosorbent assay for detection of *Clostridium botulinum* toxin type A. Jpn J Med Sci Biol 31:81–85

Notermans S, Dufrenne J, Kozaki S (1979) Enzyme-linked immunosorbent assay for detection of *Clostridium botulinum* type E toxin. Appl Environ Microbiol 37:1173–1175

Notermans S, Dufrenne J, Kozaki S (1982a) The relation between toxicity and toxin-related-antigen contents of *Clostridium botulinum* types C and D cultures as determined by mouse bioassay and ELISA. Jpn J Med Sci Biol 35:203–211

Notermans S, Hagenaars AM, Kozaki S (1982b) The enzyme-linked immunosorbent assay (ELISA) for the detection and determination of *Clostridium botulinum* toxins A, B, and E. Methods Enzymol 84:223–238

O'Brien T, Johnson LH 3rd, Aldrich JL et al (2000) The development of immunoassays to four biological threat agents in a bidiffractive grating biosensor. Biosens Bioelectron 14:815–828

Ogert RA, Brown JE, Singh BR et al (1992) Detection of *Clostridium botulinum* toxin A using a fiber optic-based biosensor. Anal Biochem 205:306–312

Ohishi I, Sakaguchi G (1980) Oral toxicities of *Clostridium botulinum* type C and D toxins of different molecular sizes. Infect Immun 28:303–309

Ozanich RM Jr, Bruckner-Lea CJ, Warner MG et al (2009) Rapid multiplexed flow cytometric assay for botulinum neurotoxin detection using an automated fluidic microbead-trapping flow cell for enhanced sensitivity. Anal Chem 81:5783–5793

Parks BA, Shearer JD, Baudys J et al (2011) Quantification of botulinum neurotoxin serotypes A and B from serum using mass spectrometry. Anal Chem 83:9047–9053

Pauly D, Kirchner S, Störmann B et al (2009) Simultaneous quantification of five bacterial and plant toxins from complex matrices using a multiplexed fluorescent magnetic suspension assay. Analyst 134:2028–2039

Pearce LB, Borodic GE, First ER et al (1994) Measurement of botulinum toxin activity: evaluation of the lethality assay. Toxicol Appl Pharmacol 128:69–77

Peck MW (2009) Biology and genomic analysis of *Clostridium botulinum*. Adv Microb Physiol 55(183–265):320

Pellett S (2012) Progress in cell based assays for botulinum neurotoxin detection. doi: 10.1007/978-3-642-33570-9-12

Phillips RW, Abbott D (2008) High-throughput enzyme-linked immunoabsorbant assay (ELISA) electrochemiluminescent detection of botulinum toxins in foods for food safety and defence purposes. Food Addit Contam 25:1084–1088

Piazza TM, Blehert DS, Dunning FM et al (2011) In vitro detection and quantification of botulinum neurotoxin type E activity in avian blood. Appl Environ Microbiol 77:7815–7822

Pier CL, Chen C, Tepp WH et al (2011) Botulinum neurotoxin subtype A2 enters neuronal cells faster than subtype A1. FEBS Lett 585:199–206

Pires-Alves M, Ho M, Aberle KK et al (2009) Tandem fluorescent proteins as enhanced FRET-based substrates for botulinum neurotoxin activity. Toxicon 53:392–399

Poli MA, Rivera VR, Neal D (2002) Development of sensitive colorimetric capture ELISAs for *Clostridium botulinum* neurotoxin serotypes E and F. Toxicon 40:797–802

Popoff MR, Walker JM (2003) Detection of toxigenic *Clostridia*. Methods Mol Biol 216:137–152

Poras H, Ouimet T, Orng SV et al (2009) Detection and quantification of botulinum neurotoxin type a by a novel rapid in vitro fluorimetric assay. Appl Environ Microbiol 75:4382–4390

Posthuma-Trumpie GA, Korf J, van Amerongen A (2009) Lateral flow (immuno)assay: its strengths, weaknesses, opportunities and threats. A literature survey. Anal Bioanal Chem 393:569–582

Potter MD, Meng JH, Kinsey P (1993) An ELISA for detection of botulinal toxin types A, B, and E in inoculated food samples. J Food Protect 56:856–861

Poulain B, Rossetto O, Deloye F et al (1993) Antibodies against rat brain vesicle-associated membrane protein (synaptobrevin) prevent inhibition of acetylcholine release by tetanus toxin or botulinum neurotoxin type B. J Neurochem 61:1175–1178

Poxton IR (1984) Demonstration of the common antigens of *Clostridium botulinum*, *C. sporogenes* and *C. novyi* by an enzyme-linked immunosorbent assay and electroblot transfer. J Gen Microbiol 130:975–981

Poxton IR, Byrne MD (1984) Demonstration of shared antigens in the genus *Clostridium* by an enzyme-linked immunosorbent assay. J Med Microbiol 17:171–176

Prévot V, Tweepenninckx F, van Nerom E et al (2007) Optimization of polymerase chain reaction for detection of *Clostridium botulinum* type C and D in bovine samples. Zoonoses Public Health 54:320–327

Rajkovic A, El Moualij B, Fikri Y et al (2012) Detection of *Clostridium botulinum* neurotoxins A and B in milk by ELISA and immuno-PCR at higher sensitivity than mouse bio-assay. Food Anal Methods 5:319–326

Raphael BH, Andreadis JD (2007) Real-time PCR detection of the nontoxic nonhemagglutinin gene as a rapid screening method for bacterial isolates harboring the botulinum neurotoxin (A–G) gene complex. J Microbiol Methods 71:343–346

Raphael BH, Luquez C, McCroskey LM et al (2008) Genetic homogeneity of *Clostridium botulinum* type A1 strains with unique toxin gene clusters. Appl Environ Microbiol 74:4390–4397

Raphael BH, Choudoir MJ, Luquez C et al (2010a) Sequence diversity of genes encoding botulinum neurotoxin type F. Appl Environ Microbiol 76:4805–4812

Raphael BH, Joseph LA, McCroskey LM et al (2010b) Detection and differentiation of *Clostridium botulinum* type A strains using a focused DNA microarray. Mol Cell Probes 24:146–153

Rasetti-Escargueil C, Jones RG, Liu Y et al (2009) Measurement of botulinum types A, B and E neurotoxicity using the phrenic nerve-hemidiaphragm: improved precision with in-bred mice. Toxicon 53:503–511

Rasetti-Escargueil C, Liu Y, Rigsby P et al (2011) Phrenic nerve-hemidiaphragm as a highly sensitive replacement assay for determination of functional botulinum toxin antibodies. Toxicon 57:1008–1016

Rasooly R, Do PM (2008) Development of an in vitro activity assay as an alternative to the mouse bioassay for *Clostridium botulinum* neurotoxin type A. Appl Environ Microbiol 74:4309–4313

Rasooly R, Do PM (2010) *Clostridium botulinum* neurotoxin type B Is heat-stable in milk and not inactivated by pasteurization. J Agric Food Chem 58:12557–12561

Rasooly R, Stanker LH, Carter JM et al (2008) Detection of botulinum neurotoxin-A activity in food by peptide cleavage assay. Int J Food Microbiol 126:135–139

Ren F, Pearton SJ (2012) Sensors using AlGaN/GaN based high electron mobility transistor for environmental and bio-applications. Phys Status Solidi C 9:393–398

Restani L, Antonucci F, Gianfranceschi L et al (2011) Evidence for anterograde transport and transcytosis of botulinum neurotoxin A (BoNT/A). J Neurosci 31:15650–15659

Rivera VR, Gamez FJ, Keener WK et al (2006) Rapid detection of *Clostridium botulinum* toxins A, B, E, and F in clinical samples, selected food matrices, and buffer using paramagnetic bead-based electrochemiluminescence detection. Anal Biochem 353:248–256

Ruge DR, Dunning FM, Piazza TM et al (2011) Detection of six serotypes of botulinum neurotoxin using fluorogenic reporters. Anal Biochem 411:200–209

Rummel A (2012) Double Receptor ancharage of botulinum neurotoxins accounts for thier exquisite neurospecificity. doi: 10.1007/978-3-642-3-570-9-4

Saiki RK, Gelfand DH, Stoffel S et al (1988) Primer-directed enzymatic amplification of DNA with a thermostable DNA polymerase. Science 239:487–491

Sakaguchi G (1982) *Clostridium botulinum* toxins. Pharmacol Ther 19:165–194

Sakaguchi G, Sakaguchi S, Kozaki S et al (1974) Cross reaction in reversed passive hemagglutination between *Clostridium botulinum* type A and B toxins and its avoidance by the sue of anti-toxic component immunoglobulin isolated by affinity chromatography. Jpn J Med Sci Biol 27:161–172

Sakamoto T, Torii Y, Takahashi M et al (2009) Quantitative determination of the biological activity of botulinum toxin type A by measuring the compound muscle action potential (CMAP) in rats. Toxicon 54:857–861

Sakuma T, Kurosaki Y, Fujinami Y et al (2009) Rapid and simple detection of *Clostridium botulinum* types A and B by loop-mediated isothermal amplification. J Appl Microbiol 106:1252–1259

Sánchez-Hernández L, Cifuentes A, Jiménez B et al (2008) Detection of *Clostridium botulinum* neurotoxin coding genes: analysis of PCR products by real time versus capillary gel electrophoresis methods. Eur Food Res Technol 227:495–502

Sapsford KE, Taitt CR, Loo N et al (2005) Biosensor detection of botulinum toxoid A and staphylococcal enterotoxin B in food. Appl Environ Microbiol 71:5590–5592

Sapsford KE, Sun S, Francis J et al (2008) A fluorescence detection platform using spatial electroluminescent excitation for measuring botulinum neurotoxin A activity. Biosens Bioelectron 24:618–625

Satterfield BA, Stewart AF, Lew CS et al (2010) A quadruplex real-time PCR assay for rapid detection and differentiation of the *Clostridium botulinum* toxin genes A, B, E and F. J Med Microbiol 59:55–64

Schiavo G, Rossetto O, Catsicas S et al (1993) Identification of the nerve terminal targets of botulinum serotypes A, D, and E. J Biol Chem 268:23784–23786

Schmidt JJ, Stafford RG, Millard CB (2001) High-throughput assays for botulinum neurotoxin proteolytic activity: serotypes A, B, D, and F. Anal Biochem 296:130–137

Scotcher MC, Cheng LW, Stanker LH (2010) Detection of botulinum neurotoxin serotype B at sub mouse LD_{50} levels by a sandwich immunoassay and its application to toxin detection in milk. PLoS ONE 5:e11047

Seddon HR (1922) Bulbar paralysis in cattle due to the action of toxicogenic bacillus, with a discussion on the relationship of the condition to forage poisoning (botulism). J Comp Pathol Therapeut 35:147–190

Sesardic D, McLellan K, Ekong TA et al (1996) Refinement and validation of an alternative bioassay for potency testing of therapeutic botulinum type A toxin. Pharmacol Toxicol 78:283–288

Sesardic D, Leung T, Gaines Das R (2003) Role for standards in assays of botulinum toxins: international collaborative study of three preparations of botulinum type A toxin. Biologicals 31:265–276

Sesardic D, Jones RG, Leung T et al (2004) Detection of antibodies against botulinum toxins. Mov Disord 19:S85–S91

Sharma SK, Eblen BS, Bull RL et al (2005) Evaluation of lateral-flow *Clostridium botulinum* neurotoxin detection kits for food analysis. Appl Environ Microbiol 71:3935–3941

Sharma SK, Ferreira JL, Eblen BS et al (2006) Detection of type A, B, E, and F *Clostridium botulinum* neurotoxins in foods by using an amplified enzyme-linked immunosorbent assay with digoxigenin-labeled antibodies. Appl Environ Microbiol 72:1231–1238

Shin NR, Yoon SY, Shin JH et al (2007) Development of enrichment semi-nested PCR for *Clostridium botulinum* types A, B, E, and F and its application to Korean environmental samples. Mol Cells 24:329–337

Shine N, Eaton L, Crawford K (2002) A continuous fluorimetric assay for high throughput screening for BoNT/A inhibitors. Naunyns Schmiedebergs Arch Pharmacol 365(Suppl. 2), Abstract R128

Shone C, Wilton-Smith P, Appleton N et al (1985) Monoclonal antibody-based immunoassay for type A *Clostridium botulinum* toxin is comparable to the mouse bioassay. Appl Environ Microbiol 50:63–67

Shriver-Lake LC, Ogert RA, Ligler FS (1993) A fiber-optic evanescent-wave immunosensor for large molecules. Sens Actuators B Chem 11:239–243

Simpson LL (1973) The interaction between divalent cations and botulinum toxin type A in the paralysis of the rat phrenic nerve-hemidiaphragm preparation. Neuropharmacology 12:165–176

Simpson LL (1974) Studies on the binding of botulinum toxin type A to the rat phrenic nerve-hemidiaphragm preparation. Neuropharmacology 13:683–691

Simpson LL (2004) Identification of the major steps in botulinum toxin action. Annu Rev Pharmacol Toxicol 44:167–193

Simpson LL, Tapp JT (1967) Actions of calcium and magnesium on the rate of onset of botulinum toxin paralysis of the rat diaphragm. Int J Neuropharmacol 6:485–492

Smith TJ, Lou J, Geren IN et al (2005) Sequence variation within botulinum neurotoxin serotypes impacts antibody binding and neutralization. Infect Immun 73:5450–5457

Stanker LH, Merrill P, Scotcher MC et al (2008) Development and partial characterization of high-affinity monoclonal antibodies for botulinum toxin type A and their use in analysis of milk by sandwich ELISA. J Immunol Methods 336:1–8
Strotmeier J, Willjes G, Binz T et al (2012) Human synaptotagmin-II is not a high affinity receptor for botulinum neurotoxin B and G: increased therapeutic dosage and immunogenicity. FEBS Lett 586:310–313
Sugiyama H (1980) Clostridium botulinum neurotoxin. Microbiol Rev 44:419–448
Sugiyama H, Oishi I, Dasgupta BR (1974) Evaluation of type A botulinal toxin assays that use antitoxin to crystalline toxin. Appl Microbiol 27:333–336
Sugiyama H, Brenner SL, Dasgupta BR (1975) Detection of *Clostridium botulinum* toxin by local paralysis elicited with intramuscular challenge. Appl Microbiol 30:420–423
Sun S, Ossandon M, Kostov Y et al (2009) Lab-on-a-chip for botulinum neurotoxin a (BoNT-A) activity analysis. Lab Chip 9:3275–3281
Sun S, Francis J, Sapsford KE et al (2010) Multi-wavelength Spatial LED illumination based detector for in vitro detection of botulinum neurotoxin A activity. Sens Actuators B Chem 146:297–306
Szabo EA, Pemberton JM, Desmarchelier PM (1992) Specific detection of *Clostridium botulinum* type B by using the polymerase chain reaction. Appl Environ Microbiol 58:418–420
Szabo EA, Pemberton JM, Desmarchelier PM (1993) Detection of the genes encoding botulinum neurotoxin types A to E by the polymerase chain reaction. Appl Environ Microbiol 59: 3011–3020
Szabo EA, Pemberton JM, Gibson AM et al (1994) Polymerase chain reaction for detection of *Clostridium botulinum* types A, B and E in food, soil and infant faeces. J Appl Bacteriol 76:539–545
Szílagyi M, Rivera VR, Neal D et al (2000) Development of sensitive colorimetric capture ELISAs for *Clostridium botulinum* neurotoxin serotypes A and B. Toxicon 38:381–389
Takahashi M, Kameyama S, Sakaguchi G (1990) Assay in mice for low levels of *Clostridium botulinum* toxin. Int J Food Microbiol 11:271–277
Takahashi H, Takakura C, Kimura B (2010) A quantitative real-time PCR method for monitoring *Clostridium botulinum* type A in rice samples. J Food Prot 73:688–694
Takeda M, Tsukamoto K, Kohda T et al (2005) Characterization of the neurotoxin produced by isolates associated with avian botulism. Avian Dis 49:376–381
Takeshi K, Fujinaga Y, Inoue K et al (1996) Simple method for detection of *Clostridium botulinum* type A to F neurotoxin genes by ploymerase chain reaction. Microbiol Immunol 40:5–11
Terilli RR, Moura H, Woolfitt AR et al (2011) A historical and proteomic analysis of botulinum neurotoxin type/G. BMC Microbiol 11:232
Theiler A, Robinson EM (1927) Der Botulismus der Haustiere. Z Infektionskr Haustiere 31: 165–220
Thomas RJ (1991) Detection of *Clostridium botulinum* types C and D toxin by ELISA. Aust Vet J 68:111–113
Torii Y, Goto Y, Takahashi M et al (2010a) Quantitative determination of biological activity of botulinum toxins utilizing compound muscle action potentials (CMAP), and comparison of neuromuscular transmission blockage and muscle flaccidity among toxins. Toxicon 55:407–414
Torii Y, Takahashi M, Ishida S et al (2010b) Quantification of potency of neutralizing antibodies to botulinum toxin using compound muscle action potential (CMAP). Toxicon 55:662–665
Torii Y, Kiyota N, Sugimoto N et al (2011) Comparison of effects of botulinum toxin subtype A1 and A2 using twitch tension assay and rat grip strength test. Toxicon 57:93–99
Umeda K, Seto Y, Kohda T et al (2009) Genetic characterization of *Clostridium botulinum* associated with type B infant botulism in Japan. J Clin Microbiol 47:2720–2728
Umeda K, Seto Y, Kohda T et al (2010) A novel multiplex PCR method for *Clostridium botulinum* neurotoxin type A gene cluster typing. Microbiol Immunol 54:308–312

van Baar BL, Hulst AG, de Jong AL et al (2002) Characterisation of botulinum toxins type A and B, by matrix-assisted laser desorption ionisation and electrospray mass spectrometry. J Chromatogr A 970:95–115
van Baar BL, Hulst AG, de Jong AL et al (2004) Characterisation of botulinum toxins type C, D, E, and F by matrix-assisted laser desorption ionisation and electrospray mass spectrometry. J Chromatogr A 1035:97–114
van Ermengem E (1897) Ueber einen neuen anaëroben Bacillus und seine Beziehungen zum Botulismus. Z Hyg 26:1–56
Varnum SM, Warner MG, Dockendorff B et al (2006) Enzyme-amplified protein microarray and a fluidic renewable surface fluorescence immunoassay for botulinum neurotoxin detection using high-affinity recombinant antibodies. Anal Chim Acta 570:137–143
Viscidi R, Laughon BE, Hanvanich M et al (1984) Improved enzyme immunoassays for the detection of antigens in fecal specimens. Investigation and correction of interfering factors. J Immunol Methods 67:129–143
Volland H, Lamourette P, Nevers M-C et al (2008) A sensitive sandwich enzyme immunoassay for free or complexed *Clostridium botulinum* neurotoxin type A. J Immunol Methods 330:120–129
Wang D, Baudys J, Kalb SR et al (2011) Improved detection of botulinum neurotoxin type A in stool by mass spectrometry. Anal Biochem 412:67–73
Warner MG, Grate JW, Tyler A et al (2009) Quantum dot immunoassays in renewable surface column and 96-well plate formats for the fluorescence detection of botulinum neurotoxin using high-affinity antibodies. Biosens Bioelectron 25:179–184
Weingart OG, Schreiber T, Mascher C et al (2010) The case of botulinum toxin in milk: experimental data. Appl Environ Microbiol 76:3293–3300
Weingart OG, Gao H, Crevoisier F et al (2012) A bioanalytical platform for simultaneous detection and quantification of biological toxins. Sensors (Basel) 12:2324–2339
Wictome M, Newton K, Jameson K et al (1999a) Development of an in vitro bioassay for *Clostridium botulinum* type B neurotoxin in foods that is more sensitive than the mouse bioassay. Appl Environ Microbiol 65:3787–3792
Wictome M, Newton KA, Jameson K et al (1999b) Development of in vitro assays for the detection of botulinum toxins in foods. FEMS Immunol Med Microbiol 24:319–323
Wilder-Kofie TD, Lúquez C, Adler M et al (2011) An alternative in vivo method to refine the mouse bioassay for botulinum toxin detection. Comp Med 61:235–242
Woudstra C, Skarin H, Anniballi F et al (2012) Neurotoxin gene profiling of *Clostridium botulinum* types C and D native to different countries within Europe. Appl Environ Microbiol 78:3120–3127
Wu HC, Huang YL, Lai SC et al (2001) Detection of *Clostridium botulinum* neurotoxin type A using immuno-PCR. Lett Appl Microbiol 32:321–325
Yamashiro S, Sano Y, Komano A et al (2007) Detection of proteinous toxins using the bio-threat alert system, part 4. Differences in detectability according to manufactural lots and according to toxin subtypes. Forensic Toxicol 25:80–83
Yoon SY, Chung GT, Kang DH et al (2005) Application of real-time PCR for quantitative detection of *Clostridium botulinum* type A toxin gene in food. Microbiol Immunol 49:505–511

Progress in Cell Based Assays for Botulinum Neurotoxin Detection

Sabine Pellett

Abstract Botulinum neurotoxins (BoNTs) are the most potent human toxins known and the causative agent of botulism, and are widely used as valuable pharmaceuticals. The BoNTs are modular proteins consisting of a heavy chain and a light chain linked by a disulfide bond. Intoxication of neuronal cells by BoNTs is a multi-step process including specific cell binding, endocytosis, conformational change in the endosome, translocation of the enzymatic light chain into the cells cytosol, and SNARE target cleavage. The quantitative and reliable potency determination of fully functional BoNTs produced as active pharmaceutical ingredient (API) requires an assay that considers all steps in the intoxication pathway. The in vivo mouse bioassay has for years been the 'gold standard' assay used for this purpose, but it requires the use of large numbers of mice and thus causes associated costs and ethical concerns. Cell-based assays are currently the only in vitro alternative that detect fully functional BoNTs in a single assay and have been utilized for years for research purposes. Within the last 5 years, several cell-based BoNT detection assays have been developed that are able to quantitatively determine BoNT potency with similar or greater sensitivity than the mouse bioassay. These assays now offer an alternative method for BoNT potency determination. Such quantitative and reliable BoNT potency determination is a crucial step in basic research, in the development of pharmaceutical BoNTs, and in the quantitative detection of neutralizing antibodies.

Keywords Botulinum neurotoxins · Potency determination · Cell-based assay · Detection assays · Neutralizing antibodies

---

S. Pellett (✉)
Department of Bacteriology, University of Wisconsin, Madison, 1550 Linden Drive, 6340 Microbial Sciences Building, Madison, WI 53706, USA
e-mail: sludwig@wisc.edu

A. Rummel and T. Binz (eds.), *Botulinum Neurotoxins*, Current Topics in Microbiology and Immunology 364, DOI: 10.1007/978-3-642-33570-9_12,

Abbreviations

NCB assay	Neuronal cell-based assay
BoNT	Botulinum neurotoxins
HC	Heavy chain
LC	Light chain
SNAP-25	Synaptosomal-associated protein 25
VAMP	Vesicle-associated membrane protein
SNARE	Soluble N-ethylmaleimide-sensitive factor attachment protein receptor
FRET	Fluorescence resonance energy transfer
FDA	Food and Drug Administration
SV2	Synaptic vesicle protein 2
MBA	Mouse bioassay
NGF	Nerve growth factor
EB	Embryoid body
mES cells	Mouse embryonic stem cells
hiPS cells	Human induced pluripotent stem cells
ELISA	Enzyme-linked immunosorbent assay
EDB	Extensor digitorum brevis
LD_{50}	Lethal dose at which 50 % of animals die
U	Mouse LD50 Unit
EC_{50}	Half maximal effective concentration
GT1b	Trisialoganglioside GT1b

Contents

1 Introduction ... 258
2 Botulinum Neurotoxins ... 259
2.1 Botulinum Neurotoxin Structure ... 259
2.2 Neuronal Cell Entry and Catalysis by BoNTs ... 260
2.3 Overview of the Detection of BoNTs ... 261
3 Neuronal Cell-Based Assays ... 262
3.1 NCB Assay Methods ... 263
3.2 Types of Cell-Based BoNT Detection Assays ... 264
3.3 Applications of NCB Assays ... 273
4 Future Perspectives and Conclusion ... 277
References ... 280

1 Introduction

Botulinum neurotoxins (BoNTs) are the most potent, naturally occurring toxins known to humankind. BoNTs are produced primarily by the gram-positive bacteria *Clostridium botulinum,* as well as by some strains of *Clostridium argentinense,* and rare strains of *Clostridium baratii and Clostridium butyricum* (Peck 2009) (Hill and

Smith 2012). BoNTs are the causative agent of botulism, which is a serious and potentially deadly neuro-paralytic human and animal disease. The toxins exert their toxic effect primarily by binding and entering peripheral cholinergic neurons and blocking acetylcholine release at neuromuscular junctions, leading to long-lasting descending paralysis (Johnson and Montecucco 2008; Schiavo et al. 2000). BoNTs are extraordinarily potent with the parenteral human lethal dose estimated to be 0.1–1 ng/kg and the oral lethal dose estimated at 1 μg/kg (Schantz and Johnson 1992; Arnon et al. 2001). This high potency, combined with the high affinity of the toxin for motor neurons and longevity of its action (up to several months), has raised serious concerns to their use as potential bioterrorism agents (Arnon et al. 2001). Remarkably, the same features have also facilitated the use of BoNTs (A and B) as extremely valuable drugs for treatment of a variety of neurological diseases as well as for cosmetic treatments. To date, BoNT/A is the most prominent serotype used in medical treatments (Truong et al. 2009; Evidente and Adler 2010) with over 1 million treatments carried out each year in the USA. Future developments of BoNTs as pharmaceuticals will no doubt utilize the specific characteristics of other BoNT sero- or subtypes in endogenous as well as recombinant BoNTs (Pickett and Perrow 2011; Cartee and Monheit 2011).

In order to establish a precise and reliable BoNT potency assay to ensure safe and consistent preparations for pharmaceutical utility, it is essential to understand the cellular biology of BoNTs and to ensure that assay considers all aspects of the BoNT intoxication process. In addition, fast, sensitive, and reliable BoNT detection platforms are desirable for research and for BoNT detection in contaminated foods, in food safety studies, and for use in the field in the case of suspected use of BoNTs for bioterrorism. Many sensitive assay platforms for BoNT detection have been developed and are applied today, with the in vivo mouse bioassay having long been regarded as the 'gold standard' (Solomon and Lilly 2001). Recent advances in cell-based assays now enable complementation or even replacement of the mouse bioassay for several applications. This chapter will first review the most important characteristics of BoNTs pertinent to assay systems, followed by a short overview of different BoNT detection methods, and an in-depth description of the current status of cell-based assays.

2 Botulinum Neurotoxins

2.1 Botulinum Neurotoxin Structure

BoNTs are classified into seven serotypes (A-G) based on immunological differences (Gimenez and Gimenez 1995), and most of the serotypes are subdivided into subtypes denoted by numbers after letters (i.e. BoNT/A1-5). At least 32 subtypes have been described based on differences in their amino acid sequences and structural models. Differences range from 35 to 70 % among BoNT serotypes and from 2.6 to 32 % among subtypes within one serotype (Smith et al. 2005; Kalb et al. 2011; Raphael et al. 2010; Macdonald et al. 2011; Hill and Smith 2012).

BoNTs are modular proteins, the structure and function of which are reviewed in detail elsewhere (Montal 2010) and in this book (Bercsenyi et el. 2012; Fischer 2012; Binz 2012). In short, all BoNTs consist of a heavy chain (HC) ($\sim$100 kDa) and a light chain (LC) ($\sim$50 kDa) linked by a disulfide bond. The first solved crystal structure was that of BoNT/A (Lacy et al. 1998). Since then the structures of BoNT/B and E, and of subdomains of those and several other serotypes and subtypes have also been reported (Swaminathan 2011). These studies indicate remarkable similarity in the individual functional domains of the BoNT serotypes, but significant differences in the organization of these domains. Unique structural features in the receptor and ganglioside binding domains as well as the catalytic domains define serotype specific binding and substrate selectivity. These structural characteristics also determine BoNT species specificity and immunological and cell trafficking properties, thus lending unique characteristics to each BoNT serotype (Swaminathan 2011). This is supported by studies showing that different BoNT serotypes bind to different protein and ganglioside receptors on neuronal cell surfaces and have different cell entry kinetics. For example, BoNT/A, D, E, and F bind to SV2 receptors with different affinities (Dong et al. 2006, 2008; Rummel et al. 2009; Peng et al. 2011), whereas BoNT/B and G bind to synaptotagmin I and II (Rummel et al. 2004; Dong et al. 2003; Nishiki et al. 1994). In addition, BoNT/B, and C bind to gangliosides with greater affinity than BoNT/A (Rummel et al. 2011). BoNT/E and BoNT/A2 have been shown to enter cells faster than BoNT/A1 (Pier et al. 2011; Keller et al. 2004).

2.2 Neuronal Cell Entry and Catalysis by BoNTs

Neuronal cell entry by BoNTs is a multi-step process, which has been reviewed in detail elsewhere (Montal 2010) and in (Rummel 2012; Bercsenyi et el. 2012; Fischer 2012). The process requires fully functional BoNT holotoxins, including both the intact heavy and light chains (HC and LC). The C-terminal domain of the HC functions in specific protein and ganglioside receptor binding at the neuronal cell surface, leading to internalization of the BoNTs into endosomes by receptor mediated endocytosis. After endocytosis, protonation in the endosome contributes to membrane insertion of HC (s) coupled to LC unfolding. This is followed by HC-chaperoned LC conduction into the cell's cytosol and subsequent release by disulfide bond reduction and its refolding in the cytoplasm of the cell (Fischer and Montal 2007a, b, Schiavo et al. 2000).

LCs of BoNTs are zinc endopeptidases that target and cleave SNARE proteins, which are core proteins involved in trafficking and release of synaptic vesicles containing neurotransmitters. The BoNT serotypes each have their own specific SNARE target and cleavage site. BoNT/A and E cleave SNAP-25 at distinct sites (Binz et al. 1994; Blasi et al. 1993a), whereas BoNT/B, D, G, and F cleave synaptobrevin 2 (VAMP-2) at distinct sites (Schiavo et al. 1992, 1993; Yamasaki et al. 1994; Schiavo et al. 1994). BoNT/C cleaves both SNAP-25 and syntaxin at unique cleavage sites (Schiavo et al. 1995; Williamson et al. 1996; Blasi et al. 1993b). In all cases, the cleavage of the SNARE protein results in a disruption of the exocytosis mechanism, and consequently a block in stimulatory neurotransmitter release, resulting in the flaccid paralysis characteristic of botulism.

2.3 *Overview of the Detection of BoNTs*

The field of botulinum toxins encompasses many areas including pharmaceutical use, biodefense, botulism diagnosis and treatment, as well as basic and applied research. The reliable detection of BoNTs is essential for all of these areas. The type and methodology of the detection assay or model used can influence the study outcome and conclusions, and it is important to understand the potential and limitations of the many different detection methods available. The first quantitative assay developed was the in vivo mouse bioassay (LD_{50} assay, MBA), in which mice are injected intraperitoneally or intravenously with dilutions of toxin and observed for signs of toxicity and death (Hatheway 1988; Schantz and Kautter 1978), and this assay has been considered the 'gold standard' for detection of BoNTs. BoNT activity is generally described in mouse LD_{50} Units [U], which is defined as the quantity of toxin required to kill 50 % of the population of injected mice. The assay is quantitative, well established, and is the only BoNT assay widely accepted by regulatory agencies of the United States for all types of samples. However, this assay has a large error (20–40 % depending on the method used and the laboratory), involves a large number of mice, incurs high costs associated with the use and housing of animals, and requires specially trained staff. Furthermore, the nature of the assay (observing mice until death) exposes the animals to pain and distress. The inclusion of a toxin standard and careful method standardization has been shown to significantly decrease the error of the mouse bioassay to approximately 15 % (Sesardic et al. 2003). However, the overall attributes of this assay still have led to a strong motivation by researchers and regulatory agencies to replace, reduce, or refine the use of animals in BoNT testing. (Adler et al. 2010). Several in vivo simulation assays, such as the hemidiaphragm assay and local injection assays (Wilder-Kofie et al. 2011; Rasetti-Escargueil et al. 2011b; Jones et al. 2006), have been developed to reduce the number of animals used (hemidiaphragm assay) and/or the suffering of animals (local injection assays). These assays are excellent research tools, but they still require the use of large numbers of animals and technical skills that make them unsuitable for medium-throughput applications such as potency determination of BoNTs or antibody screening.

Several in vitro assays for BoNT detection have also been developed, and a comprehensive review of these is presented in the preceding (Dorner et al. 2012) by Dr. Brigitte Dorner. The in vitro BoNT detection assays can be split into two groups: (1) Immunological detection methods and (2) Endopeptidase assays. While some of these assays are as or more sensitive than the MBA, they also have important restrictions. In general, immunological detection assays such as ELISA detect active as well as inactive BoNTs, and do not differentiate between holotoxin and reduced toxin in which the HC and LC are separated. Endopeptidase assays detect proteolytic activity of LCs (in reduced or nonreduced toxin), but do not require fully functional holotoxin. The buffer components or contaminating proteases have the potential to lead to falsely positive results or erroneous quantification of toxin. This problem is exemplified in the recent finding that human serum albumin itself, a commonly used excipient in pharmaceutical preparations of BoNTs, has an enzymatic activity similar to BoNT/A and C (Jones et al. 2011). A combination of specific immunological detection of the receptor-binding domain and an endopeptidase assay were recently suggested as an alternative to surmount these issues and appears to correlate with the in vivo MBA for BoNT/A (Liu et al. 2012). However, this assay is still restricted in the testing parameters of BoNT function, and extensive testing is required to ensure that potential damage to other functional regions of the toxin such as the ganglioside binding pocket, the translocation domain, or toxin aggregation do not lead to errors in potency determination.

An alternative BoNT detection method that requires the integrity of all steps of BoNT intoxication and thereby only detects fully functional holotoxin is cell-based assays, which are discussed in detail below.

3 Neuronal Cell-Based Assays

Neuronal cell-based assays (NCB assays) provide a model for BoNT detection that requires all steps of the cellular intoxication including cell surface binding, endocytosis, translocation of the LC into the cellular cytosol, and enzymatic activity of the LC on SNARE substrates. Within the past 5–7 years, cell assays have been developed that exceed the sensitivity of the mouse bioassay (Pellett et al. 2010, 2011; Whitemarsh et al. 2012; McNutt et al. 2011; Nuss et al. 2010; Kiris et al. 2011). Such cell-based assays now offer a sensitive model for the testing of fully functional BoNTs, as is necessary in the potency testing of pharmaceutical and research preparations of BoNTs. This opens the possibility of using cell-based assays as an alternative detection method to the MBA. In fact in fall of 2011, Allergan Inc., the distributor of the pharmaceutical BoNT preparation onabotulinumtoxin A (Botox®), announced that the Food and Drug Administration (FDA) for the first time had approved a cell-based assay for potency testing of its product. However, details about this assay have not been released.

In addition to BoNT detection, NCB assays provide an alternative assay to the MBA for the sensitive and specific detection of neutralizing antibodies to BoNTs (Whitemarsh et al. 2012; Pellett et al. 2007). NCB assays also have utility as a research model to study many characteristics of BoNTs, including binding to the neuronal cell surface, cell entry, intracellular catalytic activity of the LC, and inhibition of neurotransmitter release by the cells, as well as other factors.

3.1 NCB Assay Methods

Irrespective of the source of cells, all NCB assays require incubation of the cells with BoNTs for a defined time period, followed by removal of the toxin and a quantitative endpoint for determining toxin activity. Among the most specific endpoints for BoNT activity is SNARE cleavage, and this endpoint can be used in any neuronal cell type. SNARE cleavage can be determined in cell lysates by Western blot (Pellett et al. 2010; Whitemarsh et al. 2012; McNutt et al. 2011) or by ELISA (Nuss et al. 2010), or in live cells by quantitative immunofluorescence methods using cleavage-specific antibodies (Kiris et al. 2011). An alternative proposed endpoint is the introduction of a sensor into neurons such as a SNARE-FRET construct, which undergoes a change in fluorescence emission after specific BoNT cleavage inside the cells (Dong et al. 2004). While this approach has been shown to work in cells transiently transfected with FRET constructs (Dong et al. 2004), only one cell line stably expressing such a sensor has been constructed so far, and this cell line is about 2–3 orders of magnitude less sensitive than the MBA (Biosentinal Inc). The sensitivity of this endpoint compared to SNARE cleavage or neurotransmitter release remains unknown.

Another important, but less specific endpoint, is inhibition of neurotransmitter release, which can be measured in primary neuronal cell cultures and stem cell-derived neurons, as well as in some continuous cell lines. Most commonly, the cells are pre-loaded with radioactively labeled neurotransmitter, followed by measurement of the released radioactivity (Benatar et al. 1997; Bigalke et al. 1978; Keller and Neale 2001; McInnes and Dolly 1990; Sheridan et al. 2005). Other methods that determine endogenous neurotransmitter release include direct measurement by HPLC or immunoassay (McNutt et al. 2011; Yaguchi and Nishizaki 2010; Welch et al. 2000), or indirect measurement detecting enzymatic breakdown products (McMahon and Nicholls 1991) or the postsynaptic currents by voltage clamping (Akaike et al. 2010). The use of a fluorescent dye as a marker of neurotransmitter release has also been suggested (Rasetti-Escargueil et al. 2011a; Tegenge et al. 2012). Another potential but not yet developed endpoint is the measurement of electrical conductance of an entire population of cultured neurons.

The incubation time of the cells with BoNT is an important factor impacting sensitivity and is cell type specific; most cells require 24–72 h incubations for maximal sensitivity. Short incubation times of only a few hours require significantly increased toxin doses to reach the desired sensitivity. Chemically, stimulating neuronal activity in cells by modifying the toxin exposure medium or buffer to contain 56–80 mM KCl and 1–2 mM $CaCl_2$ can dramatically increase the BoNT uptake speed in some cells to 2 to several minutes (Pier et al. 2011; Keller et al. 2004; Whitemarsh et al. 2012). After removal of extracellular toxin, several hours of incubation is then required to achieve efficient SNARE cleavage, which continues to increase for 24–48 h. While useful in research applications, this approach does not reach the same sensitivity as a 24 or 48 h toxin exposure and requires at least 100-times more toxin. This indicates that both toxin binding/cell entry and SNARE proteolysis are rate limiting factors.

When exposing cultured cells to BoNTs, it is important to not only control the toxin concentration, which sometimes is the only reported parameter, but also the total volume of medium added to the cells and the specific activity of the toxin. All of these factors determine sensitivity. For example, 200 μl of a 1 pM solution of BoNT/A provides four times more toxin to the same number of cells as 50 μl of the same solution, and research in our laboratory has indicated that this affects SNAP-25 cleavage (Pellett et al., unpublished data). In addition, the specific activity of the toxin is an extremely important factor that needs to be considered when using cells (unless the NCB assay is used in a standardized and controlled format to determine specific activity), as only fully functional BoNT is able to enter the cells. Thus, the same molar quantity of a toxin preparation containing 95 % of fully functional holotoxin will have a much greater activity than a toxin preparation in which 50 % of the toxin is inactive through mishandling or degradation. Additional components included in the reaction mix such as proteins, excipients salts, metals, sugars, detergents, and many other factors may also affect the outcome of a cell-based assay. Thus, experiments using cell-based assays should always describe the toxin source and characteristics, specific activity, exposure volume, and toxin concentration in molar amounts, mass, and U/well, and the complete reaction composition.

3.2 Types of Cell-Based BoNT Detection Assays

While cell-based assays allow for a dramatic reduction in animal testing and enable BoNT potency determination, it is important to understand the characteristics and limitations of the various cell-based assays. All cell-based assays present a 'closed system'. Furthermore, they do not determine effects of pharmacokinetic properties that might play important roles in different in vivo assay (e.g., MBA) such as distribution, clearance, diffusion, transport, and other parameters. Therefore, BoNT studies conducted in cell-based assays need to be interpreted with these restrictions in mind, and some studies may require additional research using specific in vivo models. Another important consideration is the cell type and its source. Continuous cell lines are relatively easy to maintain and inexpensive compared to primary and stem cell-derived neurons, but most are derived from cancerous cells and may have altered gene expression profiles that affect their BoNT sensitivity and cellular biology. Primary neurons require the sacrifice of some animals and skilled personnel for preparation, but are relatively inexpensive. These cells can be obtained from different sources (e.g., spinal cord, dorsal root ganglion, hippocampal, or cortical neurons) and different species, which may affect the outcome of the studies. In addition, primary cells do not constitute a pure neuronal population, but rather are mixed cultures of neurons and supportive cells such as glial cells. Stem cells can be differentiated to different types of neuronal subpopulations, although the current state of stem cell technology limits the neuronal subpopulations and purity that can be achieved. As BoNT research advances to analyze different BoNT sero- and subtypes and recombinant BoNTs, cell models offer a powerful research model and assay platform if the characteristics of the model are carefully considered. Differences among cell models are discussed in the following sections.

3.2.1 Continuous Cell Lines for BoNT Detection

Several continuous neuronal cell lines derived from cancer cells have been used to detect BoNTs. The most commonly used cell lines are neuro-2a (mouse neuroblastoma) and PC12 (rat pheochromocytoma) cells (Schiavo et al. 1993; Yowler et al. 2002; Benatar et al. 1997). Both cell types are relatively insensitive to BoNTs, requiring a long incubation time for 2–3 days and large quantities of BoNTs (in the nM range) to achieve adequate SNARE cleavage. The sensitivity of neuro-2a cells was first reported to be in the range of 0.67–6 nM (Yowler et al. 2002), while other studies have reported 50 % SNAP-25 cleavage with approx. 30 nM BoNT/A (or about 30,000 U) (Dong et al. 2006; Pier et al. 2011; Fischer and Montal 2007a; Conway et al. 2010). Sensitivity can be increased to about 2 nM (~2,000 U) by over-night pre-incubation of the cells with 25–50 μg/ml of the ganglioside GT1b (Pier et al. 2011; Yowler et al. 2002).

For PC12 cells, inhibition of acetylcholine release by BoNT/A was reported with approximate EC_{50}s of ~0.02 nM in cells differentiated with nerve growth factor (NGF) (Ray 1993), while noradrenaline release at 120 nM was reported in undifferentiated cells (Shone and Melling 1992). In our laboratory, 50 % cleavage of SNAP-25 was achieved in undifferentiated PC12 cells at ~2,000 Units (~2.5 nM) after 48 h BoNT/A exposure (unpublished data). Plating of the cells onto collagen coated plates and pre-loading with 25 μg/ml of GT1b in serum-free medium increased the sensitivity to an EC_{50} of 400 U (~500 pM). Sensitivity was increased even further to ~100–200 U (EC_{50}) when the cells were exposed to toxin in a modified neurobasal medium (cell stimulation medium) containing 56 mM KCl and 2.2 mM $CaCl_2$ (Invitrogen custom medium) for 24 h, independent of GT1b pre-incubation (Pellett et al., unpublished data). Preloading PC12 cells with the ganglioside GT1b sensitized the cells to BoNT/B, allowing detection of VAMP2 cleavage with 50 nM BoNT/B (Dong et al. 2003).

There are other important characteristics of the PC12 and neuro2A cell lines with regard to BoNT testing. Besides the obvious morphological differences, PC12 cells are from rat origin and express only the SV2A isoform as receptor, whereas neuro-2a mouse cells express only the SV2C isoform (Dong et al. 2006). Both cell lines have unique ganglioside profiles. Neurite sprouting can be induced in PC12 cells by addition of NGF or pituitary adenylate cyclase-activating polypeptide (PACAP) (Ravni et al. 2006), while neuro-2a cells are neuronal growth factor insensitive. While BoNT/A analysis in neuro-2a cells used SNAP-25 cleavage as an endpoint, PC12 cells can release neurotransmitter, and thus inhibition of neurotransmitter release can serve as an additional endpoint in these cells.

Other cell lines used in BoNT research include the human neuroblastoma cell line SH-SY5Y (Purkiss et al. 2001). This cell line detected SNARE cleavage by several BoNT serotypes with significantly different potencies, and the EC_{50}s ranged from 0.54 nM for BoNT/C to 300 nM for BoNT/F. Further research on pre-differentiating the SH-SY5Y cell line by addition of retinoic acid for 5 days followed by culture in medium containing BDNF markedly increased the BoNT/A sensitivity to 100 pM as determined by a neurotransmitter release assay. Pre-incubation of the cells with GT1b further increased sensitivity to 35 pM for BoNT/A (Rasetti-Escargueil et al. 2011a; Sesardic and Das 2007). BoNT/C and D/C can also sensitively be detected in neurons differentiated from the mouse embryonal carcinoma cell line P19. Differentiation of these cells in the presence of 0.5 μM RA for 4 days followed by a 4 day maturation period in serum-free medium results in a population of neurons with cholinergic characteristics (Parnas and Linial 1995), and a sensitivity to BoNT/C and C/D similar to that in primary hippocampal cells (EC_{50} of 217 pM (syntaxin cleavage) and 310 pM (SNAP-25-cleavage), and 58 pM VAMP2 cleavage, respectively) (Tsukamoto et al. 2012).

The human neuroblastoma derived BE(2) M17 cell line shows SNAP-25 cleavage by BoNT/A with an EC_{50} of about 1 nM, which was slightly improved by pre-incubation of the cells with the ganglioside GT1b. Norepinephrine release was inhibited by 10 nM BoNT/A (Lee et al. 2008). In addition, BoNT/C inhibited norepinephrine release with an approximate EC_{50} of $\sim$1 nM, but BoNT/B, D, E, F, and G did not significantly inhibit norepinephrine release at 10 nM concentrations (Lee et al. 2008). The human neuronally committed teratocarcinoma cell line NT2, which can be differentiated to fully functioning postmitotic neurons, was recently tested as a model for BoNT/A intoxication (Tegenge et al. 2012). After a 1–3 h exposure of cells to 1 nM BoNT/A, transmitter release was significantly inhibited. However, results showing long-term exposure of these cells and effects of other factors have not been published.

While no definitive data have been released or published yet, patent descriptions indicate that Allergan has developed a cell-based assay for the onabotulinumtoxin A (Botox®) product using a clonal cell line derived from the human neuroblastoma SiMa cell line (Zhu et al. 2010). This cell line appears to have an EC_{50} of about 1 pM for BoNT/A if pre-differentiated in serum-free medium containing GT1b, which is of comparable sensitivity to the mouse bioassay.

Other cell lines that have been tested for BoNT detection include the human neuroblastoma cell line SK-N-SH, which requires addition of 50 μg/ml of GT1b for the detection of low nM amounts of BoNT/A (Yowler et al. 2002). Rat neuroblastoma B35 cells and the mouse hypothalamus N-44 cells do not express sufficient levels of SNAP-25 to be useful in a BoNT/A assay using SNAP-25 cleavage as an endpoint. The level of SNAP-25 did not change in B35 cells upon differentiation by serum deprivation. Whereas differentiation of N-44 cells by NGF treatment induced SNAP-25 expression, the cells were resistant to BoNT/A treatment at low nM amounts (Pellett et al., unpublished data). The human cortical neuronal cell line HCN1 was disregarded for BoNT assay since this cell line grows extremely slow and is difficult to propagate. The mouse motor neuron like cell line NSC-34 expresses high levels of SNAP-25 and was differentiated to motor neuron like morphology by serum deprivation, but was insensitive to BoNT/A at low nM amounts (Kiris et al. 2011) and Johnson lab, unpublished data). The known sensitivities of different neuronal cell lines to BoNT serotypes (subtypes 1, respectively) are summarized in (Table 1).

3.2.2 Primary Cells for BoNT Detection

Primary cells have been used as model systems for BoNT detection for 35 years or longer. Research on primary cells has helped immensely in our understanding of the biology and characteristics of BoNTs, including SNARE protein cleavage, inhibition of neurotransmitter release, duration of action, and cell entry mechanisms. A variety of primary cells derived from rat, mouse, or chicken embryos have been used (Keller et al. 2004; Pellett et al. 2007, 2010; Sheridan et al. 2005; Stahl et al. 2007; Bigalke et al. 1985). Among these are cortical neurons (Pier et al. 2011; Stahl et al. 2007; Saadi et al. 2012), spinal cord neurons (Keller et al. 2004; Pellett et al. 2007, 2010; Bigalke et al. 1985), hippocampal neurons (Verderio et al. 2007; Dong et al. 2007), and dorsal root ganglion cells (Welch et al. 2000; Stahl et al. 2007; Duggan et al. 2002). Early experiments in primary neurons indicated much greater sensitivity than in continuous cell lines; however, these assays did not have sensitivity comparable to the mouse bioassay. In the past few years, improvements in cell culture materials and methods have enabled primary neuronal cell assays to reach sensitivity similar to or even exceeding that of the mouse bioassay, with EC_{50} values as low as 0.3 mouse LD_{50} Units of BoNT/A in a 96-well format (Pellett et al. 2010, 2011; Whitemarsh et al. 2012).

One of the major advances is the culture of primary neurons in serum-free medium optimized for neurons (e.g., Neurobasal medium supplemented with B27 and glutamax, all from Invitrogen) (Pellett et al. 2007; Xie et al. 2000; Kivell et al. 2001; Brewer 1995; Brewer et al. 1993). This medium avoids the necessity of adding cytostatic compounds to prevent overgrowth of nonneuronal cells and results in robust and stable primary neuron cultures that can be maintained for weeks to months. In addition, plating matrices such as collagen, laminin, and matrigel are now widely available, facilitating the surface attachment of primary cells and likely also contributing to the maturation process of the cells after plating (Kohno et al. 2005; Uemura et al. 2010). Preparation of primary cells generally requires some skills and practice but can be cost-effective (e.g., one pregnant rat can yield 5–6 96-well plates of spinal cord cells), and in our experience primary cells yield very reproducible results in quantitative BoNT detection. Figure 1 outlines the steps involved in preparing and using primary rat/mouse spinal cord cells for BoNT detection. In addition, several companies now offer tissue pieces such as isolated rat or mouse embryonic spinal cords, hippocampi, or cortical pieces. While significantly more expensive, these sources allow for rapid and

Table 1 Approximate BoNT sensitivities of continuous cell lines (approximate EC_{50})

BoNT serotype/cell line	A	B	C	D	E	F	G	Reference
Neuro-2a	~3–>30 nM SC	–	–	–	–	–	–	(Pier et al. 2011; Yowler et al. 2002)
Neuro-2a + GT1b	~2 nM SC (~2,000 U) SC	–	–	–	–	–	–	(Pier et al. 2011; Yowler et al. 2002)
PC12	~120 nM NR ~2,000 U SC	–	–	–	–	–	–	(Shone and Melling 1992); Johnson lab, unpublished data
PC12, differentiated with NGF	~0.02 nM NR	–	–	–	–	–	–	(Ray 1993)
PC12 + GT1b	~400 U SC	50 nM (~50,000 U)	–	–	–	–	–	Johnson lab, unpublished data; (Dong et al. 2003)
PC12, with cell stimulation in serum-free medium	~ 100–200 U SC	–	–	–	–	–	–	Johnson lab, unpublished data
SK-N-SH	>6 nM	–	–	–	–	–	–	(Yowler et al. 2002)
SK-N-SH + GT1b and in serum-free medium	~2 nM	–	–	–	–	–	–	(Yowler et al. 2002)
SH-SY5Y	5.6 nM NR 24 nM SC	42 nM NR	0.54 nM NR	2.6 nM NR	–	>300 nM NR	–	(Purkiss et al. 2001)
SH-SY5Y, differentiated	100 pM NR	–	–	–	–	–	–	(Rasetti-Escargueil et al. 2011a)
SH-SY5Y, differentiated and + GT1b	35 pM NR	–	–	–	–	–	–	(Rasetti-Escargueil et al. 2011a)
M17 cells	~1 nM SC ≥10 nM NR	–	–	–	–	–	–	(Lee et al. 2008)
NT2, differentiated	n.d. (see text)	–	–	–	–	–	–	(Tegenge et al. 2012)
P19, differentiated	–	–	217 pM SC 310 pM SyC	58 pM VC[a]	–	–	–	(Tsukamoto et al. 2012)
SiMa, differentiated and + GT1b	~1 pM SC	–	–	–	–	–	–	(Zhu et al. 2010)

U: Mouse LD_{50} Units
NR: Neurotransmitter release, SC: SNAP-25 cleavage, SyC: syntaxin cleavage, VC: VAMP2 cleavage
[a] This study used the mosaic BoNT/CD

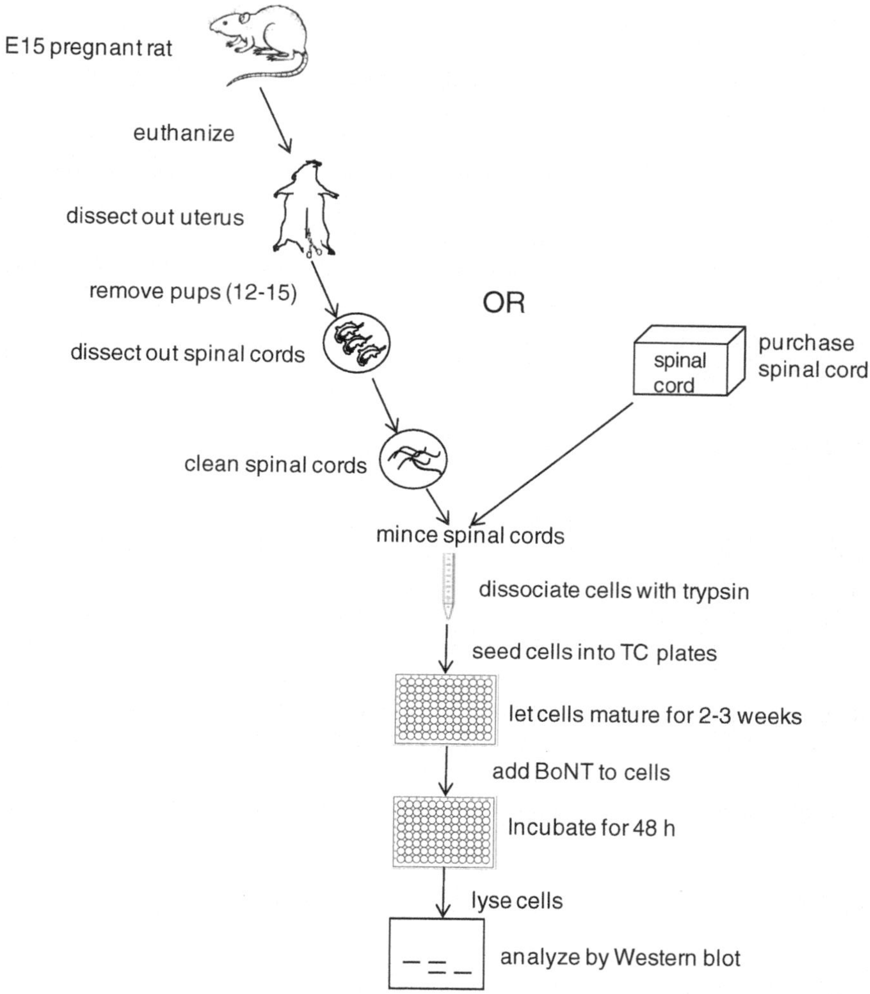

Fig. 1 Steps involved in preparing and using primary rat/mouse spinal cord cells for BoNT detection by Western bot. The uterus containing the pups of an E15 pregnant Sprague–Dawley rat or an E13 pregnant mouse (e.g., NIH Swiss) are removed and placed in dissection medium (Hanks balanced salt solution, 10 mM HEPES, 20 mM glucose. The pups are dissected out, and spinal cords removed and cleaned of all membranes and ganglia. The spinal cords are minced, and cells dissociated by trypsinization using 10 % TrypLE express (Invitrogen) in dissection medium, followed by trituration in 1 ml of culture medium (Neurobasal with B27 and glutamax) pre-warmed to 37 °C. Live cells are counted by trypan exclusion assay, and 75,000 cells/well are plated into 96-well TPP plates (MidSci) that have been coated with type I rat tail collagen (BD Biosciences), or matrigel (BD Biosciences) according to manufacturer's instructions. The cells are allowed to mature at 37 °C in a humidified 5 % CO_2 atmosphere for at least 18 days with bi-weekly changes of culture medium before they are used in the toxin assay. For the toxin assay, toxin dilutions are prepared in culture medium and added to the cells. After a specified incubation time, the toxin is aspirated, and the cells lysed in 75 μl of 1× LDS lysis buffer (Invitrogen). The samples are separated using 12 % Bis–Tris gels and MES running buffer (Invitrogen), and analyzed by Western blot using a monoclonal anti-SNAP-25 antibody (Synaptic Systems, Germany) and densitometry. The details of this method are described in (Pellett et al. 2007, 2010). The methods for preparation and use of other primary cells are similar

simple culture of primary neurons by scientists with minimal experience in primary cell preparation.

Unlike continuous cell lines, which differ in their sensitivity to BoNT serotypes based on unique expression of surface receptors, ganglioside profiles, and other often not well-defined characteristics, primary cells present normal neurons which can be used to study all BoNT serotypes and subtypes. In fact, primary cells are ideal models to study differences between BoNT sero- and subtypes in the respective neuronal subpopulations. Spinal cord cells or motor-neuron-enriched spinal cord cells would be expected to be the most relevant and sensitive BoNT cell model based on the neuromuscular symptomatology of botulism. However, other primary cells such as cortical neurons, hippocampal cells, and dorsal root ganglion cells also seem to be exquisitely sensitive to BoNTs and have been used as study models (Pier et al. 2011; Welch et al. 2000; Stahl et al. 2007; Saadi et al. 2012; Verderio et al. 2007; Dong et al. 2007; Duggan et al. 2002).

While most primary neuronal cell models are very sensitive to BoNT/A, some cell models, such as primary rat spinal cord cells, differ in their sensitivity to other BoNT serotypes. This can be explained by species-specific differences in BoNT receptors and/or SNARE proteins, as well as by other species- or cell type-specific characteristics. It is important to remember that the specific activity of all BoNT serotypes is currently determined in the mouse bioassay, and thus do not necessarily reflect the activity in other species including humans. This is exemplified by the differential potency of pharmaceutical BoNT/A and B in humans (Dressler and Benecke 2007; Flynn 2004), which is likely due to a variation in the amino acid sequence between mouse and human synaptotagmin II (the BoNT/B receptor) (Strotmeier et al. 2012; Peng et al. 2012). Thus, when using primary neurons, both the animal source and the neuronal subpopulation should be taken into careful consideration.

3.2.3 Stem Cell-Derived Neurons for BoNT Detection

Embryonic stem (ES) cells are pluripotent cells derived from the inner part of a blastocyst. ES cells can be propagated and frozen similar to continuous cell lines and can be differentiated into any cell type, including neurons. Mouse ES (mES) cells can be differentiated into neurons by addition of retinoic acid. Addition of purmorphamine or sonic hedgehog further directs this differentiation to motor neurons (Zhang 2006; Wichterle and Peljto 2008). This differentiation takes about 2 weeks, and the resulting cells are nearly as sensitive as primary neurons to all BoNT serotypes, and have equal sensitivity to the MBA (Pellett et al. 2011; McNutt et al. 2011; Kiris et al. 2011). However, differentiation of mES cells to neurons, and especially to motor neurons, requires significant experience and practice working with stem cells. Figure 2a outlines the major steps involved in mES cells differentiation to neurons, including propagation of undifferentiated mES cells, differentiating the cells into embryoid bodies (EBs), inducing neuronal differentiation of the EBs, and dissociation and plating of the differentiated neurons, followed by maturation (Fig. 2). Many variables can affect the outcome of

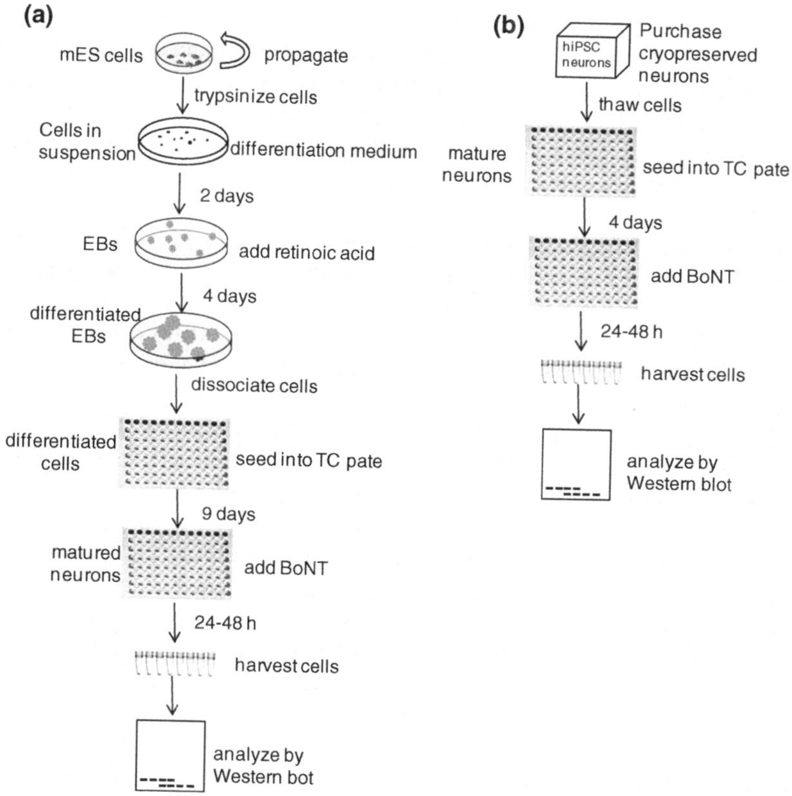

Fig. 2 Steps involved in preparing and using mouse embryonic stem cell (mES) derived neurons **a** or purchased human induced pluripotent stem (hiPS) cell-derived neurons **b** for BoNT detection by Western blot. **a**: The mES cells are maintained in a basic maintenance medium containing human leukemia inhibitory factor (LIF) to avoid cell differentiation, such as DMEM/F12, 15 % fetal bovine serum, 1× nonessential amino acids, sodium pyruvate (1 mM), glutamax, 2-mercaptoethanol (0.1 mM) and LIF, (1,000 units/ml) on 6-well plates that have been coated for 1 h at 37 °C with 0.1 % gelatin. Medium is replaced every 2 days, and cells are split by trypsin–EDTA digestion. The cells can be propagated and cryopreserved at this stage. For differentiation to neurons, $\sim$2–4 $\times$ 10^6 cells are transferred to a 100 mm Petri dish in differentiation medium such as MES/SR (same as MES, but with 15 % Knockout serum replacement instead of FBS and no LIF) in order to induce embryoid body (EB) formation. After 2 days 0.5–5 μM retinoic acid is added, and for motorneurons 1 μM purmorphamine is also added. After 4 days, the EBs are digested with accutase or TrypLE, and 75,000 cells/well are seeded into 96-well plates coated with 1 μg/well of laminin in culture medium (such as Neurobasal supplemented with B27, 1 μM cAMP, and 20 ng/ml GDNF) to maintain neurons. The neurons are allowed to mature in culture for 9 days before use in toxin assays. For details of this method see (Pellett et al. 2011; McNutt et al. 2011; Kiris et al. 2011). **b**: The hiPS cells derived neurons are purchased as already differentiated and cryopreserved cells. The cells are thawed and seeded into 96-well plates at a density of 30–40,000 cells/well (Whitemarsh et al. 2012). For the toxin assay, toxin dilutions are prepared in culture medium and added to the cells. After a specified incubation time, the toxin is aspirated, and the cells lysed in 50 μl of 1× LDS lysis buffer. The samples are separated using 12 % Bis–Tris gels and MES running buffer, and analyzed by Western blot using a monoclonal anti-SNAP-25 antibody and densitometry

neuronal differentiation, and thus sensitivity to BoNTs, including the quality of the materials, concentration of retinoic acid, time of retinoic acid addition and incubation, the dissociation of EBs, the plating matrix, medium composition, and so on. Even with standardization of the protocol, factors such as the source, storage and age of medium components, and cell handling practice can influence the differentiation success. This makes implementation of a BoNT detection assay using mES cells derived neurons challenging in a (nonstem cell focused) research lab and probably also in a commercial setting.

Differentiation of human embryonic stem (hES) cells to neurons is even more involved and requires at least 6 weeks. In addition, the differentiation rate of hES cells is significantly lower than that of mES cells (50 vs. 90 %) (Zhang 2006; Hu and Zhang 2010). Furthermore, there is considerable ethical controversy in the use of hES cells for various purposes. Probably for this reason, a BoNT detection assay using hES cell-derived neurons has not been developed.

Major advances in stem cell research during the past decade have enabled the conversion of normal adult somatic cells (e.g., skin or blood cells) to stem cells by inducing expression of a small set of genes that is otherwise silenced in these cells (Takahashi et al. 2007; Yu et al. 2007). The resulting cells are pluripotent-like ES cells and can be propagated and frozen in the undifferentiated state. These human-induced pluripotent stem (hiPS) cells can be differentiated into mature human neurons much like ES cells. In addition, methods have now been developed for cryopreservation of differentiated neurons or partially differentiated neurons from hiPS cells for commercial purposes. Thus, such neurons are now commercially available, enabling the use of high quality human neurons for BoNT detection assays in most lab settings as well as in the pharmaceutical industry while requiring only basic knowledge about tissue culture techniques (e.g., seeding of the neurons into tissue culture plates and medium changes for every 2–3 days, Fig. 2b). Importantly, commercially well-prepared cells undergo quality control tests, ensuring consistency among different cell lots within the defined parameters used for quality control standards of the company.

Human iPS cells derived neurons have been recently shown to be a highly sensitive cell model for BoNT detection with an EC_{50} around 0.3 mouse LD_{50} Unit (U) for BoNT/A (Whitemarsh et al. 2012). The hiPS cells derived neurons have also shown utility in detection of neutralizing antibodies to BoNTs (Whitemarsh et al. 2012). These cells have much potential and relevance for research and commercial detection of BoNTs and antibodies, including their human origin, and the fact that these cells resemble normal somatic neurons (unlike cancerous cells in continuous cell lines). Interestingly, the EC_{50}s in these neurons differed for different serotypes of BoNTs (BoNT/A: ~0.3 U, BoNT/C: ~0.4 U, BoNT/E: ~1.8, BoNT/B: ~16 U). Limited human studies confirm serotype specific differences in humans with a 40-fold decreased activity of BoNT/B compared to BoNT/A (Johnson and Montecucco 2008; Eleopra et al. 2004; Eleopra et al. 1998). This highlights the importance of considering species specificity when assaying different BoNT serotypes, which are currently all assayed in mice. The steps involved in the use of cryopreserved hiPS cells derived neurons for BoNT detection is shown in (Fig. 2b).

3.3 *Applications of NCB Assays*

The high sensitivity, specificity, reproducibility, and low error of NCB assays using primary or stem cell-derived neurons enable the introduction and further development of standardized cell-based assays to complement and to eventually replace the mouse bioassay for many uses. The most important applications requiring an assay that measures all steps of the intoxication process include potency determination of research and pharmaceutical BoNTs, quantitative detection of neutralizing antibodies, and use as research models. Other applications may include the sensitive detection of BoNTs in foodstuffs or in field samples, although these applications would require a combination of the cell-based assay with a BoNT isolation and purification step. The following paragraphs discuss considerations for using cell-based assays for potency determination of pharmaceutical BoNTs, neutralizing antibody detection, and as a research model.

3.3.1 Potency Determination of Pharmaceutical BoNTs

BoNTs are extremely valuable drugs for treatment of a variety of neurological diseases as well as for cosmetic treatments. To date, BoNT/A and to a lesser extent BoNT/B are being employed as pharmaceuticals (Cartee and Monheit 2011; Dressler and Benecke 2007). The number of treatments is steadily increasing with currently over 1 million treatments carried out each year in the United States alone. A reliable and well-established assay for BoNT potency testing is required to ensure safe and consistent preparations for pharmaceutical utility. This assay should test for fully functional holotoxins in a quantitative and reproducible manner, taking into consideration all steps of cellular intoxication (e.g., cell binding, endocytosis, translocation, and intracellular enzymatic activity) (Adler et al. 2010; Sesardic and Das 2007). Until recently, the MBA was the only assay approved and used for this purpose. However, the drawbacks of the MBA have led to a strong impetus by researchers and regulatory agencies to find alternatives for the potency determination of pharmaceutical preparations of BoNTs (Adler et al. 2010). Cell-based assays are especially well suited for this application, because they test for fully functional BoNTs that undergo all steps of cell intoxication are specific, and exceed the sensitivity of the mouse bioassay. Since Allergan's announcement in fall of 2011 that the FDA for the first time had approved a cell-based assay for the potency testing of its product (Botox®), the interest surrounding cell-based assays has increased markedly among commercial producers of pharmaceutical BoNTs. Research has shown that both primary cells and stem cell-derived neurons have the required sensitivity. However, the need for repeated new preparations of primary cells and the quality control issues involved makes it unlikely that this cell model will be established in a commercial setting. On the other hand, stem cell-derived neurons appear to be an ideal cell system for a potency assay, as they provide normal neurons (as opposed to cancerous cell lines), and neurons derived from hiPS cells are species specific. In order to introduce these cell assays for potency determination of pharmaceutical BoNTs, they will need to be optimized and standardized for each product, and cross-validated against the MBA. It cannot be assumed that a pharmaceutical preparation of BoNT in excipient and possibly containing complex proteins will yield the same EC_{50} values as a purified 150 kDa BoNT/A product that is stored as a concentrated stock and diluted in cell culture medium. The excipient composition as well as the potential presence of complex (or other) proteins may affect the sensitivity of a NCB assay. Thus, assay conditions need to be optimized for each specific use (product) to ensure required sensitivity and accuracy. For example, commonly used stabilizers such as human serum albumin likely increase sensitivity, and saline commonly used to re-suspend lyophilized BoNT in vials will need to be sufficiently diluted in order to avoid salt effects on the cells. After such optimization studies are completed and a toxin standard has been validated, the cell assay has the potential to fully replace the MBA for BoNT potency determination of pharmaceutical BoNTs, although periodic control assays using the MBA may be advisable. Unlike in the MBA, which is highly dependent on proper injection techniques and mice strains, the methods used in a NCB assay can readily be standardized among different laboratories. Thus, replacement of the MBA with a carefully optimized and standardized NCB assay has the potential to significantly reduce the intra-laboratory error currently observed during BoNT potency evaluations (Sesardic et al. 2003). In addition, using a human specific cell model will allow for species-appropriate potency evaluation of different BoNT sero-or subtypes. It is well known that significantly more BoNT/B mouse LD_{50} Units are required to achieve the same therapeutic effect as with BoNT/A in humans (Dressler and Benecke 2007; Flynn 2004). Recent evidence suggests that the predominant mechanism underlying this is species-specific differences in the BoNT/B and G cell surface receptor synaptotagmin II between humans and chimpanzees versus rats and mice (Strotmeier et al. 2012; Peng et al. 2012). Determining pharmaceutical BoNT Units derived from human neurons EC_{50} values (e.g., human neuron EC_{50} Unit) will likely aid physicians in converting from one BoNT product to another without the confusion of having to use a different number of Units among products.

3.3.2 Detection of Neutralizing Antibodies to BoNTs

Despite their effective use of BoNT/A and BoNT/B in clinical applications, a major adverse effect has been the formation of neutralizing antibodies that make patients refractory to treatment (Dressler et al. 2010; Borodic 2007; Muller et al. 2009; Dressler 2004). The estimates for the percentage of patients developing neutralizing antibodies vary and are dependent on the toxin dose, indication, number of repeat injections, and the commercial product. In some estimates, as many as 5–12 % of patients who received repeated BoNT/A treatments for spasticity developed resistance to treatments due to the presence of circulating neutralizing serum antibodies (Muller et al. 2009; Cordivari et al. 2006; Kessler and Benecke 1997). Currently, patients are not routinely monitored for antibody formation during their treatment regime (Sesardic et al. 2004). In vitro methods, such as ELISA, are sensitive and specific but do not currently differentiate between neutralizing and nonneutralizing antibodies. The MBA can specifically detect neutralizing antibodies, but requires a large number of mice and is not sufficiently sensitive to detect low levels of antibodies in some treatment-resistant patients. A currently used alternative test system is the mouse phrenic nerve hemidiaphragm assay (Dressler et al. 2002, 2005; Rasetti-Escargueil et al. 2009, 2011b), which quantitatively detects neutralizing antibodies without performing live-animal experiments and is also commercially available for patients. In addition, local injection assays, such as remote point injection in the forehead (Borodic et al. 1995) or the EDB (extensor digitorum brevis) assay, in which BoNT/A is 'test' injected into the EDB muscle of the patient and electrophysiological measurements, are taken before and after the injection (Cordivari et al. 2006; Kessler and Benecke 1997). While the latter assays are inexpensive, do not require the use of animals, and correlate with clinical observations of BoNT/A effectiveness, they require the injection of BoNTs into a patient for testing purposes only and are not practical.

NCB assays offer a sensitive alternative that is specific for neutralizing antibodies, provided a qualified cell model as described above is used. As proof in point, a cell-based assay using primary rat spinal cord cells was shown in a pilot study to be about 10 times more sensitive than the mouse bioassay in detection of neutralizing antibodies and correlated 100 % with patient outcome (Pellett et al. 2007). Similarly, neurons derived from hiPS cells have been shown to be about 10 times more sensitive than the MBA for neutralizing antibody detection (Whitemarsh et al. 2012). Future studies are needed to optimize and standardize a cell-based assay for this purpose and to confirm correlation of patient nonresponse to BoNT treatments and detection of neutralizing antibodies. Factors that may affect a cell assay for antibodies include the quality and sterility of the serum, which are critical to avoid cell damage, the concentration of serum and toxin used in the assay, and the incubation time and temperature of toxin with antibody or serum prior to cell exposure. When choosing a cell model, it is also important to consider that the BoNT sensitivity of the cells will also determine the neutralizing antibody detection sensitivity as more antibodies will be needed to neutralize greater amounts of toxin. Additionally, consideration must be given to the serum source and the endpoint used. In clinical trials using animals, the serum should be collected in a sterile manner and drugs that potentially may cause cytotoxicity or affect the serum quality by inducing red cell lysis (e.g., pentobarbital) should be avoided as a means of euthanasia before serum collection. If a nonspecific endpoint such as inhibition of neurotransmitter release is used instead of a SNARE cleavage endpoint, potential direct effects of the serum on this endpoint must also be evaluated.

3.3.3 Research Applications

Research on BoNTs requires the determination of BoNT specific activity. At present, this is generally determined using the MBA, but as discussed above specific toxicity determination can now be performed in cell-based assays. In addition, many research projects often address important biological properties of BoNTs such as cell binding and trafficking mechanisms. Different cell models may provide advantages and disadvantages to each specific research application. Continuous cell lines lack sensitivity but might provide characteristics desirable for a specific purpose, such as the expression of a particular surface marker. Cell lines can be manipulated to transiently or stably express a desired protein or marker, or to increase, decrease, or abolish expression levels of a particular gene. Studies demonstrating use of such techniques have, for example, led to the identification of the receptor SV2, with SV2C being the favored isoform, and polysialiogangsliosides as the receptors for BoNT/A (Dong et al. 2006; Yowler et al. 2002), strengthening the 'dual receptor' model that necessitates BoNT binding of both, a protein and a ganglioside receptor for cell entry (Montecucco 1986). While such techniques are now also possible in stem cells or stem cell-derived neurons, the methods used in those cells are much more difficult or cost-prohibitive for many studies.

Other research applications such as cellular entry and trafficking studies benefit most from using a research model that most closely mimics the behavior of BoNT in normal human cells. Since the symptoms of botulism indicate that BoNTs primarily affect cholinergic neurons, primary cell models such as mouse or rat primary spinal cord cells could present good models. In fact, many important discoveries about botulinum toxin activity have been made in primary cell models, such as for example, the much longer duration of the BoNT/A LC activity in neurons compared to the BoNT/E LC (Keller et al. 1999). However, these models are not human specific, and potential species-specific differences of the cells have to be considered when interpreting these studies. For example, the differences in human/primate and mouse/rat synaptotagmin II (Strotmeier et al. 2012; Peng et al. 2012) likely result in BoNT/B entering human neurons via the weaker synaptotagmin I receptor. Besides explaining the difference in BoNT/A and B potency in humans, this also has important implications with regard to neuron specificity of BoNT/B, considering the differential expression of synaptotagmin I and II in the central nervous system (Ullrich et al. 1994).

Neurons derived from hiPS cells offer a human specific alternative; however, the currently available neurons resemble forebrain neurons rather than neurons from the peripheral nervous system. Neurons derived from mES cells can be differentiated to mixed central nervous system neurons or motor-neuron enriched cultures, and both neuronal populations appear to be equally sensitive to BoNTs (Pellett et al. 2011; McNutt et al. 2011; Kiris et al. 2011). While all of these cell models have limitations that need to be considered, they are sensitive to BoNTs, and therefore are good models for research applications such as the screening of new BoNT antitoxins or inhibitors. The neurons derived from hiPS cells are about

10 times more sensitive for fast (5 min) BoNT cell entry during chemical cell stimulation compared to primary neurons (Whitemarsh et al. 2012), and thus are especially well suited for experiments such as screening and analyses of intracellular BoNT inhibitors. In these experiments, cells are exposed to BoNT prior to addition of potential inhibitors, which avoids potential secondary effects of inhibitors on BoNT cell entry (Fischer et al. 2009).

While cell models may not entirely replace in vivo studies, they provide an excellent research model to study the biological properties of BoNTs and inhibitors in a 'closed' neuronal model. Thus, in a hypothetical situation, a specific BoNT inhibitor may not be stable in vivo and be cleared out of the system, but may effectively enter neuronal cells and inhibit LC activity inside the cells. Such an inhibitor would be missed if only screened in animals, but identification in cell assays and modifications to improve the stability in vivo might yield a new and important drug for treatment of botulism. Equally important, cell assays allow the screening of a moderately large number of inhibitors or antitoxins before testing in animals and thereby significantly reduces the number of animals used.

4 Future Perspectives and Conclusion

Throughout the years, cell-based assays have found many applications in the field of botulinum toxins, aiding in the biologic and molecular understanding of these important toxins. The recent introduction of highly sensitive and human specific cell models enables novel areas of basic study, as well as the use of cell-based assays for commercial uses in BoNT potency determination and neutralizing antibody screening. Future developments of pharmaceutical BoNTs include the use of other BoNT sero-or subtypes and recombinantly altered BoNTs, as well as novel indications (Pickett and Perrow 2011; Cartee and Monheit 2011; Rummel et al. 2011). In addition, improved vaccines and antitoxins are being developed, and the search for an effective cure or treatment of botulism beyond antitoxin, which has to be administered within the first 24–48 h after toxin exposure, continues. Research efforts continue to identify new information on the biologic and molecular function of BoNT sero-and subtypes. NCB assays offer a unique study model that can be used to examine cell binding, entry, translocation, and intracellular trafficking and catalytic activity, and are an essential tool in these developments. The most promising NCB assays for replacement of the MBA use neurons derived from stem cells, in particular hiPS cells. Such neurons are difficult to prepare and either require collaboration with one of the several laboratories around the world that are routinely preparing them, or need to be purchased. While currently very expensive, future research in the stem cell area and competition in the market will likely lead to lowered prizes of such neurons. So far, hiPS cell-derived neurons have been tested as a model for BoNT research and potency detection from only one source (Cellular Dynamics International, WI) (Whitemarsh et al. 2012), but other companies are starting to offer similar products

Table 2 Features of cell based assays

Assay features	Continuous cell lines	Primary cells	mESC derived neurons	hiPS cell-derived neurons
BoNT/A sensitivity [EC_{50}]	≥2.5–>100 U[a]	~0.3–1 U	~0.8–1 U	~0.3 U
Dose–response	Cell line dependent	Sigmoidal, 3–4 log range	Sigmoidal, ~2.5 log range	Sigmoidal, ~1.5 log range
Reproducibility	Good	Dependent on cell preparation	Dependent on cell differentiation	Good
Cell culture set up	Easy	Difficult	Difficult	Easy
Hands-on time for cell preparation	Minimal	Moderate	Most	Minimal if purchased
Cell maturation/culture	1 Day–2 weeks	~2–3 weeks	~2 weeks	Days–2 weeks
Cell characteristics	Cancerous, cell line specific	Mixed cultures of well-defined origin	Mixed cultures, primarily neurons	Mostly neurons (≥95 %), forebrain neurons
Species	Various	Various	Mouse	Human
Cost/96-well plate	≤$10	$50–$300	~$50–$100	$1,500 (list price) if purchased differentiated

mESC derived neurons: Neurons differentiated from mouse embryonic stem cells
hiPS cell-derived neurons: Already differentiated and cryopreserved neurons from human induced pluripotent stem cells
U: Mouse LD_{50} Units
[a] The EC_{50} of 2.5 was estimated based on (Zhu et al. 2010)

(Globalstem, MD, ReproCELL, CA). In addition, neurons representing neuronal subpopulations other than the currently available forebrain neurons will likely become available in the future and will enable comparative studies of different BoNT sero- and subtypes in different neuronal subpopulations. This is exemplified by companies that are developing neurons containing a large percentage of cholinergic neurons derived from hiPS cells, as well as neuronal stem cells that can be differentiated into mature neurons of different phenotypes within 2 weeks. Eventually, more affordable cell models for research purposes are primary cells or mES cells derived neurons. Table 2 provides a summary of different cell models and their important features.

One important aspect of NCB assays is the endpoint used. The Western blot method is an excellent method for research purposes as it is easy to set up in any laboratory, does not require special equipment, provides visual results, and is applicable to any cell type. In a commercial high-throughput setting, however, this method is laborious, and using Western blots for quantitative purposes in general can introduce error. This problem is reduced by comparing cleaved versus uncleaved SNAP-25 bands for BoNT/A and E, and in fact data indicate that the error of this method is still smaller than that of the MBA (Pellett et al. 2010). Nevertheless, the error rate for other serotypes that are assayed by the disappearance of the VAMP2 band compared to a loading standard such as syntaxin is much greater (Whitemarsh et al. 2012). Other detection methods, such as ELISA of cleaved and unlceaved SNARE proteins in cell lysates, are being developed and are more suitable for commercial applications (Nuss et al. 2010; Zhu et al. 2010). These methods will likely improve accuracy, speed, and costs of NCB assays and are also applicable to any cell type. Neurotransmitter release is cell type specific, and while it is an important endpoint to explore in research especially in mES and hiPS cells derived neurons, is not useful by itself for commercial purposes because it is relatively nonspecific. The stable introduction of a sensor into a cell line that enables an easy and fast read out, such as a fluorescent or FRET sensor (Dong et al. 2004) would be of tremendous value for both research and commercial applications. Creating such a cell line can be easiest achieved in continuous cell lines (see BoCell assay, Biosentinal Inc.,), but is also possible (although much more difficult and expensive) in mES and hiPS cells cell lines that can then be differentiated to neurons. Sensitivity and species specificity of the cells should be considered before construction of such a cell model.

The mouse bioassay has been the ‘gold standard’ for BoNT detection since the beginnings of BoNT research. The recent and ongoing development of NCB assays including systems using normal human neurons opens the possibility to largely replace the mouse bioassay for BoNT potency determination and neutralizing antibody detection, and to study BoNTs in species-specific neuronal cell models. This has the potential to lead to important new insights on the behavior and effect of BoNT sero-and subtypes on specific human neurons, which will aid in the design of novel and improved BoNT-based pharmaceuticals as well as botulism countermeasures.

Acknowledgments I thank Dr. Eric A Johnson and Regina Whitemarsh for critical review of the manuscript. This work was supported by the National Institutes of Health (contract numbers 1R01AI095274-01A1 and 1R21AI082826-01A2).

References

Adler S, Bicker G, Bigalke H et al (2010) The current scientific and legal status of alternative methods to the LD50 test for botulinum neurotoxin potency testing. The report and recommendations of a ZEBET expert meeting. Altern Lab Anim 38:315–330

Akaike N, Ito Y, Shin MC, Nonaka K, Torii Y, Harakawa T, Ginnaga A, Kozaki S, Kaji R (2010) Effects of A2 type botulinum toxin on spontaneous miniature and evoked transmitter release from the rat spinal excitatory and inhibitory synapses. Toxicon 56:1315–1326

Arnon SS, Schechter R, Inglesby TV et al (2001) Botulinum toxin as a biological weapon: medical and public health management. JAMA 285:1059–1070

Benatar MG, Willison HJ, Vincent A (1997) Lack of effect of Miller Fisher sera/plasmas on transmitter release from PC12 cells. J Neuroimmunol 80:1–5

Bercsenyi K, Giribaldi F, Schiavo G (2012) The elusive compass of clostridial neurotoxins: deciding when and where to go? doi:10.1007/978-3-642-33570-9_5

Bigalke H, Dimpfel W, Habermann E (1978) Suppression of 3H-acetylcholine release from primary nerve cell cultures by tetanus and botulinum-A toxin. Naunyn Schmiedebergs Arch Pharmacol 303:133–138

Bigalke H, Dreyer F, Bergey G (1985) Botulinum A neurotoxin inhibits non-cholinergic synaptic transmission in mouse spinal cord neurons in culture. Brain Res 360:318–324

Binz T (2012) Clostridial neurotoxin light chains: devices for SNARE cleavage mediated blockade of neurotransmission. doi:10.1007/978-3-642-33570-9_7

Binz T, Blasi J, Yamasaki S et al (1994) Proteolysis of SNAP-25 by types E and A botulinal neurotoxins. J Biol Chem 269:1617–1620

Blasi J, Chapman ER, Link E et al (1993a) Botulinum neurotoxin A selectively cleaves the synaptic protein SNAP-25. Nature 365:160–163

Blasi J, Chapman ER, Yamasaki S et al (1993b) Botulinum neurotoxin C1 blocks neurotransmitter release by means of cleaving HPC-1/syntaxin. EMBO J 12:4821–4828

Borodic G (2007) Botulinum toxin, immunologic considerations with long-term repeated use, with emphasis on cosmetic applications. Facial Plast Surg Clin North Am 15:11–16

Borodic GE, Duane D, Pearce B et al (1995) Antibodies to botulinum toxin. Neurology 45:204

Brewer GJ (1995) Serum-free B27/neurobasal medium supports differentiated growth of neurons from the striatum, substantia nigra, septum, cerebral cortex, cerebellum, and dentate gyrus. J Neurosci Res 42:674–683

Brewer GJ, Torricelli JR, Evege EK et al (1993) Optimized survival of hippocampal neurons in B27-supplemented neurobasal, a new serum-free medium combination. J Neurosci Res 35:567–576

Cartee TV, Monheit GD (2011) An overview of botulinum toxins: past, present, and future. Clin Plast Surg 38:409–426

Conway JO, Sherwood LJ, Collazo MT et al (2010) Llama single domain antibodies specific for the 7 botulinum neurotoxin serotypes as heptaplex immunoreagents. PLoS ONE 5:e8818

Cordivari C, Misra VP, Vincent A et al (2006) Secondary nonresponsiveness to botulinum toxin A in cervical dystonia: the role of electromyogram-guided injections, botulinum toxin A antibody assay, and the extensor digitorum brevis test. Mov Disord 21:1737–1741

Dong M, Liu H, Tepp WH et al (2008) Glycosylated SV2A and SV2B mediate the entry of botulinum neurotoxin E into neurons. Mol Biol Cell 19:5226–5237

Dong M, Richards DA, Goodnough MC et al (2003) Synaptotagmins I and II mediate entry of botulinum neurotoxin B into cells. J Cell Biol 162:1293–1303

Dong M, Tepp WH, Johnson EA et al (2004) Using fluorescent sensors to detect botulinum neurotoxin activity in vitro and in living cells. Proc Natl Acad Sci U S A 101:14701–14706

Dong M, Tepp WH, Liu H et al (2007) Mechanism of botulinum neurotoxin B and G entry into hippocampal neurons. J Cell Biol 179:1511–1522

Dong M, Yeh F, Tepp WH et al (2006) SV2 is the protein receptor for botulinum neurotoxin A. Science 312:592–596

Dorner MB, Schulz KM, Kull S (2012) Complexity of botulinum neurotoxins: challenges for detection technology. doi:10.1007/978-3-642-33570-9_11

Dressler D (2004) Clinical presentation and management of antibody-induced failure of botulinum toxin therapy. Mov Disord 19(Suppl 8):S92–S100

Dressler D, Benecke R (2007) Pharmacology of therapeutic botulinum toxin preparations. Disabil Rehabil 29:1761–1768

Dressler D, Bigalke H (2002) Botulinum toxin antibody type A titres after cessation of botulinum toxin therapy. Mov Disord 17:170–173

Dressler D, Lange M, Bigalke H (2005) Mouse diaphragm assay for detection of antibodies against botulinum toxin type B. Mov Disord 20:1617–1619

Dressler D, Wohlfahrt K, Meyer-Rogge E et al (2010) Antibody-induced failure of botulinum toxin a therapy in cosmetic indications. Dermatol Surg 36(Suppl 4):2182–2187

Duggan MJ, Quinn CP, Chaddock JA et al (2002) Inhibition of release of neurotransmitters from rat dorsal root ganglia by a novel conjugate of a Clostridium botulinum toxin A endopeptidase fragment and Erythrina cristagalli lectin. J Biol Chem 277:34846–34852

Eleopra R, Tugnoli V, Quatrale R et al (2004) Different types of botulinum toxin in humans. Mov Disord 19(Suppl 8):S53–S59

Eleopra R, Tugnoli V, Rossetto O et al (1998) Different time courses of recovery after poisoning with botulinum neurotoxin serotypes A and E in humans. Neurosci Lett 256:135–138

Evidente VG, Adler CH (2010) An update on the neurologic applications of botulinum toxins. Curr Neurol Neurosci Rep 10:338–344

Fischer A (2012) Synchronized chaperone function of botulinum neurotoxin domains mediates light chain translocation into neurons. doi:10.1007/978-3-642-33570-9_6

Fischer A, Montal M (2007a) Crucial role of the disulfide bridge between botulinum neurotoxin light and heavy chains in protease translocation across membranes. J Biol Chem 282:29604–29611

Fischer A, Montal M (2007b) Single molecule detection of intermediates during botulinum neurotoxin translocation across membranes. Proc Natl Acad Sci U S A 104:10447–10452

Fischer A, Nakai Y, Eubanks LM et al (2009) Bimodal modulation of the botulinum neurotoxin protein-conducting channel. Proc Natl Acad Sci U S A 106:1330–1335

Flynn TC (2004) Myobloc. Dermatol Clin 22:207–211 vii

Gimenez DF, Gimenez JA (1995) The typing of botulinal neurotoxins. Int J Food Microbiol 27:1–9

Hatheway CL (1988) Botulism. In: Balows A, Hausler WH, Ohashi M, Turano MA (eds) Laboratory diagnosis of infectious diseases: principles and practice, vol 1. Springer, New York, pp 111–133

Hill KK, Smith TJ (2012) Genetic diversity within *Clostridium botulinum* serotypes, botulinum neurotoxin gene clusters and toxin subtypes. doi:10.1007/978-3-642-33570-9_1

Hu BY, Zhang SC (2010) Directed differentiation of neural-stem cells and subtype-specific neurons from hESCs. Methods Mol Biol 636:123–137

Johnson EA, Montecucco C (2008) Chapter 11 Botulism. In: Engel Andrew G (ed) Handbook of clinical neurology, vol 91. Elsevier, pp 333–368

Jones RG, Alsop TA, Hull R et al (2006) Botulinum type A toxin neutralisation by specific IgG and its fragments: a comparison of mouse systemic toxicity and local flaccid paralysis assays. Toxicon 48:246–254

Jones RG, Liu Y, Halls C et al (2011) Release of proteolytic activity following reduction in therapeutic human serum albumin containing products: detection with a new neoepitope endopeptidase immunoassay. J Pharm Biomed Anal 54:74–80

Kalb SR, Baudys J, Egan C et al (2011) Different substrate recognition requirements for cleavage of synaptobrevin-2 by Clostridium baratii and Clostridium botulinum type F neurotoxins. Appl Environ Microbiol 77:1301–1308

Keller JE, Cai F, Neale EA (2004) Uptake of botulinum neurotoxin into cultured neurons. Biochemistry 43:526–532

Keller JE, Neale EA (2001) The role of the synaptic protein snap-25 in the potency of botulinum neurotoxin type A. J Biol Chem 276:13476–13482

Keller JE, Neale EA, Oyler G et al (1999) Persistence of botulinum neurotoxin action in cultured spinal cord cells. FEBS Lett 456:137–142

Kessler KR, Benecke R (1997) The EBD test–a clinical test for the detection of antibodies to botulinum toxin type A. Mov Disord 12:95–99

Kiris E, Nuss JE, Burnett JC et al (2011) Embryonic stem cell-derived motoneurons provide a highly sensitive cell culture model for botulinum neurotoxin studies, with implications for high-throughput drug discovery. Stem Cell Res 6:195–205

Kivell BM, McDonald FJ, Miller JH (2001) Method for serum-free culture of late fetal and early postnatal rat brainstem neurons. Brain Res Brain Res Protoc 6:91–99

Kohno K, Kawakami T, Hiruma H (2005) Effects of soluble laminin on organelle transport and neurite growth in cultured mouse dorsal root ganglion neurons: difference between primary neurites and branches. J Cell Physiol 205:253–261

Lacy DB, Tepp W, Cohen AC et al (1998) Crystal structure of botulinum neurotoxin type A and implications for toxicity. Nat Struct Biol 5:898–902

Lee JO, Rosenfield J, Tzipori S et al (2008) M17 human neuroblastoma cell as a cell model for investigation of botulinum neurotoxin A activity and evaluation of BoNT/A specific antibody. The Botulinum J 1:135–152

Liu YY, Rigsby P, Sesardic D et al (2012) A functional dual-coated (FDC) microtitre plate method to replace the botulinum toxin LD (50) test. Anal Biochem 425:28–35

Macdonald TE, Helma CH, Shou Y et al (2011) Analysis of Clostridium botulinum serotype E strains by using multilocus sequence typing, amplified fragment length polymorphism, variable-number tandem-repeat analysis, and botulinum neurotoxin gene sequencing. Appl Environ Microbiol 77:8625–8634

McInnes C, Dolly JO (1990) Ca2(+)-dependent noradrenaline release from permeabilised PC12 cells is blocked by botulinum neurotoxin A or its light chain. FEBS Lett 261:323–326

McMahon HT, Nicholls DG (1991) Transmitter glutamate release from isolated nerve terminals: evidence for biphasic release and triggering by localized Ca2+. J Neurochem 56:86–94

McNutt P, Celver J, Hamilton T et al (2011) Embryonic stem cell-derived neurons are a novel, highly sensitive tissue culture platform for botulinum research. Biochem Biophys Res Commun 405:85–90

Montal M (2010) Botulinum neurotoxin: a marvel of protein design. Annu Rev Biochem 79: 591–617

Montecucco C (1986) How do tetanus and botulinum toxins bind to neuronal membranes? Trends Biochem Sci 11:314–317

Muller K, Mix E, Adib Saberi F et al (2009) Prevalence of neutralising antibodies in patients treated with botulinum toxin type A for spasticity. J Neural Transm 116:579–585

Nishiki T, Kamata Y, Nemoto Y et al (1994) Identification of protein receptor for Clostridium botulinum type B neurotoxin in rat brain synaptosomes. J Biol Chem 269:10498–10503

Nuss JE, Ruthel G, Tressler LE et al (2010) Development of cell-based assays to measure botulinum neurotoxin serotype A activity using cleavage-sensitive antibodies. J Biomol Screen 15:42–51

Parnas D, Linial M (1995) Cholinergic properties of neurons differentiated from an embryonal carcinoma cell-line (P19). Int J Dev Neurosci 13:767–781

Peck MW (2009) Biology and genomic analysis of Clostridium botulinum. Adv Microb Physiol 55(183–265):320

Pellett S, Du ZW, Pier CL et al (2011) Sensitive and quantitative detection of botulinum neurotoxin in neurons derived from mouse embryonic stem cells. Biochem Biophys Res Commun 404:388–392

Pellett S, Tepp WH, Clancy CM et al (2007) A neuronal cell-based botulinum neurotoxin assay for highly sensitive and specific detection of neutralizing serum antibodies. FEBS Lett 581:4803–4808

Pellett S, Tepp WH, Toth SI et al (2010) Comparison of the primary rat spinal cord cell (RSC) assay and the mouse bioassay for botulinum neurotoxin type A potency determination. J Pharmacol Toxicol Methods 61:304–310

Peng L, Berntsson RP, Tepp WH et al. (2012) Botulinum neurotoxin D-C uses synaptotagmin I/II as receptors and human synaptotagmin II is not an effective receptor for type B, D-C, and G toxins. J Cell Sci. March 2012 epub ahead of print

Peng L, Tepp WH, Johnson EA et al (2011) Botulinum neurotoxin D uses synaptic vesicle protein SV2 and gangliosides as receptors. PLoS Pathog 7:e1002008

Pickett A, Perrow K (2011) Towards new uses of botulinum toxin as a novel therapeutic tool. Toxins (Basel) 3:63–81

Pier CL, Chen C, Tepp WH et al (2011) Botulinum neurotoxin subtype A2 enters neuronal cells faster than subtype A1. FEBS Lett 585:199–206

Purkiss JR, Friis LM, Doward S et al (2001) Clostridium botulinum neurotoxins act with a wide range of potencies on SH-SY5Y human neuroblastoma cells. Neurotoxicology 22:447–453

Raphael BH, Choudoir MJ, Luquez C et al (2010) Sequence diversity of genes encoding botulinum neurotoxin type F. Appl Environ Microbiol 76:4805–4812

Rasetti-Escargueil C, Jones RG, Liu Y, Sesardic D (2009) Measurement of botulinum types A, B and E neurotoxicity using the phrenic nerve-hemidiaphragm: improved precision with in-bred mice. Toxicon 53:503–511

Rasetti-Escargueil C, Machado CB, Preneta-Blanc R et al (2011a) Enhanced sensitivity to botulinum type A neurotoxin of human neuroblastoma SH-SY5Y cells after differentiation into mature neuronal cells. The Botulinum J 2:30–48

Rasetti-Escargueil C, Liu Y, Rigsby P et al (2011b) Phrenic nerve-hemidiaphragm as a highly sensitive replacement assay for determination of functional botulinum toxin antibodies. Toxicon 57:1008–1016

Ravni A, Bourgault S, Lebon A et al (2006) The neurotrophic effects of PACAP in PC12 cells: control by multiple transduction pathways. J Neurochem 98:321–329

Ray P (1993) Botulinum toxin A inhibits acetylcholine release from cultured neurons in vitro. In Vitro Cell Dev Biol Anim 29A:456–460

Rummel A (2012) Double receptor anchorage of botulinum neurotoxins accounts for their exquisite neurospecificity. doi:10.1007/978-3-642-33570-9_4

Rummel A, Hafner K, Mahrhold S et al (2009) Botulinum neurotoxins C, E and F bind gangliosides via a conserved binding site prior to stimulation-dependent uptake with botulinum neurotoxin F utilising the three isoforms of SV2 as second receptor. J Neurochem 110:1942–1954

Rummel A, Karnath T, Henke T et al (2004) Synaptotagmins I and II act as nerve cell receptors for botulinum neurotoxin G. J Biol Chem 279:30865–30870

Rummel A, Mahrhold S, Bigalke H et al (2011) Exchange of the H (CC) domain mediating double receptor recognition improves the pharmacodynamic properties of botulinum neurotoxin. FEBS J 278:4506–4515

Saadi RA, He K, Hartnett KA et al (2012) SNARE-dependent upregulation of potassium chloride co-transporter 2 activity after metabotropic zinc receptor activation in rat cortical neurons in vitro. Neuroscience 210:38–46

Schantz EJ, Kautter DA (1978) Standardized assay for Clostridium botulinum toxins. J Assoc Official Anal Chem 61:96–99

Schantz EJ, Johnson EA (1992) Properties and use of botulinum toxin and other microbial neurotoxins in medicine. Microbiol Rev 56:80–99

Schiavo G, Benfenati F, Poulain B et al (1992) Tetanus and botulinum-B neurotoxins block neurotransmitter release by proteolytic cleavage of synaptobrevin. Nature 359:832–835
Schiavo G, Malizio C, Trimble WS et al (1994) Botulinum G neurotoxin cleaves VAMP/synaptobrevin at a single Ala–Ala peptide bond. J Biol Chem 269:20213–20216
Schiavo G, Matteoli M, Montecucco C (2000) Neurotoxins affecting neuroexocytosis. Physiol Rev 80:717–766
Schiavo G, Shone CC, Bennett MK et al (1995) Botulinum neurotoxin type C cleaves a single Lys-Ala bond within the carboxyl-terminal region of syntaxins. J Biol Chem 270:10566–10570
Schiavo G, Shone CC, Rossetto O et al (1993) Botulinum neurotoxin serotype F is a zinc endopeptidase specific for VAMP/synaptobrevin. J Biol Chem 268:11516–11519
Sesardic D, Jones RG, Leung T et al (2004) Detection of antibodies against botulinum toxins. Mov Disord 19(Suppl 8):S85–S91
Sesardic D, Das RG (2007) Alternatives to the LD50 assay for botulinum toxin potency testing: strategies and progress towards refinement, reduction and replacement. Special Issue 14: 581–585
Sesardic D, Leung T, Gaines Das R (2003) Role for standards in assays of botulinum toxins: international collaborative study of three preparations of botulinum type A toxin. Biologicals 31:265–276
Sheridan RE, Smith TJ, Adler M (2005) Primary cell culture for evaluation of botulinum neurotoxin antagonists. Toxicon 45:377–382
Shone CC, Melling J (1992) Inhibition of calcium-dependent release of noradrenaline from PC12 cells by botulinum type-A neurotoxin. Long-term effects of the neurotoxin on intact cells. Eur J Biochem 207:1009–1016
Smith TJ, Lou J, Geren IN et al (2005) Sequence variation within botulinum neurotoxin serotypes impacts antibody binding and neutralization. Infect Immun 73:5450–5457
Solomon HM, Lilly T (2001) Clostridium botulinum. In: BAM bacteriological analytical manual U.S. Food and Drug Administration, Chapter 17, 8th edn. Revision A
Stahl AM, Ruthel G, Torres-Melendez E et al (2007) Primary cultures of embryonic chicken neurons for sensitive cell-based assay of botulinum neurotoxin: implications for therapeutic discovery. J Biomol Screen 12:370–377
Strotmeier J, Willjes G, Binz T et al (2012) Human synaptotagmin-II is not a high affinity receptor for botulinum neurotoxin B and G: increased therapeutic dosage and immunogenicity. FEBS Lett 586:310–313
Swaminathan S (2011) Molecular structures and functional relationships in clostridial neurotoxins. FEBS J 278:4467–4485
Takahashi K, Tanabe K, Ohnuki M et al (2007) Induction of pluripotent stem cells from adult human fibroblasts by defined factors. Cell 131:861–872
Tegenge MA, Bohnel H, Gessler F et al (2012) Neurotransmitter vesicle release from human model neurons (NT2) is sensitive to botulinum toxin A. Cell Mol Neurobiol 32:1021–1029
Truong DD, Stenner A, Reichel G (2009) Current clinical applications of botulinum toxin. Curr Pharm Des 15:3671–3680
Tsukamoto K, Arimitsu H, Ochi S et al. (2012) P19 embryonal carcinoma cells exhibit high sensitivity to botulinum type C and D/C mosaic neurotoxins. Microbiol Immunol July 2012 epub ahead of print
Uemura M, Refaat MM, Shinoyama M et al (2010) Matrigel supports survival and neuronal differentiation of grafted embryonic stem cell-derived neural precursor cells. J Neurosci Res 88:542–551
Ullrich B, Li C, Zhang JZ, McMahon H et al (1994) Functional properties of multiple synaptotagmins in brain. Neuron 13:1281–1291
Verderio C, Grumelli C, Raiteri L et al (2007) Traffic of botulinum toxins A and E in excitatory and inhibitory neurons. Traffic 8:142–153
Welch MJ, Purkiss JR, Foster KA (2000) Sensitivity of embryonic rat dorsal root ganglia neurons to Clostridium botulinum neurotoxins. Toxicon 38:245–258

Whitemarsh RC, Strathman MJ, Chase LG et al (2012) Novel application of human neurons derived from induced pluripotent stem cells for highly sensitive botulinum neurotoxin detection. Toxicol Sci 126:426–435

Wichterle H, Peljto M (2008) Differentiation of mouse embryonic stem cells to spinal motor neurons. Curr Protoc Stem Cell Biol Chapter 1: Unit 1H.1.1–1H.1.9

Wilder-Kofie TD, Luquez C, Adler M et al (2011) An alternative in vivo method to refine the mouse bioassay for botulinum toxin detection. Comp Med 61:235–242

Williamson LC, Halpern JL, Montecucco C et al (1996) Clostridial neurotoxins and substrate proteolysis in intact neurons: botulinum neurotoxin C acts on synaptosomal-associated protein of 25 kDa. J Biol Chem 271:7694–7699

Xie C, Markesbery WR, Lovell MA (2000) Survival of hippocampal and cortical neurons in a mixture of MEM+ and B27-supplemented neurobasal medium. Free Radic Biol Med 28: 665–672

Yaguchi T, Nishizaki T (2010) Extracellular high K+ stimulates vesicular glutamate release from astrocytes by activating voltage-dependent calcium channels. J Cell Physiol 225:512–518

Yamasaki S, Baumeister A, Binz T et al (1994) Cleavage of members of the synaptobrevin/ VAMP family by types D and F botulinal neurotoxins and tetanus toxin. J Biol Chem 269:12764–12772

Yowler BC, Kensinger RD, Schengrund CL (2002) Botulinum neurotoxin A activity is dependent upon the presence of specific gangliosides in neuroblastoma cells expressing synaptotagmin I. J Biol Chem 277:32815–32819

Yu J, Vodyanik MA, Smuga-Otto K et al (2007) Induced pluripotent stem cell lines derived from human somatic cells. Science 318:1917–1920

Zhang SC (2006) Neural subtype specification from embryonic stem cells. Brain Pathol 16: 132–142

Zhu H, Wang J, Jacky BPS et al (2010) Cells useful for immuno-based botulinum toxin serotype A activity assays. US-Patent 12/722801 (Allergan Inc) CA, USA

Transforming the Domain Structure of Botulinum Neurotoxins into Novel Therapeutics

John Chaddock

Abstract Botulinum neurotoxins are comprised of multiple identifiable protein domains. Recent advances in understanding the relationships between domain structure and neurotoxin function have provided a number of opportunities to engineer innovative therapeutic proteins that utilise the neurotoxins and neurotoxin domains. For example, recent insights into the properties of the catalytic, translocation and binding domains open up opportunities to develop botulinum neurotoxins with enhanced properties of selectivity, potency and duration of action. In parallel, the broad scope for utilisation of the individual domains is becoming clearer as significant advancements are made to exploit the unique biology of the catalytic and translocation domains. These opportunities and the status of their development will be reviewed in this chapter.

Keywords Targeted secretion inhibitors · Protein engineering · Retargeting · Pain · Acromegaly

Abbreviations

BoNT	Botulinum neurotoxin
HC	Heavy chain of BoNT
H_C	C-terminal domain of the Heavy chain of BoNT
H_{CC}	C-terminal sub-domain of the H_C
H_N	N-terminal domain of the heavy chain of BoNT
LC	Light chain domain of BoNT
LH_N	Fragment of BoNT comprising the LC and H_N domains
LH_N/C_1	Fragment of BoNT comprising the LC and H_N domains of BoNT/C_1
SNAP-25	Synaptosomal-associated protein of 25 kDa

J. Chaddock (✉)
Syntaxin Ltd., Units 4-10, The Quadrant, Abingdon, Oxon OX14 3YS, UK
e-mail: John.chaddock@syntaxin.com

A. Rummel and T. Binz (eds.), *Botulinum Neurotoxins*, Current Topics in Microbiology and Immunology 364, DOI: 10.1007/978-3-642-33570-9_13,

SNARE	Soluble N-ethylmaleimide–sensitive factor attachment protein receptor
SV2	Synaptic vesicle glycoprotein 2
TeNT	Tetanus neurotoxin
TSI	Targeted secretion inhibitors
VAMP	Vesicle associated membrane protein

Contents

1 Introduction 288
2 Understanding the Domain Structure of Botulinum Neurotoxins Leads to New Opportunities for Development of Therapeutics 289
3 Botulinum Neurotoxin: A Platform for Protein Engineering 290
3.1 Attractive Properties of the H_C Domain 291
3.2 Attractive Properties of the LC Domain 292
4 The Potential of Neurotoxins Created from Mixed Serotype Botulinum Neurotoxin Domains 293
5 Improvements to Botulinum Neurotoxin that Could have Therapeutic Potential 294
6 Developing the Binding Domain as a Delivery Vehicle 295
7 Development of Therapeutics Utilising the BoNT Endopeptidase Domain as a Warhead 297
8 Conclusions 303
References 304

1 Introduction

Botulinum neurotoxin (BoNT), the most potent toxin known to man, is also widely used in the clinical setting. Indeed, the clinical use of BoNT products has increased the market size three-fold since the start of the century (to over $1.6 billion in 2010). This growth has been achieved with preparations of material that, though defined sufficiently robustly to meet Regulatory requirements, are nevertheless a complex mixture of active BoNT and associated proteins. These preparations are derived from the native clostridial bacterial host and, as such, are representative products of the natural protein evolution and selection. What excites many researchers in the BoNT field is the realisation that the domain structure of BoNTs provides an opportunity to harness specific functional properties of the toxins for alternative applications. The purpose of this chapter is to consider the properties of BoNT domains and to summarise the current status and the huge potential for the development of a new range of biotherapeutics based on this novel protein class.

This chapter will (1) briefly review the scientific advancements that have been achieved in understanding the relationship between BoNT structure and biological function; (2) acknowledge the broad repertoire of proteins within the seven serotypes (BoNT/A-G) and multiple subtypes of BoNTs; (3) consider the opportunities to modify the amino acid building blocks of each domain that comprises the neurotoxin, and the impact this can have and (4) review the progress made towards expanding the medical applications of the toxin through protein engineering.

Chapters elsewhere in this book will provide detail on concepts that this chapter will necessarily cover only in brief. Therefore, where it is helpful to do so, there are references to other sections throughout the text.

2 Understanding the Domain Structure of Botulinum Neurotoxins Leads to New Opportunities for Development of Therapeutics

In 1897, when Professor Emile-Pierre-Marie van Ermengem identified an anaerobic bacterium that he termed *Bacillus botulinus* (subsequently classified as *Clostridium botulinum*) that released a potent toxin that caused the symptoms associated with food-borne botulism, the structure of such a material was unknown. In fact, it took a further 50 years of experimentation, to 1946, before pure crystalline botulinum toxin of serotype A (BoNT/A) was obtained. Over the subsequent 20 years, the level of understanding of the multi-protein complex nature of BoNT increased, but it was not until the publication of the first X-ray structure of BoNT/A (Lacy et al. 1998) that the sophisticated level of understanding that we have today truly started to be achieved. Interestingly, the clarity achieved by understanding the structure of BoNT at an atomic level was achieved 20 years after the first clinical use of BoNT (Scott 1980).

Lacy's seminal work established the current view of BoNT structural organisation and provided a platform for further serotype and domain studies. Indeed, subsequent research has extensively studied the primary, secondary, tertiary and quaternary structure of the BoNT family of proteins. These investigations have facilitated an in-depth awareness of the molecular architecture of this exciting class of proteins and have enabled correlation of structural differences that lead to distinguishing functions. With respect to our understanding of BoNT molecular architecture, the gross 3-dimensional arrangements of BoNT/A and BoNT/B were determined in 1998 (Lacy et al. 1998) and 2000 (Swaminathan and Eswaramoorthy 2000), respectively and were demonstrated to be essentially similar; a three-domain protein in which the domains are arranged in a linear fashion with the translocation domain in the middle flanked by the binding and catalytic domains. While noting the overall similarity in tertiary structure, it should, however, be remembered that BoNT/A and BoNT/B are comprised of receptor-binding domains that have dissimilar binding targets, and light chain (LC) domains that cleave different soluble N-ethylamide-sensitive factor attachment protein receptor (SNARE) substrates. With these data available, it would have been reasonable to predict that the overall molecular architecture of BoNTs would be similar and, at the domain level, this has proved to be the case. However, in 2008 the tertiary structure of BoNT/E was determined (Kumaran et al. 2008) and, though the individual domains showed a high degree of structural similarity with the A and B cases, there is a major organisational difference between the holotoxin domains. Whereas the three domains are arranged in a linear organisation in A and B, in serotype E the catalytic and binding domains are arranged on the same side of the translocation domain and consequently share interactions that would be absent in A or B. The modified organisation of BoNT/E has been proposed by Kumaran et al. to correlate with the observed more rapid onset of action of BoNT/E; BoNT/E being in a 'translocation ready' conformation. These observations remind the researcher that nature has developed a number of approaches to creating multimeric proteins and all predictions are there to be challenged!

Discussed in detail in (Hill and smith 2012), it is now well known that there are multiple clusters of protein within a particular serotype that can be identified as particular subtypes. When amino acids sequences for the various subtypes within serotype families are assessed, for example, as illustrated in the work of Hill et al. (2007), it is noted that the amino acid similarity between subtypes is generally greater than 65 %. The fact that there are definable subtype families within a particular serotype, most of which have not yet been studied in any form of secondary or tertiary structural detail, acknowledges that the field has some way to go to have a complete picture of BoNT protein structure–function.

Structural data from crystallographic studies are now available for (1) the individual catalytic domains (LC) of serotypes A, B, C, D, E, F and G; (2) the binding domain (H_C) of A, B, C, D, F and G; (3) the di-chain LH_N species comprising the LC and the H_N domain of serotypes A and B and; (4) the holotoxins BoNT/A, BoNT/B and BoNT/E. It should be appreciated that, even with this level of advancement, the current level of understanding is not yet fully comprehensive. For example, the structural understanding of the holotoxins is far from complete with respect to BoNT subtypes and there is much yet to be understood about the H_N translocation domain. Nevertheless, understanding the molecular architecture of the individual domains and their inter-relationships within the macromolecule has provided opportunities for a number of paths for application of protein engineering strategies. Opportunities to utilise the individual domain components of BoNT (the LC, the H_N, the H_C domains or subcomponents/combinations of each) are discussed below.

3 Botulinum Neurotoxin: A Platform for Protein Engineering

As described in the section above, what is now known is that the scope of naturally occurring BoNTs is vastly greater than just seven variants. Within the seven major serotypes of toxin are represented currently 32 variants, called subtypes (Hill and smith 2012). Thus, this pool of evolved toxins represents a rich library of proteins that have known and potentially different properties and characteristics.

Using the BoNT/A serotype family to illustrate the point, it has been recently demonstrated that measurable functional differences exist between the subtypes.

For example, it has been reported that the A2 subtype of BoNT/A has faster onset than the current clinical BoNT/A products based on A1 (Pier et al. 2011). The mechanism of this faster onset is thought to lie in an increased efficiency of cell entry that is independent of ganglioside binding. Also, the A3 and A4 subtypes of BoNT/A have been observed to possess different catalytic properties to the A1 and A2 subtypes, with the A3 subtype exhibiting reduced efficacy for cleavage of SNAP-25 but good efficacy for a short peptide artificial substrate (SNAPtide) and the A4 subtype exhibiting significantly poorer cleavage for native SNAP-25 or artificial substrates (Henkel et al. 2009). Please see (Binz 2012) of this book for further discussion. Further study of subtype biology, in particular focusing on the LC catalytic properties, the H_C domain membrane binding and interaction properties and the pH dependency of the H_N domains, will greatly enhance the possibilities for creation of innovative protein domains utilising the broad library that nature has evolved to either directly, or through the construction of hybrids, or through engineering.

Two of the most meaningful structural features for protein engineering are therefore the H_C-mediated binding event and the LC-mediated substrate cleavage event. As part of setting the scene for a later discussion of the therapeutic opportunities afforded by these two domain-based activities, the following section summarises the attractive properties of each domain in turn.

3.1 Attractive Properties of the H_C Domain

The H_C domain has evolved to selectively bind to motorneurons, although not all BoNTs bind to the target neuron using precisely the same mechanism. Summarising a significant body of research, BoNT/B and G use gangliosides and the intraluminal domain of synaptotagmin as dual receptors, whilst BoNT/A, D, E use gangliosides and the intraluminal domain of synaptic vesicle glycoprotein 2 (SV2). Please also refer (Rummel A 2012) for further details. A protein binding domain has, however, not yet been identified in BoNT/C suggesting the BoNT/C uses ganglioside only, or an as yet unidentified proteinaceous component (Strotmeier et al. 2011). Recently, it has become evident that very careful understanding of the side-chain contacts between the heavy chain and the respective motorneuron surface entity is necessary. For example, Strotmeier et al. (2012) and Peng et al. (2012) have noted the significant differences between the affinity of BoNT/B and synaptotagmin II expressed in rodents and man; specifically, the affinity of BoNT/B for human synaptotagmin II is lower than in rodents due to the presence of a unique amino acid mutation. By binding to intraluminal domains of synaptic vesicle proteins, the H_C domain therefore locates the SNARE-cleaving LC on the outside surface of its target cell, utilising cellular features that maximise the uptake into intracellular compartments that will aid transport of the LC ‘warhead’ into the cytosol.

Exploring the domain structures in more detail, Rummel et al. (2004) identified a ganglioside binding cavity within the C-terminal sub-domain of the H_C domain (H_{CC}) of BoNT/A and/B defined by the conserved motif H…SXWY…G. Modification of residues within this site modified both the ganglioside binding affinity and toxicity of the neurotoxin. By switching the H_{CC} domain as well as H_C-fragment of BoNT/B into BoNT/A, Rummel et al. (2011) were able to enhance the potency of BoNT/A fourfold. Interestingly, according to EP1786832, specific modifications to the ganglioside binding site can both reduce binding and toxicity, or enhance it by up to threefold relative to wild-type toxin. This provides a rational basis for engineering mutated BoNT with modified, particularly enhanced, potency as improved clinical products.

In summary, therefore, the H_C effectively hijacks existing cellular processes to facilitate uptake into a sensitive cell; maximising the opportunities for the toxin to target its cellular substrate. As such, the H_C is a good candidate to supply a binding capability to neuronally focused therapeutics cargos. Additionally, the multiple binding entity nature of the H_C domain allows for modification and engineering additional binding features into the existing protein; thereby enabling design of novel targeted approaches. The potential to use the H_C domain in the design of new therapeutics is described in a later section.

3.2 Attractive Properties of the LC Domain

The second major opportunity for protein engineering is the LC domain. As described elsewhere, the LC has evolved to be a highly specific metalloendoprotease that cleaves one or more members of the SNARE protein family. The breadth of LC cleavage activities encompasses three different types of SNARE protein but is not fully comprehensive across all SNAREs. Of the approximately 40 SNAREs known to be expressed in the human genome, only a fraction are known to be substrates for BoNT LCs. No doubt this is partly a result of the complex mechanism of SNARE substrate binding which requires catalytic site alignment of the substrate and interaction between the substrate and the LC at sites distant to the catalytic site (Binz 2012). Though effective at providing the specificity exhibited by LCs, this complex requirement for substrate binding means that the LC is a complex protein engineering platform. Nevertheless, increased structural knowledge has led to a much greater understanding of the specific interactions necessary for efficient substrate cleavage and has afforded the opportunity for LC modification. Such knowledge has also led to greater appreciation of the relative promiscuity of some of the serotypes, for example BoNT/C, which has less stringent requirements for substrate binding (Jin et al. 2007).

A good example of what can be achieved by protein engineering approaches is the progress made to overcome one of the major limitations to using clostridial endopeptidases to cleave SNARE proteins and inhibit vesicle trafficking in non-neuronal cells. It is generally understood that BoNTs have evolved to target neuronal sub-populations and have therefore developed a suite of appropriate cellular targeting approaches. Similarly, the LCs have developed to effect cleavage of neuronal SNAREs. Taking the example of the BoNT-sensitive SNAP-25, SNAP-25 is restricted in its expression to neurons and the ubiquitously expressed homologue, SNAP-23, is not, in man, a substrate for any of the known BoNT serotypes. This means that BoNT/A, /C and /E, which are all SNAP-25 cleaving serotypes, are ineffective at inhibiting secretion and vesicular trafficking in non-neuronal cells. Understanding the structure-function relationship of BoNTs and their substrates has recently led to advances that have taken the first steps to overcome such limitations, with the reported mutation of a BoNT/E LC, so it possesses the capability to cleave human SNAP-23 (Chen and Barbieri 2009). Understanding the precise intra- and inter-molecular interactions necessary for cleavage of SNAP-25 by LC/E was critical to the development of mutated LC/E that exhibited the correct side chain architecture to achieve the goal of SNAP-23 cleavage. Chen and Barbieri were able to demonstrate inhibition of IL8 and mucin release from an appropriately stimulated HeLa cell. The potential to deliver a mutated LC of serotype E and impact SNAP-23 mediated vesicle trafficking is therefore a real possibility and broadens the spectrum of utility of the BoNTs, particularly when harnessed in combination with the botulinum fragment retargeting technologies described later in this chapter.

4 The Potential of Neurotoxins Created from Mixed Serotype Botulinum Neurotoxin Domains

In some instances, BoNTs isolated from natural isolates can exhibit LC and HC domains that bear significant sequence similarity to domains of dissimilar serotypes. For example, *C. botulinum* D strain OFD05 comprises the sequence of a LC and H_N domain of serotype D genetically fused to the H_C of serotype C. In a further example, as reported by Arndt and colleagues in 2006 (Arndt et al. 2006), it would appear that the BoNT/A2 subtype comprises domains that are similar to the LC of BoNT/A1 and the HC of BoNT/A3. Such natural chimera provides a further layer of variation onto the simple subtype variants that were discussed earlier, and provide scope for considering further expansion of properties. Given the primary and tertiary structural homology between the serotypes, the observations of natural chimeric variants are perhaps not unexpected. From an engineering perspective, however, it provides some level of validation that combining the domains of heterologous serotypes could be achievable and may lead to novel proteins with unique properties. In a limited number of reported cases, this has indeed been achieved. For example, Wang et al. (2008) have described the creation of chimera of BoNT/A and /E in which the H_C domain of one serotype was expressed recombinantly fused to the LC and H_N (LH_N) of the other serotype. When assessed for key functional characteristics, it was noted that the translocation and neuronal specificity properties of the hybrid were clearly differentiated: translocation reflected the behaviour of the parent LH_N domains, whilst neuronal specificity was influenced by the identity of the H_C. Such a gross domain swapping approach is clearly able to harness the inherent biological properties of the domains to create novel entities with unique properties, and is the most straightforward approach to BoNT modification. Furthermore, Wang et al. (2012) have extended this hybrid approach to replace the 50 kDa H_C binding domain of one serotype with the equivalent from another. For example, in the situation where the H_C/A is replaced with the H_C/B, the resultant hybrid induced a more prolonged neuromuscular paralysis in mice than the parent BoNT/A. The authors propose that the prolonged neuromuscular paralysis was a direct result of enhanced uptake of the LC/A via the more abundant acceptor for BoNT/B. In the complementary switch, where the H_C/A is inserted into the BoNT/B structure, the resultant molecule retained neuronal SNARE cleavage ability but also enabled cleavage of VAMP within fibroblast-like synoviocytes that lack SNAP-25 and would ordinarily be resistant to the SNARE-cleavage ability of BoNT/A. Rummel and colleagues have provided additional evidence for the domain swap engineering approach and, in an elegant series of studies, demonstrated improvements to potency that arise from incorporation of the H_{CC} or H_C domain of one serotype into another. For example, the potency of a BoNT/A-based full-length hybrid containing the H_{CC} B domain (termed AAAB; letters represent the serotype origin of the four domains) was quadrupled as compared with wild-type BoNT/A. Analogously, exchange of the H_C fragment (AABB) yielded a neurotoxin with fourfold higher potency (Rummel et al. 2011).

Such approaches to engineering of the native BoNT structure through domain switching have been further described by Dolly as the basis for the construction of 'BoTIMs' (full-length BoNTs incorporating catalytic-inactive LC/A) (Dolly et al. 2011). By recombinantly combining BoTIMs incorporating LC/E domains, a hybrid protein was constructed that utilised components within the LC/A element to extend the intracellular persistence of the LC/E, and therefore the duration of action of LC/E-induced SNAP-25 cleavage (Wang et al. 2008). Dolly proposed that the LC/E-induced cleavage of SNAP-25 would be advantageous for specific conditions, for example in the treatment of various pain states, including chronic pain.

5 Improvements to Botulinum Neurotoxin that Could have Therapeutic Potential

As described, an understanding of the relationship between BoNT structure and function provides opportunities for rational design of improvements into the molecule. Such improvements could be conceived to include modified catalytic activity, binding activity or properties that have a therapeutic impact, e.g., immunogenicity or to modify the longevity of the catalytic domain to impact on duration of action. The opportunities afforded by site-specific modification to engineer alternative cleavage properties into the LC have already been discussed. Turning the attention to the C-terminal end of the holotoxin, the increased understanding of the mode of binding provides a route to the creation of novel BoNTs with potentially improved properties. Recently, Rummel et al. (2011) applied the knowledge of relative binding affinities of the BoNT H_C domain in the design of hybrid H_C of greater potency than the parent. For example, by incorporating the H_{CC} domain of BoNT/B into the BoNT/A backbone, the greater binding affinity of the B sub-domain leads to a quadrupling of the whole molecule potency. Focusing very specifically on the H_{CC} binding domain, it has also been possible to create site-specific mutants that are therefore minimally modified, but which impart improved binding properties into the new protein. Such domain swapping and mutagenesis approaches therefore have the potential to create novel BoNTs with higher potency which could lead to lower doses, and consequently reduced off target effects and increased tolerance to the proteins. Alternatively, increased binding affinity could lead to the opportunity to use increased doses and potentially extend the duration of action of the clinical benefit, assuming a highly selective binding activity that negated the impact on non-target cells. As proposed earlier, further investigation into the properties of the BoNT sub-types may well provide additional opportunities in this area.

The discussion above, referencing the impact of modifications to the H_{CC} domain, illustrates one of the fundamental goals of clinically focused toxin research: how to improve the efficacy of the product, minimise off target effects and thereby maximise the therapeutic window. With the toxins exhibiting such a potent activity, maximising the efficacy of the material while minimising the potentially severe side effects is clearly an important strategy. Elucidating key structural elements of the toxin that impact on function has the potential to further expand the therapeutic opportunties of BoNTs. One such example is provided by the progress made to understand the impact of di-leucine motifs (Wang et al. 2011) and ubiquitination signals (Tsai et al. 2010) on the duration of action of the toxin. Indeed, the biological advantages of the hybrid molecules created by Wang et al. are in a major part proposed to be due to the transfer of such motifs within the various BoNT domains. The possibility therefore exists to create bespoke proteins that utilise domains of BoNTs that have been manipulated at the gene level to include domain or site-specific mutations that afford the novel proteins with new biological properties.

6 Developing the Binding Domain as a Delivery Vehicle

BoNTs represent a unique opportunity to develop for delivery of therapeutic molecules, particularly biological molecules, into nerve cells via their H_C binding and the beneficial properties of the in-built membrane translocation function. For these reasons, full-length BoNTs lacking a functional endopeptidase have been proposed as preferred delivery tools (Wang et al. 2008; Bade et al. 2004). Such constructs retain the evolved connectivity and functionality of BoNTs whilst being non-toxic by virtue of the inactivated endopeptidase domain.

Considering what applications there may be for using BoNT heavy chains, or fragments thereof, enzyme replacement studies such as to treat lysosomal storage diseases and treatment of oxidative injury have been considered. The first report of the cytoplasmic delivery of a cargo protein using BoNT HC was by Weller et al. (1991) through a study in which tetanus toxin (TeNT) LC was delivered into phrenic nerve by disulphide attachment to the HC of BoNT/A. This demonstrated that the BoNT holotoxin was capable of translocating another protein, albeit in this case a highly related protein.

Taking the concept one stage further into full recombinant expression, Bade et al. (2004) demonstrated the ability of BoNT/D to deliver a range of cargo proteins to neurons and achieve enzymatic activity in the neuronal cytosol. These studies also demonstrated that unfolding of the cargo protein was necessary for translocation into the nerve cell, and cargo proteins that were insufficiently flexible in their conformation were not well transported.

Further examples of delivery have been reported. As biochemical and structural studies have shown that TeNT has a similar modular arrangement of functional domains as BoNT, Pickett and Perrow (2011) notes that advances in TeNT engineering may also be applied to BoNT. For example, the binding domains of both BoNT and TeNT have been utilised to deliver DNA to target cells and enhance targeting of other transfection methods. The TeNT H_C fragment has been conjugated to polylysine, which has a high capacity to bind DNA, allowing transfection of a range of neuronal cell lines (Box et al. 2003), albeit using the diphtheria toxin translocation domain to effect intracellular membrane transfer. Recent studies with recombinantly produced BoNT domains show that proteins can be assembled by non-chemical linking, using tagging with helical motifs from the family of SNARE proteins. Such a strategy may potentially be exploited to use the BoNT-binding domain to deliver future therapeutics or other cargo into neurons (Darios et al. 2010). Indeed, Ferrari and colleagues have recently reported in vivo studies that explored the functionality of a BoNT (termed 'BiTox') that had been reconstituted using the SNARE protein based 'protein stapling' technology (Ferrari et al. 2011). In another approach to coupling a cargo to the toxin, Drachman et al. (2010) notes the early stage research into using streptavidin associated BoNT/H_C to deliver biotinylated viral vectors into motor nerve terminals. One important feature of BoNTs with respect to their application as delivery vehicles is the advantages provided by the translocation domain. In the examples reported above, the presence of the translocation domain is an essential feature. Indeed, in the work of Box et al., where the H_c domain of tetanus neurotoxin is used for the binding function, the translocation domain of diphtheria toxin was included to facilitate crossing of the endosomal vesicle membrane.

In addition to utilising clostridial neurotoxin binding domains to deliver cargos that will affect a particular cellular property that is unrelated to the toxin itself, there is also the hypothetical opportunity to affect toxin-related biology.

For example, Kuo et al. (2011) have postulated the use of BoNT or BoNT heavy chain fragments to deliver 'targeted F box' agents to BoNT intoxicated neurons as a potential therapeutic for BoNT poisoning. The F-box strategy is hypothesised to be successful, because the targeted F box agents cause increased ubiquitination and accelerate turnover of the targeted BoNT/A protease within neurons. Indeed, Kuo reported a camelid heavy chain-only V(H) (VHH) domain specific for a BoNT protease fused to an F-box domain recognised by an intraneuronal E3-ligase is effective at accelerating turnover of the targeted BoNT/A in neurons (Kuo et al. 2011). Although elegant in design, the strategy critically requires targeted delivery to the correct neuronal population and BoNT (or fragments of) would seem to continue to be the best candidate to achieve this. However, it should be acknowledged that a BoNT/A intoxication is related to the state of the neuron; in an highly active neuron, BoNT/A uptake is potentiated. The opposite is therefore true: an intoxicated neuron has significantly reduced neurotransmitter release activity, and as a consequence also has reduced SV2 presentation. With a mechanism of action linked to binding and internalisation, it is important to consider that BoNT-intoxication results in self-limited uptake of similarly targeted species, such that any targeted therapeutics for BoNT poisoning will have to overcome such a naturally restricted process.

In summary, various cargos have been proposed for delivery by suitable non-toxic BoNT-based vehicles for treatment of diseased neuromuscular junctions or to enhance motor neuron function. These include anti-neurotoxin therapies, anti-neurotropic viral treatments, neuronal enzyme replacement, ion channel modulators, neurotrophic factor receptor modulators and protein replacement for hereditary or autoimmune presynaptic disorders. Whilst an attractive opportunity supported by model system studies, there are not yet any clinical applications of drug delivery using a BoNT-based vehicle and more needs to be understood regarding uptake rates and in structural terms for the translocation domain, the kinetics of translocation and how this impacts on its limitations.

7 Development of Therapeutics Utilising the BoNT Endopeptidase Domain as a Warhead

Understanding the discrete modular organisation of the LC, H_N and H_C domains within BoNT has encouraged a number of engineering approaches that utilise the domains in novel molecule construction. A good example of this approach is provided in the design of new therapeutics based on the LH_N BoNT fragment comprising the LC and H_N domains. Referred to as targeted secretion inhibitors (TSI), this novel approach to the development of therapeutics harnesses the power of the endopeptidase domain to modify the intracellular processes of the target cells and inhibit secretion (Foster and Chaddock 2010). This approach therefore aims to extend the therapeutic application of BoNT-derived activities away from

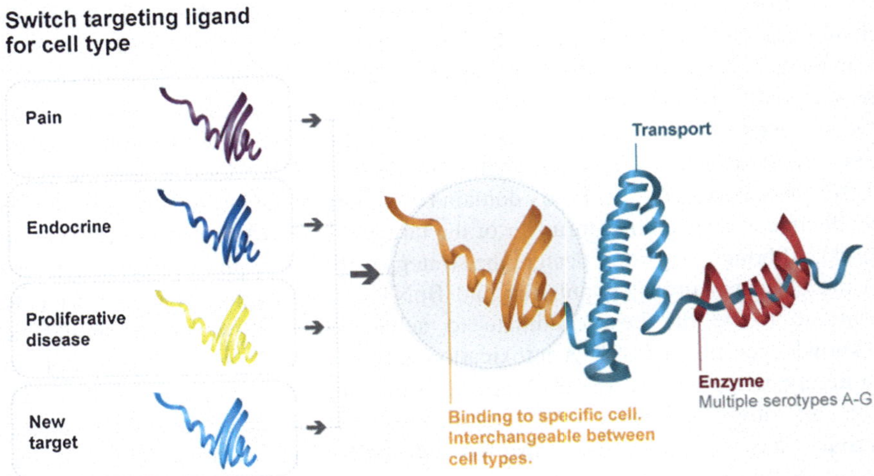

Fig. 1 Schematic representation of botulinum neurotoxin structure in relation to engineered targeted secretion inhibitor (*TSI*) proteins

the limitation of simply neuromuscular junction targeting to the wider field of specific cellular targeting.

In contrast to the cell-ablation type technologies, in which cytotoxic payloads are delivered to cells (often of a cancerous nature) by peptide or antibody targeting approaches, the TSI strategy utilises the BoNT LC mechanism of action to manipulate cell processes (inhibition of secretion and membrane receptor/channel presentation) rather than lead to cell death. By not incorporating the BoNT H_C domain into the TSI protein, but replacing the binding function with a chosen peptide or protein, TSI are able to target BoNT endopeptidase domains to non-neuronal cells not targeted naturally by BoNTs. A range of such novel recombinant proteins have been created and it has proved possible to deliver the LC into cell types not sensitive to native BoNTs, cleave the relevant SNARE protein and thereby inhibit secretion from a range of otherwise resistant cells. This has enhanced the utilisation of the BoNT domains far beyond that achievable by natural selection.

TSI are a new drug design platform of multidomain proteins with an overall architecture as illustrated in Fig. 1. In common with the architecture of BoNT, the first TSI component is the LC domain of one of the BoNT serotypes. This provides the TSI with a SNARE cleavage capability that is dependent on the LC chosen but is also open to modification to facilitate wider SNARE cleavage. The second component of the TSI is the BoNT H_N domain which provides the intracellular translocation ability for the LC. The third TSI component is a binding domain which could be derived from BoNT (was neuromuscular targeting to be required) but is more commonly specified by inclusion of a peptide or a protein that interacts with a receptor of choice on the target cell.

TSI were initially created by chemical conjugation techniques; in the first instance, a chemical conjugate of nerve growth factor (NGF) and the LH_N/A fragment of BoNT/A was demonstrated to result in cleavage of SNAP-25 and inhibition of noradrenaline release from PC12 cells (Chaddock et al. 2000a). Subsequently, a chemical conjugate of wheat–germ agglutinin (WGA) and the LH_N/A fragment was observed to deliver the endopeptidase into neuronal and non-neuronal cell types with a consequent cleavage of SNAP-25 and inhibition of secretion (Chaddock et al. 2000b). This latter example confirmed the scope of TSI beyond the neuron. Following establishment of proof of principle for the retargeting approach, a therapeutically relevant application was developed utilising a conjugate of *Erythrina cristagalli* lectin and LH_N/A (ECL-LH_N/A) to target nociceptive afferents with the intention of inhibiting the release of neurotransmitter from the nociceptors and consequently inhibition of pain. ECL-LH_N/A was demonstrated to inhibit both substance P and glutamate release from cultured embryonic dorsal root ganglion neurons. Furthermore, intrathecally administered ECL-LH_N/A significantly reduced the nociceptive inputs to convergent dorsal horn neurons by primary sensory afferents of the C-fibre and Aδ types, whereas there was little or no effect on sensory inputs from Aβ-fibres (Duggan et al. 2002). Intrathecal ECL-LH_N/A also resulted in prolonged withdrawal latency in a 'hot-plate' model of acute thermal pain (Chaddock et al. 2004). Retention of the duration of action of the parent BoNT LC is a key feature of TSI and, though it is possible to create modified LCs to have different durations of action, it is also extremely encouraging that the extended durations of action of BoNTs can be retained in a novel therapeutic construct. This means that recombinant proteins based upon this approach are particularly suitable for treating chronic diseases.

Although satisfactory for demonstration of principle, there are several drawbacks to using chemical conjugation to produce therapeutic proteins: of most concern being the inevitable heterogeneous mixture of species that is created using this process and the difficulties of controlling such a process within the manufacturing environment. To progress such a strategy of retargeted BoNT fragments, it is necessary to develop a recombinant platform that would underpin the TSI approach.

Recombinant expression of a catalytically active, stable LH_N fragment of BoNT/A was reported in 2002 (Chaddock et al. 2002), followed soon afterwards with successful preparation of LH_N/B and LH_N/C_1 (Sutton et al. 2005). In all cases, the recombinant LH_N fragments had dramatically reduced toxicity (by a factor of 10^6) because of the lack of the necessary H_C domain with which to bind to acceptors on the neuronal surface. The recombinant LH_N proteins were demonstrated to be stable, catalytically active and effective at intra-cellular cleavage of the target SNAREs. Recently, recombinant LH_N/A and LH_N/B have been reported to retain the structure of the equivalent domains in the intact BoNT protein (Masuyer et al. 2009, 2011a, 2011b). Moreover, the LH_N structure appears to be resilient to incorporation of additional peptides at various locations within the primary sequence, and therefore be a robust platform on which to build multidomain, multifunctional TSI. Figure 2 illustrates the significant structural similarity between LH_N/A and BoNT/A which is

Fig. 2 Tertiary structure of the LH_N fragment of BoNT/A (**a**) and BoNT/A holotoxin (**b**). LC is depicted in *red*, H_N domain in *green*, H_{CN} in *blue* and H_{CC} in *orange*

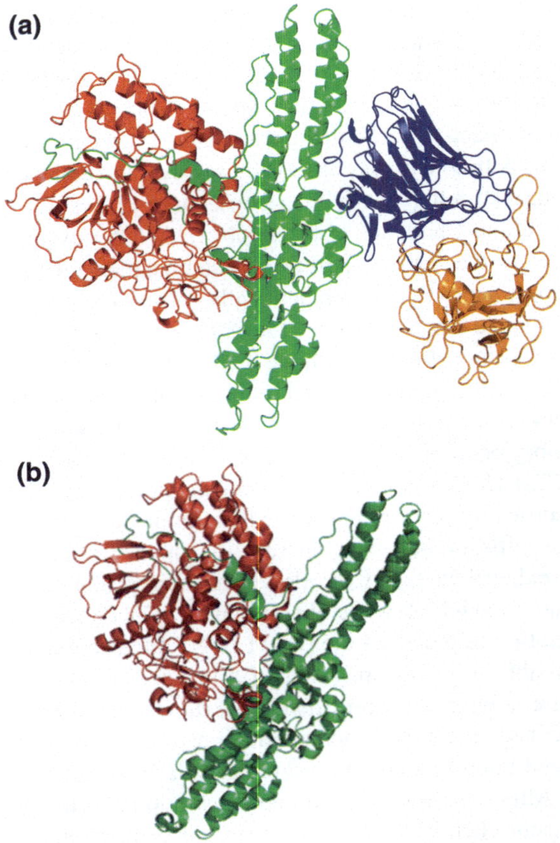

also illustrated in the observed root-mean-squared deviation (r.m.s.d) of 0.009 angstroms bond length and 1.3° bond angle.

Having demonstrated the robustness of the LH_N fragment to recombinant expression and to modification, the stage was set for the development of a range of modified LH_N incorporating new receptor targeting domains. It is now possible to envisage a toolbox of DNA fragments encoding protein domains that could be spliced together to form a series of bespoke recombinant proteins that utilise the power of the endopeptidase and translocation domains of BoNT coupled to the receptor targeting specificity of a peptide or protein. Such a fully recombinant TSI would have the capability to find a specific cell type of interest (through the engineered ligand-receptor binding activity), to facilitate entry of the LC into the cytosol of the cells, and then to inhibit secretion by a SNARE-cleavage-dependent mechanism. Hence TSI.

The first report (Foster et al. 2006) of a fully recombinant TSI fusion protein comprised the LH_N-fragment of BoNT/C_1 (LH_N/C_1) and epidermal growth factor (EGF); a protein of an overall mass of approximately 106 kDa. Such a molecule combined an ability to bind EGF-receptors with the syntaxin-cleavage activity of the BoNT/C LC; coupled with the function of intracellular membrane translocation afforded by the H_N/C. Such a novel construct therefore had the potential to inhibit secretory events from cells expressing EGF receptors. Creating fully recombinant proteins that selectively deliver LC into a specified cell represents a tremendous opportunity to develop a new range of therapeutic proteins that inhibit secretion from cells involved in a wide variety of diseases.

Syntaxin Ltd., have reported that the TSI platform can lead to a portfolio of engineered molecules that have the potential to lead to novel biopharmaceuticals for use in a wide range of diseases. Harnessing the LC endopeptidase and H_N translocation activities, a series of innovative proteins have been created that are reported to have potential in the treatment of pain, endocrine disease (acromegaly) and cancer. As reported at the 93rd meeting of The Endocrine Society (ENDO2011) recombinant TSI proteins comprising modified versions of the growth hormone receptor hormone (GHRH) peptide covalently fused to the N-terminus of the H_N domain and the LC of BoNT/D have been created that specifically target the somatotroph cells of the pituitary with a view to establish new opportunities for the treatment of conditions such as Acromegaly (Somm et al. 2011). The TSI strategy would be a novel approach to treatment of such conditions and molecules have been designed to inhibit the excessive release of growth hormone from pituitary adenomas. It is proposed that inhibition of growth hormone release would result in a decrease in the growth signals, thereby alleviating some of the symptoms of this debilitating condition. Studies presented at ENDO2011 report that engineered TSI decrease the circulating levels of growth hormone for many days when assessed in various pre-clinical studies. Interestingly, such effects are achieved after only a single bolus of test material (rather than continuous infusion) and the material was administered systemically (i.v.) rather than being limited to local administration at the site of therapeutic benefit. Such a novel administration route for a molecule derived from BoNT indicates the significant progress that has been made to (1) engineer out unwanted toxic effects that would be inherent in BoNT; (2) engineering in a high level of cell specificity whilst; (3) retaining the powerful biochemical mechanisms of substrate cleavage and membrane translocation.

Utilising a variation of the TSI strategy to explore the utility of the LC into additional therapeutic areas, Yeh et al. (2011) reported that antibody decorated LH_N/B can be taken up by human bone marrow derived macrophages to result in inhibition of TNF-α release and a decrease in macrophage recruitment in in vivo studies. These studies therefore further demonstrate the potential for inhibiting secretion from target cells and impacting on processes of medical value.

The development of the TSI platform from a concept to a pre-clinical reality is a significant step forward. Moreover, many of the opportunities afforded by the TSI approach have the potential to meet clinical needs in situations where BoNT would be clinically ineffective, either due to the significant off-target safety liability of the neurotoxin or due to the lack of sufficient efficacy in a cell (for example, due to the lack of a BoNT receptor). Though details of targeting domains are not always available, natural peptides and proteins are candidates for achieving effective receptor-ligand interactions and so ensuring that the TSI locates the target cell. It is possible that not all natural ligands are suitable for incorporation into a recombinant TSI protein; however, the growing field of protein scaffolds and recombinant antibody technologies are possible routes to expanding the breadth of targeting domains. Many of these scaffolds have pre-existing tertiary structural data available and their binding interfaces have been extensively studied. The opportunity therefore exists to combine the BoNT and scaffold protein structural information and facilitate optimal engineering of the products.

No discussion of the potential for engineering novel recombinant therapeutics based on bacterial protein backbones would be complete without consideration of potential immunogenicity. This is one of the most common issues raised when considering the possible side-effect profile of such engineered materials. It is well acknowledged that the rate of non-responders to BoNT treatment in the clinic that results from the development of an anti-drug antibody response is relatively small (only a few percent). One reason for this could be the exquisite potency of the BoNT therapeutic, meaning that vanishingly small doses of material are required to achieve a significant biological effect. Additionally, the mode of action of the toxin being dependent on relatively rapid cell surface exposure followed by a significant longevity of action inside the cell, minimises exposure of the patient to the drug and so minimises the opportunity for the immune system to recognise the BoNT. Additionally, the extended period of time between applications is of benefit in decreasing exposure to the immune system. However, in the case of proteins engineered to incorporate significant stretches of BoNT protein, one can hypothesise that potential immune reaction to the protein would be a concern. Data on this important safety/efficacy topic are awaited; however, it is reasonable to assume that maintaining the potency, efficacy and duration of action of such novel agents will ensure that such a potential problem is minimised.

Encouragingly, the story of the development of retargeted BoNT fragments comprising the LC has progressed significantly further than pre-clinical studies. It is well known that the medical applications of BoNTs are expanding widely and the number of approved indications is increasing; some of the more recent ones being US FDA approval of BOTOX® (onabotulinumtoxin A) for the treatment of urinary incontinence in adults with neurological conditions including multiple sclerosis and spinal cord injury and the emerging opportunities in certain types of migraine. Despite the continued development of the BoNT markets, the use of BoNT is naturally limited by its mechanism of cell targeting and, though the TSI platform has the potential to significantly expand such opportunities, clinical proof of this has been awaited. In the first quarter of 2011 Syntaxin Ltd., announced that its partner Allergan had initiated two Phase II trials to evaluate the safety and efficacy of its re-targeted endopeptidase drug candidate AGN-214868. The Phase II trials are focused on patients with post herpetic neuralgia (PHN; ClinicalTrials.gov Identifier NCT01129531) administering 3 and 16 μg of AGN-214868 and Idiopathic Overactive Bladder and Urinary Incontinence (ClinicalTrials.gov Identifier NCT01157377) injecting 0.5–18 μg of AGN-214868. With the initiation of Phase II trials, the TSI technology platform reached a significant point of

Table 1 Examples of the versatility of BoNT domains

The binding domain as a delivery vehicle		The endopeptidase domain as an effector function	
Cargo	Vehicle	Cargo	Vehicle
LC of TeNT	HC of BoNT/A	LH_N/C_1	EGF
LC of BoNT/A	BoNT/D	LH_N/D	Growth hormone releasing hormone
LC of BoNT/A	BoNT/Di	LH_N/A	*E.cristagalli* lectin
Di-hydrofolate reductase	BoNT/D	LH_N/A	NGF
LC of BoNT/E	BoNT/Ai	LH_N/A	Wheat germ agglutinin
DNA	H_C of TeNT	BoNT/A,B,F	Transferrin
Viral vector	H_C of BoNT/A	BoNT/B, LH_N/B	Antibody
Quantum dots	H_C of BoNT/A		

"i" signifies the use of a mutated BoNT that is catalytically inactive

development since the AGN-214868 candidate was discovered under the collaboration using Syntaxin's proprietary discovery platform. The PHN trial will provide the first clinical data for a retargeted BoNT fragment.

In summary, therefore, the BoNT LC represents a warhead that can be delivered to target cells via the cognate BoNT receptor or via a wide range of ligand-receptor interactions to result in cleavage of intracellular SNAREs and inhibition of cellular secretory events. The LC is amenable to mutagenesis and, as a minimum, the SNARE-substrate range could be extended through site specific modification. As an example of harnessing the potential of such a strategy, Syntaxin Ltd's, TSI platform has the potential to deliver a wide range of exciting new approaches to the treatment of multiple conditions.

8 Conclusions

The BoNT engineering field has progressed significantly in the past few years. One major contributory factor is the significant developments in understanding of the structure of BoNT and the discovery that BoNT domains are amenable to protein engineering. Furthermore, understanding the functional roles of specific elements of the BoNT domains, and having the tools to assay for function, has provided an impetus to design of new BoNT-based therapeutics as a complementary strategy to extend the market opportunities for existing current natural products (Table 1).

It is well understood that BoNT/A is a product of major therapeutic and commercial success that can be widely used to treat various neurological and neuromuscular conditions. The therapeutic success of the BoNTs results from two structure-function derived outputs: (1) the specific and potent inhibition of neurotransmitter release from peripheral cholinergic neurons that results from the neuronal binding event and the unique substrate cleavage activity and; (2) the LC-derived duration of action that can often be measured in months. The clinical utility of the neurotoxins is, however, severely constrained, both by their limited range of target cells and narrow therapeutic window. Strategies to overcome such constraints and to identify new opportunities for utilisation of the BoNT mechanism of action are exemplified in the creation of hybrid BoNTs, modification of the binding domain to increase potency and, with the broadest application, the development of the TSI platform. The TSI approach is already progressing in the clinical setting and pre-clinically for a range of indications not treatable with neurotoxin products. Demonstration of the clinical success of new materials consisting of BoNT domains will be a significant step forward as it will validate the opportunity to explore a wide range of new therapeutic spaces. Harnessing the properties of the neurotoxins' protein domains in novel recombinant proteins will lead to the creation of completely new classes of biologics. As one advanced illustration, those based on the LC activity which thereby are designed to inhibit secretion, have the potential to treat chronic conditions like chronic pain, which currently have few effective treatments.

Hopefully, such clinical and pre-clinical progress will also inspire a further focus of effort onto maximising the potential of this exciting class of proteins.

References

Arndt JW, Jacobson MJ, Abola EE, Forsyth CM, Tepp WH, Marks JD, Johnson EA, Stevens RC (2006) A structural perspective of the sequence variability within botulinum neurotoxin subtypes A1–A4. J Mol Biol 362:733–742

Bade S, Rummel A, Reisinger C, Karnath T, Ahnert-Hilger G, Bigalke H, Binz T (2004) Botulinum neurotoxin type D enables cytosolic delivery of enzymatically active cargo proteins to neurones via unfolded translocation intermediates. J Neurochem 91(6):1461–1472

Box M, Parks DA, Knight A, Hale C, Fishman PS, Fairweather NF (2003) A multi-domain protein system based on the HC fragment of tetanus toxin for targeting DNA to neuronal cells. J Drug Target 11(6):333–343

Carter AT, Paul CJ, Mason DR, Twine SM, Alston MJ, Logan SM, Austin JW, Peck MW (2009) Independent evolution of neurotoxin and flagellar genetic loci in proteolytic *Clostridium botulinum*. BMC Genomics 10:115

Chaddock JA, Purkiss JR, Duggan MJ, Quinn CP, Shone CC, Foster KA (2000a) A conjugate composed of nerve growth factor coupled to a non-toxic derivative of *Clostridium botulinum* neurotoxin type A can inhibit neurotransmitter release in vitro. Growth Factors 18(2):147–155

Chaddock JA, Purkiss JR, Friis LM, Broadbridge JD, Duggan MJ, Fooks SJ, Shone CC, Quinn CP, Foster KA (2000b) Inhibition of vesicular secretion in both neuronal and nonneuronal cells by a retargeted endopeptidase derivative of *Clostridium botulinum* neurotoxin type A. Infect Immun 68(5):2587–2593

Chaddock JA, Herbert MH, Ling RJ, Alexander FCG, Fooks SJ, Revell DF, Quinn CP, Shone CC, Foster Ka (2002) Expression and purification of catalytically active, non-toxic endopeptidase derivatives of *Clostridium botulinum* toxin type A. Protein Expr Purif 25:219–228

Chaddock JA, Purkiss JR, Alexander FC, Doward S, Fooks SJ, Friis LM, Hall YH, Kirby ER, Leeds N, Moulsdale HJ, Dickenson A, Green GM, Rahman W, Suzuki R, Duggan MJ, Quinn CP, Shone CC, Foster KA (2004) Retargeted clostridial endopeptidases: inhibition of nociceptive neurotransmitter release in vitro, and antinociceptive activity in in vivo models of pain. Mov Disord 19(Suppl 8):S42–S47

Chen S, Barbieri JT (2009) Engineering botulinum neurotoxin to extend therapeutic intervention. Proc Natl Acad Sci USA 106(23):9180–9184

Darios F, Niranjan D, Ferrari E, Zhang F, Soloviev M, Rummel A, Bigalke H, Suckling J, Ushkaryov Y, Naumenko N, Shakirzyanova A, Giniatullin R, Maywood E, Hastings M, Binz T, Davletov B (2010) SNARE tagging allows stepwise assembly of a multimodular medicinal toxin. Proc Natl Acad Sci USA 107(42):18197–18201

Dolly JO, Wang J, Zurawski TH, Meng J (2011) Novel therapeutics based on recombinant botulinum neurotoxins to normalize the release of transmitters and pain mediators. FEBS J 278(23):4454–4466

Drachman DB, Adams RN, Balasubramanian U, Lu Y (2010) Strategy for treating motor neuron diseases using a fusion protein of botulinum toxin binding domain and streptavidin for viral vector access: work in progress. Toxins (Basel) 2(12):2872–2889

Duggan MJ, Quinn CP, Chaddock JA, Purkiss JR, Alexander FC, Doward S, Fooks SJ, Friis LM, Hall YH, Kirby ER, Leeds N, Moulsdale HJ, Dickenson A, Green GM, Rahman W, Suzuki R, Shone CC, Foster KA (2002) Inhibition of release of neurotransmitters from rat dorsal root ganglia by a novel conjugate of a *Clostridium botulinum* toxin A endopeptidase fragment and *Erythrina cristagalli* lectin. J Biol Chem 277(38):34846–34852

Ferrari E, Maywood ES, Restani L, Caleo M, Pirazzini M, Rossetto O, Hastings MH, Niranjan D, Schiavo G, Davletov B (2011) Re-assembled botulinum neurotoxin inhibits CNS functions without systemic toxicity. Toxins (Basel) 3(4):345–355

Foster KA, Chaddock JA (2010) Targeted secretion inhibitors—innovative protein therapeutics. Toxins 2(12):2795–2815

Foster KA, Adams EJ, Durose L, Cruttwell CJ, Marks E, Shone CC, Chaddock JA, Cox CL, Heaton C, Sutton JM, Wayne J, Alexander FC, Rogers DF (2006) Re-engineering the target specificity of Clostridial neurotoxins—a route to novel therapeutics. Neurotox Res 9(2–3):101–107

Henkel JS, Jacobson M, Tepp W, Pier C, Ea Johnson, Barbieri JT (2009) Catalytic properties of botulinum neurotoxin subtypes A3 and A4. Biochemistry 48:2522–2528

Hill KK, Smith TJ, Helma CH, Ticknor LO, Foley BT, Svensson RT, Brown JL, Johnson EA, Smith LA, Okinaka RT, Jackson PJ, Marks JD (2007) Genetic diversity among botulinum neurotoxin-producing clostridial strains. J Bacteriol 189(3):818–832

Jin R, Sikorra S, Stegmann CM, Pich A, Binz T, Brunger AT (2007) Structural and biochemical studies of botulinum neurotoxin serotype C1 light chain protease: implications for dual substrate specificity. Biochemistry 46(37):10685–10693

Kumaran D, Eswaramoorthy S, Furey W, Navaza J, Sax M, Swaminathan S (2009) Domain organization in *Clostridium botulinum* neurotoxin type E is unique: its implication in faster translocation. J Mol Bio 386(1): 233–245

Kuo CL, Oyler GA, Shoemaker CB (2011) Accelerated neuronal cell recovery from botulinum neurotoxin intoxication by targeted ubiquitination. PLoS One 6(5):e20352

Lacy DB, Tepp W, Cohen AC, DasGupta BR, Stevens RC (1998) Crystal structure of botulinum neurotoxin type A and implications for toxicity. Nat Struct Biol 5(10):898–902

Masuyer G, Thiyagarajan N, James PL, Marks PM, Chaddock JA, Acharya KR (2009) Crystal structure of a catalytically active, non-toxic endopeptidase derivative of *Clostridium botulinum* toxin A. Biochem Biophys Res Commun 381(1):50–53

Masuyer G, Beard M, Cadd VA, Chaddock JA, Acharya KR (2011a) Structure and activity of a functional derivative of *Clostridium botulinum* neurotoxin B. J Struct Biol 174(1):52–57

Masuyer G, Stancombe P, Chaddock JA, Acharya KR (2011b) Structures of engineered *Clostridium botulinum* neurotoxin derivatives. Acta Crystallogr, Sect F: Struct Biol Cryst Commun 67(Pt 12):1466–1472

Peng L, Berntsson RP, Tepp WH, Pitkin RM, Johnson EA, Stenmark P, Dong M (2012) Botulinum neurotoxin D-C uses synaptotagmin I/II as receptors and human synaptotagmin II is not an effective receptor for type B, D-C, and G toxins. J Cell Sci 125(13): 3233–3245

Pickett A, Perrow K (2011) Towards new uses of botulinum toxin as a novel therapeutic tool. Toxins 3:63–81

Pier CL, Chen C, Tepp WH, Lin G, Janda KD, Barbieri JT, Pellett S, Johnson EA (2011) Botulinum neurotoxin subtype A2 enters neuronal cells faster than subtype A1. FEBS Lett 585(1):199–206

Rummel A, Mahrhold S, Bigalke H, Binz T (2004) The HCC-domain of botulinum neurotoxins A and B exhibits a singular ganglioside binding site displaying serotype specific carbohydrate interaction. Mol Microbiol 51(3):631–643

Rummel A, Mahrhold S, Bigalke H, Binz T (2011) Exchange of the H(CC) domain mediating the double receptor recognition improves the pharmacodynamic properties of botulinum neurotoxin. FEBS J 278(23):4506–4515

Scott AB (1980) Botulinum toxin injection into extraocular muscles as an alternative to strabismus surgery. J Pediatr Ophthalmol Strabismus 17(1):21–25

Somm E, Martinez A, Marks P, Toulotte A, Bonnet N, Ferrari S, Huppi P, Jones R, Aubert M (2011) SXN101742, a botulinum toxin-derived targeted secretion inhibitor (TSI) inhibits GH synthesis and secretion: a new concept for the management of acromegaly. Endocr Rev 32:P1–P413

Strotmeier J, Gu S, Jutzi S, Mahrhold S, Zhou J, Pich A, Eichner T, Bigalke H, Rummel A, Jin R, Binz T (2011) The biological activity of botulinum neurotoxin type C is dependent upon novel types of ganglioside binding sites. Mol Microbiol 81(1):143–156

Strotmeier J, Willjes G, Binz T, Rummel A (2012) Human synaptotagmin-II is not a high affinity receptor for botulinum neurotoxin B and G: increased therapeutic dosage and immunogenicity. FEBS Lett 586(4):310–313

Sutton JM, Wayne J, Scott-tucker A, Brien SMO, Marks PMH, Alexander FCG, Shone CC, Chaddock JA (2005) Preparation of specifcally activatable endopeptidase derivatives of *Clostridium botulinum* toxins type A, B, and C and their applications. Protein Expr Purif 40:31–41

Swaminathan S, Eswaramoorthy S (2000) Crystallization and preliminary X-ray analysis of *Clostridium botulinum* neurotoxin type B. Acta Crystallogr D Biol Crystallogr 56(Pt 8): 1024–1026

Tsai YC, Maditz R, Kuo CL, Fishman PS, Shoemaker CB, Oyler GA, Weissman AM (2010) Targeting botulinum neurotoxin persistence by the ubiquitin-proteasome system. Proc Natl Acad Sci USA 107(38):16554–16559

Wang J, Meng J, Lawrence GW, Zurawski TH, Sasse A, Bodeker MO, Gilmore MA, Fernandez-Salas E, Francis J, Steward LE, Aoki KR, Dolly JO (2008) Novel chimeras of botulinum neurotoxins A and E unveil contributions from the binding, translocation, and protease domains to their functional characteristics. J Biol Chem 283(25):16993–17002

Wang J, Zurawski TH, Meng J, Lawrence G, Olango WM, Finn DP, Wheeler L, Dolly JO (2011) A dileucine in the protease of botulinum toxin A underlies its long-lived neuroparalysis: transfer of longevity to a novel potential therapeutic. J Biol Chem 286(8):6375–6385

Wang J, Zurawski TH, Bodeker M, Meng J, Boddul S, Aoki KR, Dolly JO (2012) Longer-acting and highly-potent chimeric inhibitors of excessive exocytosis created with domains from botulinum neurotoxin A and B. Biochem J 444(1): 59–67

Weller U, Dauzenroth ME, Gansel M, Dreyer F (1991) Cooperative action of the light chain of tetanus toxin and the heavy chain of botulinum toxin type A on the transmitter release of mammalian motor endplates. Neurosci Lett 122(1):132–134

Yeh FL, Zhu Y, Tepp WH, Johnson EA, Bertics PJ, Chapman ER (2011) Retargeted clostridial neurotoxins as novel agents for treating chronic diseases. Biochemistry 50(48):10419–10421

Botulinum Toxin: Application, Safety, and Limitations

Hans Bigalke

Abstract Botulinum neurotoxin type A (BoNT/A), despite its high toxicity, is approved for therapy of many neurological (e.g., dystonia, spasticity) and non-neurological (e.g., achalasia, hyperhidrosis) disorders. Its mode of action is well understood. This has led to more and more indications (e.g., pain, gastrointestinal and urologic disorders), in which the toxin can reduce disturbing symptoms. In general the application is safe (pharmacological index 20–100, depending on indication). Few unwanted reactions may occur. In worst cases BoNT treated patients may develop neutralizing antibodies. These patients are excluded from further treatment. A more recently approved second serotype (BoNT/B) could be effective in those secondary non-responders, however, due to less potency in humans higher doses have to be applied leading to an only transient successful treatment. Other serotypes as BoNT/A and B, e.g., BoNT/C should be approved as medicines.

Keywords OnabotulinumtoxinA · IncobotulinumtoxinA · AbobotulinumtoxinA · RimabotulinumtoxinB · Therapeutic index

Abbreviations

BoNT	Botulinum neurotoxin
NNT	Number needed to treat
SNAP-25	Synaptosomal associated protein
LERS	Lambert-Eaton-Rooke-Syndrome
MG	Myasthenia gravis
Ona	OnabotulinumtoxinA
Abo	AbobotulinumtoxinA
Inco	IncobotulinumtoxinA
Rima	RimabotulinumtoxinB

H. Bigalke (✉)
Institute of Toxicology, Medical School of Hannover, 30625 Hannover, Germany
e-mail: bigalke@toxogen.de

A. Rummel and T. Binz (eds.), *Botulinum Neurotoxins*, Current Topics in Microbiology and Immunology 364, DOI: 10.1007/978-3-642-33570-9_14,

Contents

1 Clinical Applications 308
2 Safety 309
3 Limitations 313
4 Conclusion 315
References 315

1 Clinical Applications

The introduction of botulinum neurotoxin (BoNT), the most poisonous drug known in clinical use, was a milestone in the therapy of many neurological disorders comparable with the invention of anti-Parkinson drugs of L-dopa type (Birkmayer and Hornykiewicz 1962). In both cases the bases of the diseases are located in the central nervous system. Both kinds of drugs do not cure the diseases, but the patients affected by the malady are relieved from the painful, bothersome, and defacing symptoms as long as they are treated with the respective medicine. Anti-Parkinson drugs have to be applied systemically at daily intervals, whereas BoNT is given locally only quarterly. In general, the latter drug keeps its efficacy during the lifetime of the patient in contrast to the former that loses efficacy while the disease proceeds.

Clostridial neurotoxins had been in the focus of a small community of scientists for decades, e.g., pharmacologists, microbiologists, and toxicologists. Astonishingly, none of them has borne the idea to exploit the well-known paralytic activity of the toxin for therapeutic purposes. Because of its extremely high toxicity the toxin was never considered to be applied therapeutically (except by Kerner in 1822), but perversely, it was produced in grand-scale to be used as a biological weapon (Bigalke and Rummel 2005).

It was a physician, the ophthalmologist Alan Scott, who has remembered Paracelsus's phrase,

> All things are poison, and nothing is without poison; only the dose permits something not to be poisonous.

In challenging experiments he tested in primates many substances known to weaken the muscle tone to implement the best working drug in the therapy of childish strabismus. Eventually, he received allowance from the FDA for a first clinical trial with botulinum toxin to treat strabismus. He found that the BoNT injection was simpler than surgery, side effects were seldom and, more important, transient, and that loss of vision had never occurred (Scott et al. 1985). Although surgery has remained the method of choice in the treatment of most cases of strabismus, his pioneering work has paved avenues for applications in other fields of medicine. In the beginning the toxin was used to return to patients their lost vision who suffered from blepharospasm. This successful treatment was the breakthrough for applications in other neurological disorders, the group of dystonia. The number needed to treat (NNT) in this kind of sickness is almost one, a number that is achieved by a few drugs only. In other words, the symptoms of dystonia almost disappear in each treated patient. During the last 2 decades, physicians explored the effect of BoNT on more disorders in which cholinergic overactivity was involved like hyperhidrosis, achalasia, hyperfunctional larynx, focal spasticity, gastrointestinal disorders, tremors, urologic disorders, and last but not least, cosmetic flaws. Whereas all these malfunctions can be attributed to disturbed cholinergic transmission the toxin is thought to help also in the management of pain and headache as migraine (Aurora 2006), however, not all patients benefit from the treatment, i.e. the NNT is increased compared to the number in cholinergic dysfunctions. The background of the mode of action in non-cholinergic disorders is not well-understood but the release of transmitters other than acetylcholine is also inhibited by BoNT (Bigalke et al. 1981; Purkiss et al. 2000) and an increased, uncontrolled release of these transmitters could be the cause of various forms of pain. The same is true for the efficacy of BoNT given in benign prostate hyperplasia (Boy et al. 2008).

A recently conducted clinical trial demonstrated efficacy of BoNT/A in the treatment of depression. This promising finding is related to changes in mental behavior following variation of facial expression caused by the toxin (Wollmer et al. 2012). Table 1 gives an overview of the diseases and symptoms reacting upon BoNTs.

Due to the complex and well organized structure of BoNT it can be disassembled into its single domains that can newly be recomposed with other ligands that target non-neuronal but secretory cells. With this approach it is possible to treat diseases that are caused by an overactivity of secretory processes, e.g., hypermegalie (Somm et al. 2012, Chaddock 2012).

Taking everything together, botulinum toxins have evolved from a single symptom drug to a multifunctional medicine that is useful in the treatment of diseases resulting from various origins (Aoki 2004). All these diseases have in common that secretory processes are disturbed. Moreover, the transport unit of the toxin may be utilized to carry other cargos into neurons, e.g., virostatics. Thus, BoNT medically evolves like the formerly antipyretic drug Aspirin that nowadays benefits patients suffering from stroke, heart attack, and even cancer (Sonnemann et al. 2008).

2 Safety

Despite their high toxicity botulinum toxins belong to the safest drugs!

Unwanted reactions resemble anticholinergic functions and may occur if the substance is accidently injected into non-affected muscles where they block the physiological transmitter release from innervating nerves. This could cause weakening of normally contracting and functioning muscles. Since the effects are fully reversible a malfunction is transient and generally without any adverse

Table 1 Indications of BoNT with respective status of approval (Jost 2007)

Disease	Indication	Status of approval Abo	Inco	Ona	Rima
Group of dystonia	Spasmus hemifacialis	A	A	A	-
	Blepharospasmus	A	A	A	-
	Several forms of cervical dystonia	A	A	A	A
	Lingual dystonia	-	-	-	-
	Laryngeal dystonia	-	-	-	-
	Dystonia of upper limb (writer's cramp, musician dystonia)	-	-	-	-
	Dystonia of lower limb	-	-	-	-
	Dystonia of the trunk (Camptocormia)	-	-	-	-
Spasticities	Upper limb (hand in consequence of a stroke)	A	A	A	-
	Upper limb (arm in consequence of a stroke)	A	A	-	-
	Lower limb (pes equinus due to infantile cerebral palsy)	A	-	A	-
	Spasticities due to various origins (Multiple sclerosis, inflammation of the brain etc.)	-	-	-	-
Urological disorders	Bladder weakness leading to incontinence	-	-	A	-
Dermatology	Hyperhidrosis	-	-	A	-
	Glabellar lines	A	A	A	-
	Facial wrincles (e.g., crow's feet)	-	-	-	-
Internal medicine	Achalasia	-	-	-	-
Various symptoms	Crocodile tears	-	-	-	-
	Migrane	-	-	A	-
	Essential tremor	-	-	-	-

Not all indications are approved for all products on the market. Some indications are not approved at all and others not in all countries, however, the scientific evidence of efficacy in the above indications is provided. The table shows the indications and in cases of approval (A) the status in Germany. Since the status of approval expands quickly the present status should be learnt by the summary of the product characteristics. The mode of action is the same in each BoNT/A containing product, thus, with respect to efficacy the products are interchangeable. Because of the diverse formulations and variation in assays, the doses vary between the BoNT products; a simple replacement of one product by the other is therefore not recommended. Moreover, patients with low titers of neutralizing antibodies generated by the complex protein containing products not only respond well to the complex free product, but also the antibody titers in some of the patients decrease despite the continuing treatment with this product (Hefter et al. 2012). The equivalence dose of the BoNT/B containing product is more difficult to determine, because it seems that the conversion factor is different for various indications (Dressler and Benecke 2004). The list is not complete, but provides an overview of the most important indications for BoNT. The doses applied may vary due to mass of the affected muscle or size of the organ. Moreover, particularly in cases of dystonia and spasticity, different muscles or groups of muscles can be involved leading to varying doses. For more details see (Jost 2007) knock-on effects. Only in special cases if a fully functional muscle is vitally important, the erroneous injection could harm the patient. Since the pivotal respiratory muscles are normally distant from any injection site respiratory failure has never occurred in consequence of a therapeutic BoNT application. However, in the case that for unjustified economique reasons unlicensed preparations are illegally used (Pickett and Mewies 2009) and high concentrated experimental BoNTs are applied, severe intoxications can occur (Chertow et al. 2006). A potential dangerous injection point is the laryngeal areal where the esophagus and trachea intersect. Weakened muscles in this area could interfere with physiological deglutition leading to asphyxia or secondary induced pneumonia. Special care should also be taken when children are involved. Few reports describe the appearance of iatrogenic botulism related to an overdosage of BoNT/A. This special form of botulism produces the same distinct clinical syndrome as food borne or wound botulism like symmetrical cranial nerve palsies followed by descending, symmetric flaccid paralysis of voluntary muscles, which may progress to respiratory failure and death (Sobel 2005; Coban et al. 2010; Crowner at al. 2007). Since correct injection sites are crucial to avoid unwanted reactions the knowledge of anatomic settings is a prerequisite for the operating doctor. In complex cases in which deeply located muscles are afflicted by uncontrolled neuronal activity the injection needle may be guided by electromyographic control. Due to the unphysiological contractions of the afflicted muscle the electrical noise increases when the needle approaches this muscle (Martos-Díaz et al. 2011). An overview of various neurological and non-neurological diseases with the corresponding muscles and the identification of correct injection points are provided by the 'Botulinum Toxin Injection Atlas' (Jost 2007) which also helps with the finding of the precise doses.

Since striated muscles are generally enwrapped by a fascia, injected toxin is almost trapped, as due to its molecular mass it can hardly penetrate the barrier. A slow distribution within the organism takes place by diffusion through the pores between capillary endothelial cells and by transport via the lymphatic vessels whereupon the toxin reaches the bloodstream and is distributed in the extracellular space that represents approximately 70 % of the total body volume. One can easily calculate that the factor of dilution is so large that systemic unwanted reaction could not occur. Only if high doses of toxin are applied can systemic effects be seen only with single fiber recordings (Lundh et al. 1977; Lange et al. 1991; Roche et al. 2008). These effects will not impair the daily life of a patient.

Facial muscles lack a fascia and similar glands are located in the skin, thus, toxin injected subcutaneously can diffuse without any barrier into adjacencies. Though it dilutes with distance from injection site, the nearby located muscles or glands may be affected and unwanted reactions may appear e.g., dry mouth, if the salivary gland is involved, ptosis, if the lid elevator is weakened, or enervation of mimic muscles if facial nerve endings are inhibited. But also these reactions are weak, transient, and in worst cases aggravating and disturbing for the patient. They are never life-threatening (Crowner et al. 2010).

A crucial parameter for the safety of a certain drug is represented by the therapeutic index and not by its absolute toxicity. A highly toxic substance could be safe when used as a drug, while a drug with low toxicity may be dangerous if the sickness relieving dose is close to the dose causing toxic effects. A few examples can support this statement. The multifunctional drug acetylsalicylacid can prevent stroke when given in low dosage (100–300 mg/d) to risk patients (NNT 70). Within the same dose range it will cause bleeding in a certain number of patients and in worst cases fatal cerebral hemorrhage. Given as pain reliever (>300 mg/d) it may provoke cancer inducing gastrointestinal ulcer. Acetaminophen/paracetamol [p-(Acetylamino) phenol] is the drug of choice in fever even for children. However, an approximately sixfold therapeutic dose will cause hepatic damage. Both medicines are available without prescription in the pharmacy. Compared with anti-Parkinson drugs BoNT demonstrates its superiority with respect to side effects. While the toxin causes in worst cases symptoms like muscle weakness and mild vegetative suppression, anti-Parkinson drugs in therapeutic dosages provoke dyskinesia, hallucinations, nausea and vomiting. Another example is insulin, a beneficial medicine in diabetes. It causes hypoglycemia if the therapeutic dose is doubled. Hypoglycemia can occur even in therapeutic dosage if patients do not eat. Hypoglycemia can lead to irreparable brain damage. Insulin is in general applied parentally by the patients themselves and probably is responsible for many cases of confusion in the elderly. Compared to these medicines BoNT is safe. Even a tenfold dose, with respect to costs an implausible fiction, given accidently intravenously is diluted to a concentration that may lead to an only slight and transient faintness of muscles, harmless for the patient (exception in children, see above).

BoNT is contraindicated during gravity (see summary of product characteristics) because the toxin passes the placenta barrier. Data are missing on the risk for the embryo but one can assume that a developing nervous system that is not protected by a blood–brain barrier may be hampered by an, even so, slight decreased transmitter release.

BoNT was shown to be transported retrogradely like tetanus neurotoxin into the spinal cord (Wiegand et al. 1976). Moreover, there is also evidence that the toxin undergoes transcytosis followed by cleavage of postsynaptic located SNAP-25 (Restani et al. 2011, Schiavo et al. 2012). In consequence the release of transmitters in the spinal cord is inhibited and similar to tetanus neurotoxin BoNT would also cause peripheral symptoms due to its spinal action. Does this behavior generate any short- or long-term risks for the patients?

BoNT reduces the inhibition of the monosynaptic gastrocnemius reflex when applied in high doses into the gastrocnemius muscle of cats (Wellhoener and Wiegand 1977). This experimental finding was observed with lethal doses only. Moreover, any spinal effect would be masked by the peripheral blockade of neuronal transmission. Even in patients suffering from severe botulism putative central effects were never described and beyond that, if patients survive, the intoxication will be cured without any residual symptoms. The latter is also true in case of tetanus neurotoxin poisoning accompanied by excessive spasms of striated muscles. Thus, short-term risks as well as long-term effects in the wake of a single application can obviously be excluded.

BoNT is given mostly over years or decades. One can speculate that sufficient amounts of toxin accumulate within the spinal cord and interrupt synaptic circuits. This could lead to a remodeling of these circuits leading to spinal disorders. However, in the first place spinal effects were only observed experimentally when high doses were applied; thus, it is unlikely that enough toxin is transcytosed to produce long-term inhibition in postsynaptic neurons. In the second place, BoNT has been in use for more than 20 years and long-term effects have never been observed (Naumann et al. 2006). Thus, it is uncertain that botulinum toxin would alter spinal conductivity.

Taken everything together, the statement, given above, bears up a critical analysis.

3 Limitations

In general, patients with local cholinergic overactivity tolerate the toxin very well. There are, however, some kinds of diseases that cause relaxation of striated muscles. The relaxation caused by the congenital myasthenia is conditioned by a gene defect, while the relaxations due to the Lambert-Eaton-Rooke-Syndrome (LERS) and the myasthenia gravis (MG) rely on an autoimmune response. Patients suffering from LERS produce antibodies against the presynaptic calcium channels leading to a reduced acetylcholine release and those suffering from MG develop antibodies against the acetylcholine receptor of the N-type leading to a low density of postsynaptic receptors. These patients should if at all be treated with special care.

There are also drugs which may interfere with BoNT by blocking the acetylcholine receptor, by reducing the intramuscular free calcium concentration and by decreasing acetylcholine release due to spinal inhibition of excitation. Thus, one has to act with caution in case the patient is treated with drugs modifying cholinergic transmission like aminoglycoside antibiotics, central or peripheral, and myotropic muscle relaxants like benzodiazepines and analogs, curare-like drugs, and dantrolene.

BoNTs represent foreign proteins for the human organism. Physiologically, the body reacts with an immune response because the application of the toxin is synonymous with a BoNT inoculation. Fortunately, it is the unrivaled potency, which allows a successful treatment in a dose range that is in general below the immune cell stimulating dose range. However, approximately 1 % (Brin et al. 2008) of patients treated with the toxin develops a secondary non-response due to neutralizing antibodies. The probability to become a non-responder depends among others on the dose applied (Göschel et al. 1997; Lange et al. 2009). Thus, the percentage of non-responders might be higher in patients suffering from torticollis spasmodicus or spasticity and lower in those with blepharospasm because different diseases demand various dosages. Nevertheless, it seems that the number of non-responders is low, however, one can roughly calculate from the worldwide revenue of BoNT/A ($ 1,500,000,000) that 4,100,000 patients are treated with BoNT/A. Thus, it appears that approximately 40,000 individuals belong to the group of the secondary non-responders. The patients affected by non-response have almost no other therapeutic option except the replacement of BoNT serotype A by another serotype (see below). In rare cases surgery may alleviate symptoms and deep brain stimulation could suppress local spasms. Most patients, however, are afflicted with their painful and bothersome disease for the rest of their lives. Thus, the development of alternative highly potent BoNT serotypes by one of the big pharmaceutical winners even as an orphan drug was more than justified to provide a therapeutic option for the unfortunate losers.

In general, the BoNT of first choice is represented by the serotype A. This serotype is distributed by three companies that offer products with different formulations and diverse preparation grades. While onabotulinumtoxinA (Ona, Botox®) and abobotulinumtoxinA (Abo, Dysport®) contain the neurotoxin and various sets of non-toxic clostridial proteins, incobotulinumtoxinA (Inco, Xeomin®) consists of the neurotoxin only (Eisele et al. 2011). All products include stabilizing agents like human serum albumin. Although the biological activity is given for each product in units measured as median lethal dose in mice only the units for onabotulinumtoxinA and incobotulinumtoxinA equal each other (Jost et al. 2005; Wabbels et al. 2011) while the units for abobotulinumtoxinA show less potency in humans (Nüssgens and Roggenkämper 1997). This discrepancy could be explained by the diverse formulations of the products and different potency assays (Wohlfarth et al. 1997). Until now, however, evidence is lacking that these disparities contribute to unwanted immune responses leading to non-response.

As mentioned above, BoNT/A could be replaced by another of the seven serotypes A–G if an immune response against type A had occurred. Only BoNT/B is approved for cervical dystonia and available for therapeutic purposes. The activity of rimabotulinumtoxinB (Rima, Neurobloc/Myobloc®) is also given in units determined in mice. Strangely, this toxin has to be applied in 40-fold higher doses when compared with BoNT/A drugs. This discrepancy is well-founded by the affinity of BoNT/B to human and rodent protein receptor synaptotagmin-II. It was demonstrated that a single point mutation in the human neuronal protein receptor synaptotagmin-II drastically decreases the affinity toward BoNT/B (Strotmeier et al. 2012, Rummel 2012). This result explains the potency difference of BoNT/A and B in treatment of humans in contrast to, e.g., potency assays in mice. This result also raises the question of whether the treatment with BoNT/B bears a higher risk to stimulate immune cells because the protein load injected is much higher compared with BoNT/A. Clinical trials have revealed that more than 30 % of all patients developed antibodies in the course of treatment with rimabotulinumtoxinB (Jankovic 2006). However, the appearance of antibodies seemed not to be accompanied by the development of a non-response. Since antibody titers were determined in the mice protection assay antibodies are supposed to be of the neutralizing type. The ineffectiveness of these antibodies in humans can be attributed to the different dosages of rimabotulinumtoxinB in mouse and man. The test was conducted with highly sensitive animals in which low titers of antibodies can be detected, titers too low to neutralize enough of the highly dosed neurotoxin. However, during long-term treatment titers increase and non-response occurs in more than 40 % of patients (Dressler et al. 2005). Since the high dose necessary for BoNT/B reflects the low affinity to the receptor protein, a promising research approach describes a site directed mutagenesis in the receptor binding domain of BoNT/B leading to a protein with increased affinity to human synaptotagmin II (Rummel et al. 2005). This improved protein would represent a true alternative to type A in non-responders as well as in responders.

4 Conclusion

Despite their high efficacy and potency, BoNT drugs are beneficial for a high number of patients suffering from painful, disturbing, and defacing symptoms. The application is safe when recommended doses are used. Only few and mild unwanted reactions may occur. In worst cases the development of neutralizing antibodies may take place. This means that due to the medical application, the painful symptoms reoccur in 1 % of patients. The approval of a drug that provokes those painful symptoms in 1 % of treated patients is justified because it suppresses the symptoms in 99 % of treatments.

References

Aoki KR (2004) Botulinum toxin: a successful therapeutic protein. Curr Med Chem 11(23): 3085–3092

Aurora S (2006) Botulinum toxin type A for the treatment of migraine. Expert Opin Pharmacother 7(8):1085–1095

Bigalke H, Rummel A (2005) Medical aspects of toxin weapons. Toxicology 214(3):210–220

Bigalke H, Heller I, Bizzini B, Habermann E (1981) Tetanus toxin and botulinum a toxin inhibit release and uptake of various transmitters, as studied with particulate preparations from rat brain and spinal cord. Naunyn Schmiedebergs Arch Pharmacol 316(3):244–251

Birkmayer W, Hornykiewicz O (1962) The L-dihydroxyphenylalanine (L-DOPA) effect in Parkinson's syndrome in man: on the pathogenesis and treatment of Parkinson akinesis. Arch Psychiatr Nervenkr Z Gesamte Neurol Psychiatr 203:560–574

Boy S, Seif C, Braun PM, Jünemann KP (2008) Botulinum toxin in the treatment of benign prostatic hyperplasia: an overview]. Urologe A 47(11):1465–1471

Brin MF, Comella CL, Jankovic J, Lai F, Naumann M (2008) CD-017 BoNTA Study Group. Long-term treatment with botulinum toxin type A in cervical dystonia has low immunogenicity by mouse protection assay. Mov Disord 23(10):1353–1360

Chaddock J (2012) Transforming the domain structure of botulinum neurotoxins into novel therapeutics. doi:10.1007/978-3-642-33570-9_13

Chertow DS, Tan ET, Maslanka SE, Schulte J, Bresnitz EA, Weisman RS, Bernstein J, Marcus SM, Kumar S, Malecki J, Sobel J, Braden CR (2006) Botulism in 4 adults following cosmetic

injections with an unlicensed, highly concentrated botulinum preparation. JAMA 296(20): 2476–2479
Coban A, Matur Z, Hanagasi HA, Parman Y (2010) Iatrogenic botulism after botulinum toxin type A injections. Clin Neuropharmacol 33(3):158–160
Crowner BE, Brunstrom JE, Racette BA (2007) Iatrogenic botulism due to therapeutic botulinum toxin a injection in a pediatric patient. Clin Neuropharmacol 30(5):310–313
Crowner BE, Torres-Russotto D, Carter AR, Racette BA (2010) Systemic weakness after therapeutic injections of botulinum toxin a: a case series and review of the literature. Clin Neuropharmacol 33(5):243–247
Dressler D, Benecke R (2004) Autonomic side effects of botulinum toxin type B therapy. Adv Neurol 94:315–320
Dressler D, Lange M, Bigalke H (2005) Mouse diaphragm assay for detection of antibodies against botulinum toxin type B. Mov Disord 20(12):1617–1619
Eisele KH, Fink K, Vey M, Taylor HV (2011) Studies on the dissociation of botulinum neurotoxin type A complexes. Toxicon 57(4):555–565
Göschel H, Wohlfarth K, Frevert J, Dengler R, Bigalke H (1997) Botulinum toxin A therapy: neutralizing and non-neutralising antibodies—therapeutic consequences. Exp Neurol 147 :96–102
Hefter H, Hartmann C, Kahlen U, Moll M, Bigalke H (2012) Prospective analysis of neutralising antibody titres in secondary non-responders under continuous treatment with a botulinum-toxin type A preparation free of complexing proteins–a single cohort 4-year follow-up study. BMJ Open 2(4) (4 Aug 2012)
Jankovic J, Hunter C, Dolimbek BZ, Dolimbek GS, Adler CH, Brashear A, Comella CL, Gordon M, Riley DE, Sethi K, Singer C, Stacy M, Tarsy D, Atassi MZ (2006) Clinico-immunologic aspects of botulinum toxin type B treatment of cervical dystonia. Neurology 67(12): 2233–2235
Jost WH (2007) Bildatlas der Botulinumtoxin-Injektion Der Medizinverlag 2007
Jost WH, Kohl A, Brinkmann S, Comes G (2005) Efficacy and tolerability of a botulinum toxin type A free of complexing proteins (NT 201) compared with commercially available botulinum toxin type A (BOTOX) in healthy volunteers. J Neural Transm 112(7):905–913
Kerner J (1822) Das Fettgift oder die F ettsäure und ihre Wirkungen auf den thierischen Organismus Tübingen Cotta 1922
Lange DJ, Rubin M, Greene PE, Kang UJ, Moskowitz CB, Brin MF, Lovelace RE, Fahn S (1991) Distant effects of locally injected botulinum toxin: a double-blind study of single fiber EMG changes. Muscle Nerve 14(7):672–675
Lange O, Bigalke H, Dengler R, Wegner F, deGroot M, Wohlfarth K (2009) Neutralizing antibodies and secondary therapy failure after treatment with botulinum toxin type A: much ado about nothing? Clin Neuropharmacol 32:213–218
Lundh H, Schiller HH, Elmqvist D (1977) Correlation between single fibre EMG jitter and endplate potentials studied in mild experimental botulinum poisoning. Acta Neurol Scand 56(2):141–152
Martos-Díaz P, Rodríguez-Campo FJ, Bances-del Castillo R, Altura-Guillén O, Cho-Lee GY, Mancha-de la-Plata M, Escorial-Hernandez V (2011) Lateral pterygoid muscle dystonia. A new technique for treatment with botulinum toxin guided by electromyography and arthroscopy. Med Oral Patol Oral Cir Bucal 16(1):e96–e99
Naumann M, Albanese A, Heinen F, Molenaers G, Relja M (2006) Safety and efficacy of botulinum toxin type a following long-term use. Eur J Neurol 13(Suppl 4):35–40
Nüssgens Z, Roggenkämper P (1997) Comparison of two botulinum-toxin preparations in the treatment of essential blepharospasm. Graefes Arch Clin Exp Ophthalmol 235(4):197–199
Pickett A, Mewies M (2009) Serious issues relating to the clinical use of unlicensed botulinum toxin products. J Am Acad Dermatol 61(1):149–150
Purkiss J, Welch M, Doward S, Foster K (2000) Capsaicin-stimulated release of substance P from cultured dorsal root ganglion neurons: involvement of two distinct mechanisms. Biochem Pharmacol 59(11):1403–1406

Restani L, Antonucci F, Gianfranceschi L, Rossi C, Rossetto O, Caleo M (2011) Evidence for anterograde transport and transcytosis of botulinum neurotoxin a (BoNT/A). J Neurosci 31(44):15650–15659

Roche N, Schnitzler A, Genêt FF, Durand MC, Bensmail D (2008) Undesirable distant effects following botulinum toxin type a injection. Clin Neuropharmacol 31(5):272–280

Rummel A (2012) Double receptor anchorage of botulinum neurotoxins accounts for their exquisite neurospecificity. doi:10.1007/978-3-642-33570-9_4

Rummel A, Weil T, Gutscaits A (2005) Carrier for targeting nerve cells: toxogen GmbH, Hannoverfiled: 26th April 2005 Application no.: DE102005019302.1; PCT/EP2006/003896

Schiavo G et al (2012) The Elusive Compass of Clostridial Neurotoxins: deciding when and where to go? doi:10.1007/978-3-642-33570-9_5

Scott AB, Kennedy RA, Stubbs HA (1985) Botulinum a toxin injection as a treatment for blepharospasm. Arch Ophthalmol 103(3):347–350

Sobel J (2005) Botulism. Clin Infect Dis 41(8):1167–1173

Somm E, Bonnet N, Martinez A, Marks PM, Cadd VA, Elliott M, Toulotte A, Ferrari SL, Rizzoli R, Hüppi PS, Harper E, Melmed S, Jones R, Aubert ML (2012) A botulinum toxin–derived targeted secretion inhibitor downregulates the GH/IGF1 axis. J Clin Invest 122(9):3295–3306.

Sonnemann J, Hüls I, Sigler M, Palani CD, le Hong TT, Völker U, Kroemer HK, Beck JF (2008) Histone deacetylase inhibitors and aspirin interact synergistically to induce cell death in ovarian cancer cells. Oncol Rep 20(1):219–224

Strotmeier J, Willjes G, Binz T, Rummel A (2012) Human synaptotagmin-II is not a high affinity receptor for botulinum neurotoxin B and G: increased therapeutic dosage and immunogenicity. FEBS Lett 586(4):310–313

Wabbels B, Reichel G, Fulford-Smith A, Wright N, Roggenkämper P (2011) Double-blind, randomised, parallel group pilot study comparing two botulinum toxin type a products for the treatment of blepharospasm. J Neural Transm 118(2):233–239

Wiegand H, Erdmann G, Wellhöner HH (1976) 125I-labelled botulinum a neurotoxin: pharmacokinetics in cats after intramuscular injection. Naunyn Schmiedebergs Arch Pharmacol 292(2):161–165

Wiegand H, Wellhöner HH (1977) The action of botulinum a neurotoxin on the inhibition by antidromic stimulation of the lumbar monosynaptic reflex. Naunyn Schmiedebergs Arch Pharmacol 298(3):235–238

Wohlfarth K, Göschel H, Frevert J, Dengler R, Bigalke H (1997) Botulinum a toxins: units versus units. Naunyn Schmiedebergs Arch Pharmacol 355(3):335–340

Wollmer MA, de Boer C, Kalak N, Beck J, Götz T, Schmidt T, Hodzic M, Bayer U, Kollmann T, Kollewe K, Sönmez D, Duntsch K, Haug MD, Schedlowski M, Hatzinger M, Dressler D, Brand S, Holsboer-Trachsler E, Kruger TH (2012) Facing depression with botulinum toxin: a randomized controlled trial. J Psychiatr Res 46(5):574–581

Index

16S rRNA comparisons, 3, 4

A
AbobotulinumtoxinA, 314
Acromegaly, 301
Actin, 100, 101
Adherens junction, 52, 54
AGN-214868, 302
Antibodies, 310, 313–315
Axonal transport, 91, 100, 101, 102, 104

B
Beltless H_N, 123
Binding, 95, 96, 98, 100, 102, 105
Bioassay (MBA), 219, 220, 225–229, 232, 234–236, 239
BiTox, 296
BoNT/Ai, 26, 28, 29, 31, 33
BoTIMs, 294
Botulinum neurotoxin gene location, 6, 7, 9–11
Botulism, 311, 312

C
Caco-2, 49, 51, 52
Carbohydrate-binding activity, 51, 53
Cell-based BoNT detection assays, 257, 264
Cell body, 100, 101, 103, 105, 106
Channel, 117, 127, 131
Chaperone, 24, 30, 124, 132
Chemical conjugation, 299
Chimera, 293
Clathrin coated pit, 92, 93, 97
Clathrin coated vesicle, 92, 97
Clathrin-dependent, 97, 103
Countermeasure, 38
Cytoplasmic dynein, 95, 100–102
Cytotoxicity, 52

D
De-ubiquitinases, 189
Di-leucine motifs, 295
Diphtheria toxin, 296
Disulfide bridge, 117, 118, 130, 131, 133, 141
Domain swap, 293
Drug discovery, 198, 201–203, 213
Dual receptor hypothesis, 96
Dynamin, 94, 97, 99
Dystonia, 307, 309, 310, 314

E
E-cadherin, 50, 52–65
E3-ligase, 189, 191
E3 ubiquitin ligase, 192
Electrospray ionization (ESI), 220, 229, 236, 237
Embryonic dorsal root ganglion neurons, 299
Endocytosis, 75, 78, 83, 93, 97, 98, 117, 125, 128, 132, 139, 141, 161, 168–170, 172, 223, 260, 262, 273
Endopep-MS, 229, 230, 237, 240, 243
Enzyme-linked immunosorbent assay (ELISA), 220, 222, 231–235, 239, 241–242

A. Rummel and T. Binz (eds.), *Botulinum Neurotoxins*, Current Topics in Microbiology and Immunology 364, DOI: 10.1007/978-3-642-33570-9,

E (*cont.*)
Epidermal growth factor, 300
Epithelial barrier, 48, 49–52, 55, 56

F
F box, 297
Food borne botulism, 45, 55, 224
Förster resonance energy transfer (FRET), 220, 228, 240, 242

G
Ganglioside, 23, 29, 33, 51, 56, 63–66, 68, 69, 71–73, 75, 80–83, 95–98, 106, 117, 118, 121, 132, 141, 171–173, 199, 202, 223, 229, 233, 260, 262, 265, 266, 270, 276, 291
Ganglioside binding site (GBS), 62, 68, 71, 72
Gastrointestinal tract, 21, 22
Group designation, 3, 5
Growth hormone receptor hormone (GHRH), 301

H
Hemagglutinin (HA), 23, 46, 47, 49–53
Hybrid, 294

I
Iatrogenic botulism, 311
Idiopathic overactive bladder, 302
IL8, 293
Immune response, 313, 314
Immunogenicity, 302
Immuno-PCR, 232, 233, 241
IncobotulinumtoxinA, 314
Infant botulism, 46, 48
Inhibitor, 118, 131, 198, 200–214
Inhibitory interneuron, 94, 95, 98, 100
Intestinal botulism, 46, 56
Ion channel, 119, 122, 160, 169

K
Kinesin, 100, 101

L
Lateral flow assays (LFA), 235, 242
LC refolding, 115, 126, 130, 133
LH_N, 83, 124, 130, 287, 290, 297, 301, 303
Lipid microdomain, 97
L-PTC, 25, 27
Lysosome, 180, 182

M
Mass spectrometry (MS), 220, 229, 236, 243
Matrix-assisted laser desorption (MALDI), 220, 229, 236, 237
Microfluidic, 106, 107
Microtubule, 100
Molecular motors, 95, 100
Monoclonal antibody (mAb), 220, 234
Motor neuron, 94, 95, 97–103, 105–107
M-PTC, 22, 24–33
Multiplex, 233, 237–239
Myosin, 100, 101

N
Nerve sprouts, 181
Nerve terminal, 97, 98
Neuroendocrine intestinal crypt cells, 51
Neuromuscular junction (NMJ), 22, 65, 68, 94, 144, 223, 225, 259, 297
Neurotoxin-associated protein (NAP), 22
Neurotrophin, 97, 103
nLoop, 30
Nontoxic neurotoxin-associated proteins (NAP), 46, 47
Nontoxic non-hemagglutinin (NTNHA), 23, 46, 47, 49

O
OnabotulinumtoxinA, 314
Oral toxicity, 46, 49
orfX, 56

P
Pain, 299
Palmitoylation, 186
Persistence, 180, 181
Phase II, 302
pH dependence, 123
pH sensor, 32, 33
Polyclonal antibody (pAb), 220, 234
Polymerase chain reaction (PCR), 220, 222, 237, 238
Polysialoganglioside, 95–98, 106, 107
Post herpetic neuralgia, 302
Potency, 313–315

Primary sensory afferents, 299
Progenitor toxin complex (PTC), 22, 38
Proteasome, 182
Protein engineering, 290
Protein receptor, 23, 33, 38, 75–83, 94, 96–98, 107, 117, 143, 171, 172, 181, 199, 202, 223, 226, 289, 314
Protein receptor binding site, 61, 75, 77, 79
Protein stapling, 296

R
Rab proteins, 97, 102, 103
Receptor binding domain, 189
Retrograde, 95, 98, 99, 101, 103–105
RimabotulinumtoxinB, 314

S
Serotype comparisons, 14
Signalling endosome, 91, 101, 102
SNAP, 143, 162
SNAP-23, 144, 149, 153, 166, 171, 292
SNAP-25, 24, 62, 79, 92, 94, 104, 105, 118, 129–131, 147, 151, 164, 166, 171, 183, 187, 200, 264, 293
SNAPtide, 291
SNARE, 22, 24, 63, 94, 97, 117, 118, 133, 141–145, 153, 166, 167, 170, 173, 181–185, 189, 191, 193, 198, 199, 202, 223, 224, 227–230, 262, 264, 266, 275, 279, 289, 291–294, 296–299, 303
Somatotroph, 301
Spasticity, 309, 310, 313
Species specificity, 54
Strabism, 308
Substance P, 299
Subtype, 13–15, 80, 84, 106, 131, 143, 153, 172, 180, 222, 225, 226, 230, 234, 237–239, 243, 259, 265, 270, 274, 279, 288–290, 293
Subtype comparisons, 15
Subtype definition, 13, 14
SV2 protein, 51, 55, 164, 169
Synaptic vesicle (SV), 61, 75, 79, 160, 161, 168, 173
Synaptobrevin (VAMP), 24, 79, 94, 142, 143, 162, 223, 260
Synaptotagmin (Syt), 163, 167, 171, 172, 314
Syntaxin, 142–144, 152, 161, 162, 164, 166, 169, 170

T
Targeted secretion inhibitors, 297
Tetanus neurotoxin, 312
Tetanus toxin, 94, 104
Therapeutic index, 307, 311
Tight junction, 52, 54
Toxin gene clusters, 11–13
TRAF2, 189
Transcytosis, 49–55, 95, 100, 103, 104, 107, 312
Translocation domain, 29, 83, 116, 187, 189, 190, 199, 202, 289, 296, 297

U
Ubiquitination, 295
Ubiquitin-proteasome, 180

V
VAMP/synaptobrevin, 54, 62, 79, 92, 142, 147, 149, 150, 162, 163, 167, 223
VHHs, 191

X
X-ray structure, 289

Z
Zinc protease, 141

The manufacturer's authorised representative in the EU is Springer Nature Customer Service Centre GmbH, Europaplatz 3, 69115 Heidelberg, Germany. If you have any concerns regarding our products, please contact ProductSafety@springernature.com

Printed and bound by CPI Group (UK) Ltd, Croydon, CR0 4YY
22/09/2026
02220948-0016